Health and Sustainability

Health and Sustainability

An Introduction

TEE L. GUIDOTTI

OXFORD
UNIVERSITY PRESS

Oxford University Press is a department of the University of Oxford. It furthers the University's objective of excellence in research, scholarship, and education by publishing worldwide.

Oxford New York
Auckland Cape Town Dar es Salaam Hong Kong Karachi
Kuala Lumpur Madrid Melbourne Mexico City Nairobi
New Delhi Shanghai Taipei Toronto

With offices in
Argentina Austria Brazil Chile Czech Republic France Greece
Guatemala Hungary Italy Japan Poland Portugal Singapore
South Korea Switzerland Thailand Turkey Ukraine Vietnam

Oxford is a registered trademark of Oxford University Press in the UK and certain other countries.

Published in the United States of America by
Oxford University Press
198 Madison Avenue, New York, NY 10016

Library of Congress Cataloging-in-Publication Data
Guidotti, Tee L., author.
Health and Sustainability : An Introduction / Tee L. Guidotti.
p. ; cm.
Includes bibliographical references and index.
ISBN 978-0-19-932533-7 (alk. paper)
I. Title.
[DNLM: 1. Ecosystem. 2. Environmental Health. 3. Environmental Pollution. WA 30.5]
RA1226
615.9′02—dc23
2014035163

This book is dedicated to Donna Marie, who sustains me.

Contents

Foreword on Health

"HEALTH AND SUSTAINABILITY" is not necessarily a familiar marriage of words (and shouldn't it be the other way around)? Surely a key reason for living sustainably and equitably is to secure the possibility of high and continuing levels of population health? Yes, but there are deeper, more interactive and multidirectional aspects to this pairing: these are aspects that have received little formal attention. Here is a book that is demanding in its rigor and can also help us see beyond the surface.

The word "sustainability" is often misunderstood. It has become cosmetically fashionable and promiscuously used in the realms of government, private enterprise, the NGO world and academia. But what, really, does "sustainability" mean—or at least, in Lewis Carroll's terms, what do serious users of the word intend it to mean? This needs careful exploration.

Meanwhile, most of us assume we know what "health" means. Well, at a superficial and personal level that may be so; we all have an opinion about our own state of health. Even so, our self-assessment works better in relation to personal physical ailments and to "feeling sick" than for gradations of mental and emotional health. But once we begin to consider human health in relation to the overarching issue of environmental and social sustainability it necessarily takes on a different complexion. We must shift to an *ecological* frame, and try to understand how the profile of health within a whole human community or population relates to the activities, vitality and functioning of the other communities of living species and associated physical processes that make up the greater ecosystem of which *Homo sapiens* is (or should be) an integral part.

Tee Guidotti emphasizes the same point: that "population health" is a much broader concept. It reflects, he writes, the difference between the health status and properties of an entire population as measured by various population-level indicators (such as average life expectancy, vaccination coverage or the socio-economic gradient of child nutritional stunting) versus the more sterile simple summation of all individuals with specific health conditions in the population (such as totting up the number of people with arthritis or heart disease). The former perspective carries us further, into understanding population-level emergent properties such as "herd immunity" against infectious epidemics, or how the unequal distribution of particular health disorders reflects, and often reinforces, differential circumstances of exposures. It thus belongs within the idiom of ecological relationships.

There is further added value in the approach taken in this book. Exploring how the achievement of a sustainable future for humankind both influences and depends on levels of population health is a topic that has largely been skirted around. The questions are not easy ones. How should we actually measure and monitor that relationship over time? And why have we not understood two key issues more clearly? First, that human well-being and health provides a sentinel outcome marker (albeit one with some built-in delay in realization) of the extent to which our modifications of practices are proving sustainable—that is, are not eroding the life-support capacity of the natural environment. And second, that this is a two-way street: population health is more than an output; it represents a human capital asset that will facilitate the reforming our social and economic structures and technological choices. Understood thus, population health is the *pivot* of sustainability.

Tee Guidotti proposes ten key sustainability "values," such as conserving resources, minimizing harm, preserving social structures, maintaining population health and ensuring future momentum. All relate to the successful sustaining of the processes of the biosphere that are the necessary and continuing foundation for health and life. To reduce this Herculean meta-question to manageable proportions the author disaggregates the settings, exposures and behaviors of human communities into a sequence of chapters that encompass the core of the "sustainability" challenge: dealing with catastrophes, pollution, ecosystem viability, workplace conditions, anxieties, cultural phenomena, socially mediated health outcomes, infectious diseases, professionalism, and decision making.

Beyond the conceptual clarity on offer here, there are of course things to be learned from the long historical experience of many earlier societies that suffered localized or regional crises arising from nonsustainable cultural practices and technologies. The driving principle of economies is consumption. Throughout history the first-order task of societies has been to harness local resources to feed, house, and clothe people; societies should also provide physical security, generate wealth and maintain social stability.

The most basic need is for food and the expansion and intensification of agriculture. Striving to keep up with growing populations (including the demands of the urban elites and their craftsmen and armies) often exceeded the sustainable limits of soil fertility, seasonal rainfall, and river flows. As yields faltered, often aggravated by century-long changes in regional climates, crises ensued, unrest escalated, and civilizations collapsed—the Sumerians, Akkadians, Harappans, Moche, Mayans, Anasazi, and others met their demise. The sequence of increasing and chronic hunger, conflict, starvation and then rebellion, deaths, and political dissolution was a recurring sign that the system was no longer sustainable.

This is a timely book, with a structured analytic exploration and discussion of the many facets of sustainability, how they relate to human population well-being, health, and social stability, and how those in turn bear on the achievement and modulators of sustainable ways of living. The task of recharting a course to a sustainable future is great, urgent, and formidable. What is needed is not only a clear understanding of the "health and sustainability" relationship but also the reassurance that achieving environmental and social sustainability has enormous consequences for human well-being, happiness, health and survival.

—Anthony J. McMichael
Australian National University

Foreword on Sustainability

AT THE EARTH Summit in 1992, world leaders embraced the concept of sustainable development. Bringing together environmental, economic, and social considerations promised much improved decisions. It was the politics of hope—quite the seductive concept. We were asked to imagine real improvements in the health of the environment, a more equitable sharing of the earth's resources, and a much-improved quality of life for more of the planet's people.

Years later any assessment of progress would find us wanting. The promise of Rio and of the Millennium Development Goals has not been fully realized. To be charitable, sustainable development remains a "work in progress." Compared to the scope of the change that is needed, we are largely tinkering in the margins. There is clearly a disconnect between what we negotiated and what we delivered.

When we look critically at the state of the world, our performance is simply not good enough. Compelling and mind-numbing statistics result in pessimistic reviews of the ecological fate of the earth. Suffice it to say that on the environmental front, any assessment concludes that there is no room for complacency. Virtually every indicator, from water shortages to air pollution, shows symptoms of a world in wobbly disequilibrium.

And one could undertake a similar and compelling analysis in other areas of concern such as social justice. This is a time of blurring sovereignty, blinding technological change, integrated economies, and growing alienation between political processes and peoples' passions. We live in a world where ideas cross borders as if they did not exist, where cyberspace is beyond national control, and where the speed and magnitude of capital flows is incredible. The horrors of continuing unrest in many parts of the world surely illuminate the extent of our interconnectedness and the fragility of a world of inequity.

We do not seem to understand how to avoid a collision between growing ecological pressures, significant challenges to social cohesion and economic expansion. So the conversation that we should be having is centered around the following questions: Why is action on the sustainable development agenda so elusive? Why are countries and companies not living up to what they have promised? Why is there a gap between policy and action? Why is the policy response so weak and hesitant? What do we need to do differently if we are to succeed on a grander scale?

We started down the path to sustainability—a deceptively simple concept—only to discover that it has many challenges in implementation. It is intrinsically holistic and interdisciplinary. It embodies complexity. It makes value judgments about equity. It is long term in character, which is quite inconsistent with the time frames of elected governments. We are finding that the politics of anticipate and prevent is much harder than of react and cure.

In the search for answers many are turning to the power of science and technology in making an immense contribution to human progress. A network of scientists has been asking some fundamental questions about the nature and purpose of science. Is science generating useful knowledge and know-how that society and its leaders want and need?

Is it directed to the solution of sustainable development problems and not just their definition? Is it a place-based science, supporting the agenda of local communities and helping them assess options? Does it empower people to make better decisions? Is it accessible to decision makers? Does it encourage interdisciplinary understanding and sharpen our response to sectoral challenges? Is it more than just an integrated understanding of biogeochemical, climatic, ecological and speciation processes but also the workings of politics and markets, social institutions, human behavior, and technological innovation?

The World Health Organization defines health in a broad manner as a state of complete physical, mental, and social well-being and not merely the absence of disease or infirmity. Slowly it is becoming apparent that the lens of sustainability with its complex interlinkages among health, ecosystem functioning, and multiple economic and social dimensions may lead to improvements in understanding the vulnerabilities in human population health. But even now in the current dialogue about the development of new Sustainable Development Goals, although ensuring healthy lives and promoting well-being is promoted as a goal, the connections are not made to other important goals to end hunger, end poverty, achieve gender equality, and promote sustained and inclusive economic growth.

The contribution of this book goes beyond a simplistic and obvious discussion of water and air quality, of climate change, nutrition, traditional medicines, and pharmaceuticals. It attempts to clarify our understanding of sustainability and health by articulating sustainability values and suggesting ways in which they can be applied consistently in the design of health systems and delivery of primary care.

The author also recognizes that there are no "right" answers to many of the ethical questions we face in examining health and sustainability. How do we accommodate the desires of the current generation while recognizing that the decisions we make now may affect the lives of our children, their children, and many generations to come? How heavily should we rely on emerging technologies? What forms of institutions and governance inspire trust and confidence? With its capacity to re-create nature and even change what it means to be human, science is now confronting us with moral dilemmas and profound choices that will require deeper global dialogue and greater systemic thinking than we have ever achieved.

Our challenges will not be met with traditional entrenched modes of thinking. Fresh and innovative perceptions need to be brought to bear. Preventing a genomics divide or responding to profound inequities and questions of ethics call upon different skills and approaches. Issues such as climate change require us to shift from the local and short-term focus to a focus that is regional and international in scope and to problems that are highly uncertain and not amenable to quick technological fixes. Agility and responsiveness need to be designed into our institutional structures just at the time when so many of them are going through a mid-life crisis. Innovation and technology development need to be encouraged, particularly to replace ecologically empty economic models, and best practices identified and replicated. It is not the scientific and technical challenges that should occupy our greatest attention. It is the challenge of attitudinal and behavioral change—both of individuals and institutions—that will prove to be formidable over time.

The challenges of alleviating poverty, building a safe and secure world and shaping globalization require our best efforts. A future of more mouths to feed, uncertain environmental conditions, and unmet development expectations should make us pause. Sustainable development remains largely theoretical for the majority of the world's people.

—E. Dowdeswell
July 2014

Preface

THIS BOOK IS dedicated to exploring the specific relationships between health and sustainability with a clear idea of what sustainability really means. It is meant to be read profitably by people with a deep interest—professional or personal—in either health or sustainability or both. Features of the book have also been designed to support its use as a text by students in either the health sciences (particularly public health) or in sustainability studies, including programs in environmental studies and environmental sciences.

My hope in writing this book is to encourage students and sustainability practitioners to study health as part of their preparation, just as they learn about ecosystem stability and environmental economics. At the same time, I would like to encourage fellow students of public health and the health sciences to think about what we do in the context of sustainability. I believe that there is an intimate relationship between sustainability and health that will only grow more important in both a professional and scientific sense. Another goal is to present health-centered careers as a viable option for students who are in environmental sciences, environmental studies, and sustainability programs.

Much of this book is derived from classes and seminars I taught at the University of Alberta and later at the George Washington University. My thinking has been shaped by being a medical doctor, of course, but I have been as much a public health physician as a specialist clinician in my long career, and so my orientation is not necessarily to clinical medicine, although I am comfortable in the realm of practice. I have been equally informed by an unusual and eclectic education received formally at the University of Southern California, the University of California at San Diego (including a profoundly helpful class at Scripps Institution of Oceanography), and Johns Hopkins, and informally (but no less gratefully) at the Los Angeles County Museum of Natural History, the Foundation for Advanced Education in the Sciences (the graduate-level teaching institute at the National Institutes of Health), as an environmental activist in my youth, and through numerous challenging projects I have done for clients as a part-time or full-time consultant. It has been profoundly shaped by the opportunity to compare and contrast numerous case studies and real-life situations, often as a consultant, particularly in Canada and the United States but also in China, Turkey, Australia, the Middle East, Africa, and elsewhere.

I have benefited from stimulating conversations throughout my life, but in the context of this book, in particular, I would like to acknowledge the direct influence and feedback of Tony McMichael, Steve Hrudey, Weiping Zhang, Colin Soskolne, Colin Butler, Garry Bowen, Sally Kane, Richard Jackson, Bernard Goldstein, David Blockstein, Nader Nassif, Trevor Hancock, Warren Bell, Elisabet Lindgren, John Last, Don Franklin, and Komali Naidoo. In particular, I am grateful to Marina S. Moses for urging me repeatedly to apply my work to sustainability and in so doing reach outside my core competency area of environmental and occupational health and medicine.

Any errors or significant omissions in the book are solely my own responsibility.

The voice of this book should normally be studiously neutral, except in passages where the weight of evidence is clear, in order to provoke thought. If the reader cannot discern my personal opinion when the topic is ambiguous, then I feel that I have succeeded in presenting the issue objectively. The biased reader will probably always assume that I am on the "other" side, and if so, I will have succeeded in challenging their assumptions. The casual reader with no opinion may assume that I have no opinion either, because I rarely spoon-feed readymade conclusions on complicated issues. I respect the reader more than that.

I was motivated to write this book precisely because I do have deep concern and hold strong convictions about both health and sustainability, separately and together. My personal views regarding sustainability include the following:

- The primary problem in sustainability is cultural. To endure from generation to generation, sustainability must relate to the permanent values of the culture, not to a persuasion or ideology.
- The best hope for achieving sustainability is technical rather than behavioral. A world in which environmental and energy sustainability are built in rather than having to be chosen anew with every economic decision is much more likely to achieve the goals and values of sustainable development.
- At the same time, technical solutions depend on a supportive culture that values sustainability and sets priorities for its application.
- People and their culture should shape technology, rather than having technology dictate to culture.
- We need a new approach in the environmental movement to projecting the vision of a sustainable future. The public may temporarily accept an

ascetic view of the future when they feel threatened, as today, but will soon fall back into prodigal consumption habits and push for unsustainable infrastructure again when the perceived crisis is over. We need to project a concept of a future that is attractive, culturally rich, and diverse and *incidentally* sustainable.

- The greatest challenge in sustainability facing us is not the shape of the alternative future, which is likely to take care of itself, but rather managing the transition through the next ten to twenty years, during which time petroleum will continue to dominate, and we will have to work with the infrastructure of the previous century.
- There will be a period of about two decades when fossil fuels will be irreplaceable in the world economy, during which time mitigating measures will be needed to see us through the transition. Oil dependency is a big problem, and its complexity needs to be understood because the transition away from oil will take time. Coal dependency, however, is an enormously greater problem in comparison and the single most important threat in the energy sector and possibly in all sustainability. On the way to eventual energy sustainability, natural gas is a transitional solution.
- The precautionary principle is applied wrongly. We use it too often for *de minimis* problems and not often enough on the major society-destabilizing issues it was designed to address.
- Sustainability has to be nonideological in order to be, well, sustainable. No one political faction has a monopoly on enlightened policy and insight. On the other hand, not every political faction has something useful to say on every given issue.

For there to be hope in achieving sustainability, there must be a will to change. For the needed change to take place, it is essential that peoples of the world believe that change is possible and likely to succeed. This means that the twin enemies of sustainability (and very likely survival) are fatalism and pessimism.

—Tee L. Guidotti

About the Author

TEE L. GUIDOTTI has had a long and diverse career as a physician, professor of public health and medicine, and international consultant. He is currently based in the Washington, DC, area but most of his professional work has been in Canada, where he retains close ties. His core field of expertise is occupational and environmental medicine. He has had a career spanning four decades in clinical medicine, public health teaching and practice, environmental protection, and research. His interests and professional opportunities have also given him unique opportunities to branch out widely into many related and distant fields.

Dr. Guidotti became interested in the environment in 1957, at age eight, when he became a Smokey the Bear Junior Forest Ranger through a program at the Burbank (California) Public Library. Much later, he was a student activist in the ecology movement in the 1960s, peaking in 1970 with the first Earth Day.

He maintained that interest through his undergraduate education at the University of Southern California, medical school at the University of California at San Diego, and a masters degree in public health from Johns Hopkins, where he also completed medical training in internal medicine, pulmonary medicine, and occupational medicine; he eventually qualified as a specialist in these fields. He also trained in medical research at the National Institutes of Health. He remains active in medicine as a consultant physician.

Dr. Guidotti spent most of his academic career at the University of Alberta, where he was the founding head of the program in occupational health and also did considerable work in environmental health and risk science. In Alberta, he had the opportunity to work closely with the oil and gas industry and other parts of the energy sector. During this period, he was one of the three founders of the Canadian Association of Physicians for the Environment, which became an important national organization.

Although his career was mostly bound up with Alberta, it began as the founding head of the occupational and environmental health division at the Graduate School of Public Health at San Diego State University. At the time of his retirement from full-time academic life, he was chair of the Department of Environmental and Occupational Health in the School of Public Health and Health Services at the George Washington University in Washington, DC, where he also taught occupational and environmental medicine. He has been a visiting professor at many institutions worldwide, and at the time of publication is serving a half-year award as Fulbright Visiting Chair in the Institute for Science, Society and Policy at the University of Ottawa.

Dr. Guidotti has written a major textbook, *The Praeger Handbook of Occupational and Environmental Medicine*, coauthored others, and has edited several important books in occupational and environmental medicine. He edited the best-selling *Canadian Guide to Health and the Environment*. He has written over 300 scientific papers, reviews, book chapters, and published reports, and numerous editorials, features, and short contributions. He is editor in chief of a historically important journal, *Archives of Environmental and Occupational Health*.

Dr. Guidotti is the recipient of many honors and has been elected to fellowships in many organizations, most of them in medicine and public health; however, he is also the only physician to have been named a Fellow of the Energy Institute (London), which is a reflection of his interest in the sector. He has served as president or in other leadership roles in various organizations, mostly in medicine and public health.

1

Health and Sustainability

HEALTH AND SUSTAINABILITY, as the latter is commonly understood, seem intuitively to go together. One naturally assumes they are related. Figuring out exactly how they go together in practice is not so simple.

The concepts of health and sustainability have become inextricably entwined in popular perception, politics, and advocacy. For such foundational ideas in modern society, the definitions of both "health" and "sustainability" can be surprisingly slippery. "Health" describes a state of well-being and well-functioning that also combines the sense of being whole and can be applied to individuals and to populations (i.e., communities). Sustainability is a concept that embraces environmental protection and includes stewardship so that resources are available equitably and to future generations. Health and sustainability go together not because they are linked in an obvious or physical way—so that improvement in one automatically means improvement in the other—but because health protection and enhancement, on the one hand, and sustainability and environmental protection, on the other, are driven by similar values.

If sustainability just meant continual persistence, then there would be no reason for health and sustainability to be linked. If it were, then throughout the existence of the human species, the most "sustained" human condition (by that dismal standard) was to be poor, short-lived, small in number, and sick by today's standards, mostly because of parasite infestations and malnutrition. There were a few long periods of decent living conditions for humans before the modern era, but from the beginning of human communities, life seems to have been burdensome and hard for all but a few privileged families. If sustainability only meant the ability for the collective human community and social order to be sustained or to abide, why would anyone assume then that health is a prerequisite of sustainability or that sustainability results in good health?

The reason health is linked to sustainability in the modern era is that in the twenty-first century sustainability does *not* mean just survival. The modern concept of sustainability incorporates values incompatible with the mere persistence of an inequitable and marginal society.

What Is Sustainability?

Sustainability is a strategy. Unlike previous environmental and social movements, the movement for sustainability cannot be fulfilled without solving many of the world's other important problems. Health issues are an obstacle to sustainability where and when health is poor and a resource where and when health is good, while freeing up energy for participation and engagement. For health, sustainability is an opportunity to correct old problems, reduce disparities among communities, and make progress in prevention. In a future sustainable society, health will also play an important role as an indicator of how well sustainability is doing. Sustainability, as a contemporary concept, has more to do with values than with "sustainable yield," which has been oversimplified to a business model that generates predictable revenues, harvest, recycling, technological innovation, or environmental preservation. The practice of sustainability flows from these values and is rooted in a society that cares about future generations and about leaving the planet as good or better than how the present generation found it. Caring for the legacy left to future generations and the integrity of the ecosystem is more important in such a society than leaving a legacy of monuments or grandeur, or an influence on military history. Sustainability is a value that society accepts: and when they do accept it, sustainability supports numerous actions and measures that may be individually small but perhaps create something more lasting.

Sustainability in this modern sense means balancing what is taken and what is given, as well as preserving carefully what is good in human life and what supports the planet; however, sustainability also means keeping a forward momentum for human progress and a constant reduction in demands placed on the resources that cannot be replaced. To achieve this, a future world would have to be very efficient, exquisitely conscious of limits, and committed to stewardship; the society it supports would have to be forward looking, extremely adept technologically, and grounded in values that commit the society to this way of life and prevent a return to careless or ruthless exploitation.

Sustainability provides a framework for achieving the same goals as traditional conservation and environmental health protection but with less conflict and rancor. It is an approach that tolerates the essential tensions inherent in environmental protection, on the one hand, and sustainable yield and business practice on the other. Yet it keeps these contradictions from impeding progress toward a viable sustainable future. To achieve this long-term sustainability, also called sustainable development, the society and the world that hosts it must all the while achieve growth from within, rather than through numbers and expansion, and allow people to reach their human potential through design and freedom rather than destruction and resource consumption.

In many ways, sustainability as a "big concept" is the progeny of the environmental movement, which in the 1960s described the threat to the planet and to human health and demanded protection for nature and for people in their communities. The environmental movement itself grew out of two earlier movements: the early-twentieth-century movement for conservation of nature and natural resources and the post–Second World War "toxics" movement against contamination, pollution, and health harm. In the 1960s these movements came together during a period of deep anxiety over the destruction of the natural environment and risk to human life, now popularly called the "environmental movement" (but more often called the "ecology movement" by activists at the time).

Sustainability offers a rare opportunity to take essentially "negative" or discouraging messages of environmental protection (in other words, how bad things are and how they are getting worse) and convert them into positive motivational messages (about how good things can be). Commitment to sustainability redefines environmental protection and turns it into an opportunity instead of a holding action.

Dictionary Definitions

To "sustain" something obviously means to carry something forward or to design it in a way that it can continue indefinitely. Therefore the conventional definition of "sustainable" simply describes an activity or product that can be used, conducted, or produced without limits in the long term. (For environmental sciences, the "long term" may be generations. For business, the "long term" may be two years. Obviously determining what may be the so-called long term is a major part of the definitional problem.)

Dictionary definitions of sustainability imply either standing still or going forward unchanged. The implication is that once a process, operation, or activity begins, it continues uninterrupted and unmodified. One might conceive of a sustainable system that operates for long periods but is static or steady state in which conditions merely persist essentially unchanged for a long time, like human life before the invention of fire. Or, one might conceive of a sustainable system that sees changes only in its nonessential features and details but in which the essential features remain the same for a long time and are reinforced rather than eroded by intrinsic tension between values and by debate (like a stable democracy).

The modern understanding of sustainability, however, is anything but static. To be meaningful in contemporary society, sustainability requires constant adaptation and change within the framework of the environment, technology, and society. Sustainability today is a dynamic concept and requires adjustment and response and ultimately constant progression. Sustainability is more apt to be akin to a steady state, one in which inputs and outputs balance but having plenty of room to innovate within a particular budget.

"Sustainable", it was observed, does not just mean something than can be sustained. By contemporary standards, some of the most "sustained" societies in world history, societies lasting thousands of years, have been poor peasant cultures with low technological capacity, at least compared to what came later. The real issue for contemporary sustainability is achieving some way of life that is sustainable, in a world that someone might actually want to live in.

The word "sustainability" is probably better understood by its practitioners and advocates as shorthand for an agenda and for an emerging consensus of what needs to be done to protect people and the planet; however, "sustainability" is also understood as allowing continuation of constructive and positive economic and social development well into the future and ideally indefinitely. "Sustainability" therefore describes a basket of different ideas and objectives, or a cluster of values that have in common a core of meaning that is more future oriented and values laden than simple environmental awareness.

Sustainability arose from the concept of "sustainable development," which was formally defined from the beginning: "Sustainable development is development that meets the needs of the present without compromising the ability of future generations to meet their own needs." Sustainability became shorthand for a cluster of ideas consistent with this "Brundtland" definition of sustainable development (to be described in greater detail later), with "sustainable yield," and the emerging concept of the "ecological footprint." The definition of sustainable development, which will be revisited later in this chapter, brings together the most essential and well-recognized elements of a definition of sustainability itself: current requirements balanced against anticipated future requirements and opportunities. It also implicitly reflects a moral or ethical dimension of sustainability, which is to balance the right of the current generation for a decent quality of life against the need for ensuring the quality of life in the future: this is so that resources are saved to support and ecological damage does constrain choices of future generations. Because of this emphasis on the future, sustainability as a concept is not value-free or neutral. It rests on a foundation of responsibility.

It is often said that the concept of "sustainability" is an elusive one and that the term has become a "buzzword" devoid of real meaning; it is also said that sustainability is always defined in the interests of the organization using it. On the other hand, there is widespread understanding that the word refers to an activity that can be continued for a long time, if not indefinitely, and that is not destructive to the environment or to the quality of human life. This must be an activity that conforms or exceeds conventional regulatory standards, in effect substituting ethical imperatives for legal requirements. These ideas, taken together, are not a bad basis for a working definition.

For the purpose of this book, sustainability will be defined the terms in the following imperfect but workable manner:

> "Sustainability" is the state achieved when an activity action, operation, enterprise, or product that 1) can be maintained over a long period, with an aspiration to last at least several generations; 2) conforms to a set of expectations and goals such that it achieves at least one or more of the following objectives and does not interfere with the others: a) enhance social progress, b) conserve resources, c) preserve the natural environment, d) enhance the built or human-made environment, e) comply with and exceed legitimate regulations and rules, f) avoid destructive influences on society, and g) be controlled by a coherent sense of ethics and social responsibility; 3) is responsive to community values; and 4) does not require constant external oversight and organized opposition to remain aligned with these goals while it endures.
>
> "Sustainable" describes an activity, action, operation, enterprise, or product that can be maintained over a long period that furthers progress toward one or more of these goals, that does not interfere with progress in achieving the other goals, and that preserves opportunities and resources for the future. This definition implies that sustainability is possible but not guaranteed by any one action.

These are not perfect definitions. They are too long, depend too much on the subsidiary definitions (such as what constitutes a "long period"), and are value laden. However, these working definitions will suffice for now, as the language evolves.

"Sustainable Development"

Many definitions of "sustainability" are derived from the definition of "sustainable development," which was introduced to the world by *Our Common Future*, the 1987 report of the World Commission on Environment and Development (WCED): "Sustainable development is development that meets the needs of the present without compromising the ability of future generations to meet their own needs." This is the basic definition for discussions of sustainability, as demonstrated above, because it is concise, unambiguous, and universally accepted. However, it is a definition of sustainable development, not sustainability as such. Although the two concepts are closely related, they are not identical.

"Sustainable development" was a concept interjected into the environmental movement in 1987 by the report of a historic commission set up by the United Nations (the World Commission on Environment and Development, WCED), which was chaired by a physician with extensive training and personal experience in public health, Dr. Gro Harlem Brundtland (b. 1939), who had been environment minister and then prime minister of Norway. (See figure 1.1.) Following her service chairing the WCED, which came to be known as the "Brundtland Commission," she subsequently became the director-general of the World Health Organization and at the time of this writing (2014) serves the United Nations as special envoy on climate change. Her influence both as a leader and as a personal role model in bringing together health and environmental protection will be evident throughout this book.

FIGURE 1.1 Dr. Gro Harlem Brundtland, physician, public health practitioner, political leader, international civil servant, and diplomat. The United Nations commission that she chaired, which came to be known as the "Brundtland Commission," defined "sustainable development."

For human societies, sustainability requires progress that differs from the way material and social progress are commonly understood—developmental rather than quantitative growth. Sustainability is therefore integrally tied to the idea of social development and environmental management based on principles that can be traced back to evolution. Soon, however, the idea of sustainable development was conflated (intentionally or not) with the older and related (but distinct) idea of "sustainable yield." The German word for "sustainable yield" has long been used in forestry: *Nachhaltigkeit*, which means "capable of being sustained over time." It was coined in 1713 by Hans Carl von Carlowitz (b. 1645–d. 1714), a visionary civil servant and forestry expert who wrote a historic text on forestry in response to a serious wood crisis and introduced the principles of "sustainable yield" to the world. In the late twentieth century, sustainable yield came to be applied indiscriminately to profits, extraction of renewable resources, and long-term planning (with a typical time horizon in finance of five years and in business accounting of two years).

The mandate of the WCED was to create a space for discussion of the global response to rapid and destructive industrial growth on a global scale, with accompanying pollution, loss of habitat, and resource consumption that threatened to degrade the environment for all people and to deny people in developing countries and their children the opportunity for progress and the right to live in a clean and safe environment. The idea of sustainable development was that both economically developed (i.e., rich) and developing (i.e., poor) countries had a stake in economic development that could be sustained for the future and would not constrain future quality of life. Having had their opportunity to build wealth and having benefited from exploitation of resources and the economic growth that came at the cost of pollution, developed countries had an even greater responsibility to curb resource depletion and pollution in the near future. Developing countries needed to grow their economies without repeating the mistakes that developed countries had already made.

The WCED concept of sustainable development was received in public discourse as the idea that economic growth should not be destructive and that rich countries should help poor countries grow in ways that do not exhaust resources or pollute. This reflected an era when foreign aid was still a major driver of economic development in developing countries, and it was assumed that rapacious foreign companies operating inside countries with weak regulation was the main problem. Since then, the roles have changed. The role of rich countries now is largely to create demand for commodities, including scarce resources, which are extracted at excessive levels and cause incentives for over-production in ways that cause pollution and ecological damage. Large multinational companies, particularly businesses dependent on their reputation in Europe or North America, are as often as not in-country benchmarks for best practices in the developing world, and are less often the rapacious exploiters of the past. Local entrepreneurs are just as often the unrestrained overproducers, polluters, and exploiters nowadays. In the past there were relatively few successful entrepreneurs in developing countries, especially in Africa. Now there are many, and more opportunities for local development to be sustainable. Poor infrastructure, restricted access to capital, few startup business opportunities, and practical barriers to entry have tended to limit the options of local entrepreneurs, even if they were aware of the benefits of sustainable development.

Applying the notion of sustainable development to developed countries was a less obvious need in 1987, but it was key to generalizing the concept. A developed country was recognized as having a responsibility to reduce its own demand for nonrenewable resources, curb pollution, and support sustainable development in developing countries. But developing countries were then thought of as relatively passive and limited by circumstances. When it became clear that both developed and wealth-holding developing countries (with China, for example, qualifying as a wealth-holding developing country) were threatening the rest of the world by contributing to climate change and driving up the price of essential commodities, the idea of sustainable development took on new relevance for developed countries. The technology revolution pointed the way to practical sustainable development for developed countries and sustainability in practice was an object lesson in how society and technology could evolve continuously. When life-cycle analysis was taken into account, it was not even absurd to speak of "sustainable mining," because mining a nonrenewable resource might be considered sustainable if recycling were considered.

First Uses of "Sustainability"

"Sustainability" is a new word. It does not even register on Google Ngram until 1975, so it is a true neologism. Unlike many neologisms, it is also a useful one. However, both the concept and the word have deep roots.

Sustainable development reflects the newer meaning of preserving future use and opportunity, but it also reflects the older and related but distinct idea of "sustainable yield." The German word for sustainable yield, as previously noted, came out of forestry: *Nachhaltigkeit*, literally meaning "holding something over for later", conveyed more than the idea of continuity in access to the resource. It also meant sustainable yield and conserving resources for future use. In the late twentieth century, sustainable yield came to be applied indiscriminately in business to profits, dividends, and revenues rather than the resources that were at the heart of the original meaning.

The word "sustainable" and "sustainability" were heavily used in the 1972 report *Limits to Growth* but mostly in the sense of avoiding imminent collapse of the economy and resource base. Although based on models originated by a pioneer in systems thinking named Jay Forrester (b. 1918), the book was written by a young husband-wife team, Dennis Meadows (b. 1942) and Donnella Meadows (b. 1941–d. 2001), all working at the Massachusetts Institute of Technology. The rest of their careers were spent defending and elaborating on the work. The book could not be dismissed as antibusiness because its sponsor, the Club of Rome, consisted of prominent industrialists led by Aurelio Peccei (b. 1908–d. 1984) of Italy, and the work was funded by the Volkswagen Foundation. Although widely decried as alarmist, the book was remarkably insightful in retrospect. Its basic message was that exponential growth was unsustainable not only in population but also in resource consumption and externalized adverse consequences and that these constraints and externalities, if not recognized and corrected, would always lead to catastrophic failure or slow strangulation and decline. The authors did not specify the time frame or make predictions for any particular country.

Seven years later, James C. Coomer (b. 1939) published *Quest for a Sustainable Society*, in which he wrote that "the sustainable society is one that lives within the self-perpetuating limits of its environment. That society . . . is not a "no growth" society . . . [but rather] a society that recognizes the limits of growth . . . [and] looks for alternative ways of growing."

The subsequent popularization of the concept of "sustainability" itself followed the introduction of the term "sustainable development" and became a sort of shorthand for a cluster of ideas consistent with the Brundtland definition, sustainable yield, and the emerging concept of the "ecological footprint" introduced by William Rees (b. 1943), who is a prominent scholar of environmental sciences.

"Sustainability" in its current usage can be thought of as operationalizing sustainable development to mean actions, institutions, and resource utilization that can continue indefinitely within planetary, local, social, and economic limits and do not deny resources or compromise opportunity for the future (including use by future generations). It also makes the idea of sustainable development universal, since it applies to both developed and developing economies.

A Cluster of Values

Most definitions of sustainability seem to include at least three of the following ten distinct elements, although most do not contain them all:

I. unlimited (at least within the foreseeable future), long-term continuity
II. no or minimal harm to the environment
III. conservation of resources for the future

IV. either no harm to or enhancement of social structures
V. maintaining health and the quality of life
VI. optimization or maximization of economic, social, and environmental performance
VII. a low risk of catastrophic disruption
VIII. compliant with regulation and best practices
IX. stewardship and a responsibility that goes beyond ethics in dealing with other people to include other species, life and some notion of Nature
X. capable of continuing progress under its own momentum and therefore stable in the short term.

These principles are numbered in Roman numerals not because they are commandments but to remove the implication that they are necessarily a hierarchy or list of priorities.

Table 1.1 describes these values more formally and attaches a name to each. This table format will be used in subsequent chapters to demonstrate how the topics fit into the overall framework of sustainability.

"Sustainability" variously implies environmental protection, nonrenewable resource conservation and reuse, minimal impact on the environment, "long-term" viability in business, social relationships and structures that are durable, a population that does not exceed the resource base necessary to provide basic necessities (carrying capacity), a cultural context that will not be subject to upheaval and disruptive change, and a good deal more. What sustainability "ought to mean" depends on the speaker, but a modern understanding would involve taking no more of resources than one needs and not depleting resources for the next generation or producing more waste than the system can handle by conversion or recycling. By any standard, contemporary industrial society is far from sustainable.

One indication of a well-recognized global consensus on how sustainability might look and goals it might seek to achieve would be the Millennium Development Goals, a set of targets set by a conference of 189 member nations of the United Nations in 2000 to achieve by 2015. Box 1.1 lists these goals.

Although the individual elements of this meaning cluster are distinct, they are not unrelated. At least notionally, the following elements support one another and together create a long-lasting, stable ecological structure and social platform for human development and environmental protection. The logic can be understood along these lines:

> The essence of sustainability is optimization or maximization of economic, social, and environmental benefits [VI] and operational performance [VI] across generations [I]. This requires conservation of resources for the future

Table 1.1 Table of sustainability values: Value cluster in the definition of sustainability

	Sustainability Value	Sustainability Element
I.	Long-term continuity	Sustainability ensures long-term continuity.
II.	Do no/minimal harm	Sustainability embraces actions that do no harm or that minimize harm to the environment.
III.	Conservation of resources	Sustainability uses a minimum of resources necessary, reuses those it can, and conserves resources for the future.
IV.	Preserve social structures	Sustainability respects diversity in people, tradition, and values of sustainability and preserves social structures compatible with those values.
V.	Maintain health, quality of life	Sustainability respects the quality of life and maintains health.
VI.	Performance optimization	Sustainability protects finite resources and optimizes efficiency for optimal or maximal economic, social, and environmental performance.
VII.	Avoid catastrophic disruption	Sustainability seeks the lowest feasible risk of catastrophic disruption and the highest possible probability of continuity and so seeks the optimum level of resilience.
VIII.	Compliant with regulation	Sustainability includes compliance with reasonable regulations and best practices.
IX.	Stewardship	Sustainability embraces stewardship and a responsibility that goes beyond ethics in dealing with other people to include other species, life and some notion of Nature.
X.	Momentum	Sustainability maintains forward momentum, and is open to constructive change but resistant to destructive or meaningless change.

[III] and protecting the ecosystem on which we all depend [II]. On a practical level, it also means not degrading ecosystems and the local environment where people live and on which they depend for economic life to the point of irreversible stress or failure [VII], and protecting people both as individuals and communities [V], other species, ecosystems, and the earth

BOX 1.1

The Millennium Development Goals (United Nations, 2000)

1) Eradicate extreme poverty and hunger
   a) Halve, between 1990 and 2015, the proportion of people whose income is less than one dollar a day.
   b) Halve, between 1990 and 2015, the proportion of people who suffer from hunger.
2) Achieve universal primary education
   a) Ensure that, by 2015, children everywhere, boys and girls alike, will be able to complete a full course of primary schooling.
3) Promote gender equality and empower women
   a) Eliminate gender disparity in primary and secondary education, preferably by 2005, and in all levels of education no later than 2015.
4) Reduce child mortality
   a) Reduce by two-thirds, between 1990 and 2015, the under-five mortality rate.
5) Improve maternal health
   a) Reduce by three-fourths, between 1990 and 2015, the maternal mortality ratio.
6) Combat HIV/AIDS, malaria, and other diseases
   a) Halt by 2015, and begin to reverse, the spread of HIV/AIDS.
   b) Halt by 2015, and begin to reverse, the incidence of malaria and other major diseases.
7) Ensure environmental sustainability
   a) Integrate the principles of sustainable development into country policies and programs and reverse the loss of environmental resources.
   b) Halve, by 2015, the proportion of people without sustainable access to safe drinking water.
   c) Achieve, by 2020, a significant improvement in the lives of at least 100 million slum dwellers.
8) Develop a global partnership for development
   a) Develop further an open, rule-based, predictable, nondiscriminatory trading and financial system.
   b) Address the special needs of the least developed countries.
   c) Address the special needs of landlocked countries and small island developing states.

d) Deal comprehensively with debt problems of developing countries through national and international measures to make debt sustainable in the long term.
e) In cooperation with developing countries, develop and implement strategies for decent and productive work for youth.
f) In cooperation with pharmaceutical companies, provide access to affordable, essential drugs in developing countries.

In cooperation with the private sector, make available the benefits of new technologies, especially information and communications.

itself from harm. [IX] To ensure that this way of living provides security from harm and does not collapse, it must be capable of continuing under its own momentum in the long term [X] and therefore must be resilient to shocks and disruption in the short term [VIII]. It must also meet the needs of people [V] because if their legitimate needs are not satisfied, they will otherwise reject it, rebel against it, bypass it, or undermine it and take every opportunity to pursue their own interests. As a practical matter, this means being compliant with regulations for the protection of the environment [VIII] but that should be the *minimum* level of performance and it implies that those regulations are wise and sound.

These ten elements, in various combinations, comprise the core of the sustainability agenda in its contemporary sense.

Other Views of Sustainability

The original definition of sustainable development introduced the notion that development should meet the needs of the present without denying future generations or people or other places the opportunity to meet their needs in the future. This spare core definition still leaves unanswered questions such as: What are legitimate needs? What are dispensable luxuries? Into this ambiguity have entered many ways of thinking about sustainability, interpretations, and shades of meaning. Some of these are spin, some are self-serving, and others represent thoughtful points of view.

The meaning of sustainability has been expanded upon in different ways depending on the interests of who defines it. For example, both government agencies and the private sector tend to describe sustainability in terms of measurable outcomes and compliance. Environmental and social scientists and advocates tend to describe sustainability as a process of continuous improvement, one that is never finished and in which gains can always be made. Advertising and commercial uses of the word tend to treat sustainability as if were a characteristic or selling feature, and equate it to "green" or "eco-friendly."

The US Environmental Protection Agency (EPA), while it does not formally define sustainability, states on its website that it is based on "a simple principle. Everything we need for our survival and well-being depends, directly, and indirectly, on our natural environment. Sustainability creates and maintains the conditions under which humans and nature exist in productive harmony, that permit fulfilling the social, economic, and other requirements of present and future generations." Note the insinuation of the word "productive," which can mean biologically productive (i.e., a healthy natural ecosystem) or economically productive (i.e., effectively and efficiently meeting human needs). The overall definition provides scope for aligning projects and regulations with EPA's goal "to make sustainability the next level of environmental protection by drawing on advances in science and technology to protect human health and the environment, and promoting innovative green business practices."

Note that the EPA definition also moves beyond environmental protection, which is at the core of the agency's mandate but is only implied the wording of the goal. The EPA definition makes the transition from avoiding harm and restoring something that is broken to emphasizing a future of promise, sufficiency, and productivity. This shift is highly significant for an agency that was created in 1970 to stop rampant environmental degradation but not to manage resources. (In the United States, that is primarily the responsibility of the Department of the Interior.) The definition opens the door leading to a new level of environmental opportunity as well as restorative management.

The business and commercial approach to sustainability can be recognized by an emphasis on the enterprise: sustainability is treated as a characteristic or commitment of the company or its product. There are at least two threads to the business approach to sustainability.

The first approach is based on the notion that business objectives and performance cannot be separated from the future availability of resources to provide goods and services, the well-being of communities in which either producers or customers or both live and work and on which the business depends, and the long-term effect of the business on the environment, which supports all human activity. This approach is sometimes called the "triple bottom line," which is a term introduced by John Elkington (b. 1949) to refer to three dimensions of performance evaluation ("people, planet, profit"): benign social impact, conserving natural capital and preventing pollution, and economic value.

The other business thread is a commitment to sustainable revenue and continuity of business operations. The sustainability report of SAP, a global company based in Germany that specializes in business management software, presents its own definition: "Sustainability means that we strive to create a better run world where organizations balance short- and long-term profitability, and holistically manage economic, environmental and social risks and opportunities." Note that in this context the emphasis is on management and the sustainability of profit and business viability but done in such a way as to avoid harm and hopefully to be constructive.

Dimensions of Sustainability

Sustainability has many dimensions, each of which flesh out the definition, and is undertaken on several different levels of commitment. The dimensions apply to the different domains in which sustainability is applied and reflect how human society works, not the natural world, which is unified as an indivisible whole. At the same time, sustainability is undertaken on different levels, from the global to the individual. Individual behavioral levels of sustainability can be identified in the choices made every day and how they are influenced by peers, family, education, culture, advertising and propaganda, and role expectations.

Commonly recognized dimensions include the following:

- Environmental sustainability, which may be divided into ecological sustainability (is the natural ecosystem stable and capable of continuing uninterrupted and evolving?) and environmental quality (does the pollution and waste generated overwhelm the "environmental services" that dispose of it?)
- Economic sustainability, which has implications for fair markets, accounting for externalities, and pricing of resources
- Business sustainability, which refers to whether a particular business or sector can continue operations and remain profitable
- Financial sustainability, which is about whether a good practice, enterprise, or property can be supported financially (in ongoing cost as well as initial capital investment) and whether an enterprise can be sustained as a business over time
- Social sustainability, which is about whether the community or culture will support the effort over a long period, ideally measured in generations. Social sustainability has elements of ethics, equity, and justice as well as acceptance and motivation
- Religious, moral or spiritual sustainability, which is about a more profound moral case for treating the earth and its creatures responsibly through stewardship, arising from a mandate from God, moral responsibilities, a one-ness with the natural world, or an ethical obligation that transcends relationships among people and creates duties and rights with species and ecological communities ("deep ecology")

Social sustainability is usually taken to include political sustainability, but as in economic and business sustainability, the distinctions reflect the long and short term, respectively, and so it is useful to separate them. Certainly political sustainability rests on values within the broader culture and society, but it is distinct because it normally operates over a much shorter term than cultural sustainability. If political and justice values are not broadly and consensually shared in a society, political support will most likely be interrupted sooner or later, and "sustainable" measures will then be reversed or abandoned.

The interface between each dimension of sustainability is as important as the role of each dimension as an individual pillar of sustainability, because dimensions only reflect how human society is organized; the world itself is actually indivisible. By one popular formulation, the relationship between environmental and economic sustainability is said to represent "viability," while that between environmental and social sustainability represents "bearability" (in the sense of tolerance of a burden), and that between social and economic sustainability represents "equitability." This formulation raises many concerns. The most fundamental relationship has to be "viability" because in theory it represents physical limits that cannot be negotiated. Much of the argument of sustainability hinges on the role of technology in pushing these limits. If the relationship between environmental and social sustainability is so onerous as to be a burden that can only be "bearable," then sustainability is indeed in deep trouble because it will then depend on accurate perception and tolerance, both of which are in short supply in human societies. This issue is explored more fully in chapter 11.

"Laws" and Principles of Sustainability

Sustainability is about mitigating or limiting the downside risk of disruption and future scarcity and protecting the benefits that one generation enjoys while allowing future generations to reuse or harvest at least as much resources as current generations. In other words, sustainability is about carrying on indefinitely without future generation paying the price. This is not the same as designing for most efficient use, greater utility, or maximizing profit, and the principles are not the same as they might be for, say, engineering or business.

Contemporary views of sustainability draw heavily on insights from the twentieth-century environmental movement, which in turn relied heavily on a popularized version of the science of ecology. Ecology, which is the systematic study of biological communities rather than individuals or species, is not a simple science with a few basic rules. Rather, it relies heavily on data and the analysis of particular ecosystems rather than deduction from broad principles. For example, one eminent biologist (Pierre Dansereau, b. 1911–d. 2011, Québec) identified no less than twenty-seven "basic" principles that apply to speciation, ecology, and community biology; he could not further reduce them to simpler laws.

Whether ecology does have laws in the literal sense is questionable. Several authors have pointed out that the closest thing to an undisputed "law" in ecology is the exponential rise in unrestricted population growth. It is grounded in the biology of replication and the knowledge that growth by replication is described by an exponential curve. This "law" is actually a *description* of a tendency, or the behavior of an idealized or abstract system, not a fundamental relationship derived from cause and effect in the way that the laws of physics apply to planetary motion because they describe the action of gravity on objects. Analogies to physical laws are therefore metaphorical. Ecology (and therefore sustainability) is too complicated to be easily described by a handful of selected principles and too local to generalize. This is in part because both the science and the field it gave rise to address the performance of a total system and resist reductionist decomposition.

Many environmental scientists have tried to simplify the ideas of ecology to a more manageable and more easily understood set of principles, usually calling them "laws" but really meaning principles for the practice of sustainability. The most successful, useful, and popular set of these principles was formulated by Barry Commoner (b. 1917–d. 2012), whose influence on the environmental movement was profound. (See figure 1.2.) Commoner reduced the essential lessons of ecology to four basic "laws," which are presented in table 1.2, with comments. If they seem familiar, it is because they have been repeated many times and have been absorbed into the culture. It should be recognized, however, that these simple principles are neither comprehensive nor applicable to ecology as a scientific discipline. They are statements of principles important mostly to guide efforts at sustainability.

Sustainability as a forward-looking design movement has also been profoundly affected by the legacy of a visionary American thinker, Buckminster Fuller (b. 1895–d. 1983). Fuller was educated at Harvard in mathematics but was largely self-taught in engineering and architecture. His work laid the foundation for adapting engineering, design, and technology to problems of sustainability. Fuller saw that conventional engineering, management practice, and design were about exploitation of resources and extracting the most "benefit" over the life cycle of a resource or product. This is distinctly different from knowing what is best for future generations.

Sustainability does not necessarily depend on technological development and affluence, but it is certainly facilitated by innovation and the availability of resources to make the investment in transition to a more efficient, less resource-consuming regime. Fuller emphasized design that required a minimum of

FIGURE 1.2 Professor, scientist, ecological theorist, environmental advocate, and peace activist Barry Commoner. Most of his career was spent at Washington University in St. Louis, Missouri. Commoner laid the broad foundations for what became the "ecology" arm of the 1960s environmental movement. He and his colleagues provided a model for responsible activism and social change through science (see chapter 2), initially through his work documenting the extent of radionuclide fallout from atmospheric nuclear weapons testing and later by documenting the complex interrelationships of technology, pollution, population, and poverty—which in turn laid the groundwork for environmental sustainability studies.

materials and a maximum in resiliency and efficiency, so that his insight that "less is more" (taken from a line in a Robert Browning poem) redefined the implicit link between function, productivity, and accessibility. In a sense, Fuller advocated the substitution of intelligence in design for matter. His signature example was the geodesic dome, although he did not invent it. The result was better performance than conventional technology, and furthermore it was more aesthetically pleasing. This central but simple idea has been elaborated on many times and in many ways. The idea of "less is more" became a guiding principle of contemporary architecture, design, and technology and is now fundamental to the application of sustainability. One can see his influence later in contemporary technology, such as in the miniaturization of electronics fused with the design

Table 1.2 The four laws of ecology, as formulated by Barry Commoner

	Commoner's Law	Restatement	Origin
1.	Everything is connected to everything else.	This is a restatement of John Muir's quote: "When we try to pick out anything by itself, we find it hitched to everything else in the universe." The "web of life" concept, that species, including humans, whether as individuals or in communities, are interconnected.	Could be construed as Newton's Third Law of Motion "every action has an equal and opposite reaction," but it is a stretch.
2.	Everything must go somewhere.	Matter cannot be destroyed, so there has to be waste, pollution, or a product for every use and transformation.	Matter and energy can neither be created or destroyed: conservation of energy and matter. (First Law of Thermodynamics)
3.	Nature knows best.	A natural system has inherent stability and can repair itself if the damage is not too great.	Teleology: assumes that nature has a plan. Actually refers to biological communities that are stable already and have sufficient resilience to restore themselves to equilibrium. Concept of "homeostasis" (Claude Bernard, 1869, named by Walter Bradford Cannon, 1926): living beings tend to regulate their internal environment; applied to ecosystem.
4.	There is no such thing as a free lunch.	To do or make something, there is always some cost and expenditure of energy required.	The entropy (disorder) of an isolated system tends always to increase. (Second Law of Thermodynamics)

Source: Barry Commoner, *The Closing Circle*. New York: Random House, 1971.

aesthetic of Steve Jobs (b. 1955–d. 2011) to create high-performing but streamlined and resource-conserving products.

The result of this legacy, and other trends, is that the movement that eventually resulted in the modern idea of sustainability took a progressive, pro-technology direction. Under other circumstances, the idea could easily have been antitechnological, even Luddite, more along the lines of a "back to the land" movement. However, that did not happen. Instead, sustainability became compatible and friendly to technological change, which was a very positive development indeed.

What is Health?

Health is integral to the cluster of values that constitute sustainability. Health relates to sustainability both directly and indirectly, both obviously and subtly. Logically, if health is to be applied to societies, ecosystems, and planets, then a "healthy" environment would be a sustainable one. Sustainability emphasizes the "health of the environment." A "healthy" person or a "healthy" society on a "sick" planet is inconceivable, because all living things are part of an indivisible whole.

However, as familiar as the idea is, health is a surprisingly elusive concept and defined for the most part by what it is not: illness. There are only two words in English for the state of not being ill: "health" (and its archaic equivalent "hale", both of which come from the same root word as "whole" and have connotations of being complete) and "well" (which just means "satisfactory," or "in good condition"). The meaning of "health" is therefore much deeper than that of "well" and has contained within it the idea of being complete as a person and also fitting into a broader society. ("Sane," which derives from the Latin *sana*, meaning healthy, is used exclusively in English for mental health.) It is perhaps reflective of history that although there are only the two words in English for the state of being well (three words, if one counts "sane"), there are many synonyms for "sick" (including "ill," "infirm," "ailing," "diseased," "laid-up," and so forth), as well as an entire vocabulary dedicated to manifestations of illness (including "feverish," "nauseous," "feeble," "impaired," "suffering," "indisposed," and so forth). This gives one a fairly clear idea of the human condition, as least as it related to health, as it was when the English language was coming together.

For many practicing health professionals, "health" remains, conceptually, a baseline state of function that is "normal" for that individual and to which a person returns between bouts of illness. In the past when the prevalence of disease and disability were high and most people were glad to be alive, that idea of health may have seemed adequate. Today, when most people in developed countries are not obviously diseased or dysfunctional and can expect to live long lives, this definition of health seems inadequate and a state defined as a double negative (not being obviously sick or unhealthy) rather than a useful concept of wellness.

For most practicing physicians, "health" remains just a baseline state of function. However, for the patient that baseline state may be anything but healthy in actual fact and future risk and certainly human potential.

Most members of the general public, asked on the spot and not given time to think the question through, would probably define "health" as a state in which the body is functioning well without disease or disability. This definition would address the current state of health of an individual and would be benchmarked against what appeared to be "normal" in appearance and function. Good health means minimal or no signs of illness, includes only non-serious, short-term diseases (such as influenza or the common cold), and suggests only minimal or short-term incapacity. Good health is more or less undifferentiated; bad health is on a continuum from inconvenient but not serious to desperately ill or moribund. Disease plays a much more important role in defining health than injury, it would seem, unless the injury is severe and disabling. In other words, when asked about health most people think of the absence of disease in an individual. A young football player with a knee injury would be considered healthy; however, an old man whose only problem was an easily treatable form of cancer (a "dread" disease, in risk perception parlance, meaning that people tend to become anxious just thinking about it) would not be. For most of us, "healthy people" are perceived to be characterized by the following:

- Free of a visible, serious or "dread" disease, such as cancer, but not necessarily of minor ailments, say, influenza, a common cold, or a sprained ankle
- Able to meet expectations for playing their normal social role
- Subject to only short-term impairment and inconvenience as when they have an injury
- Mentally balanced, such that they do not experience constant anxiety or cause it in others
- Capable of appropriate responses to life's small traumas, whether they are minor infections, wound healing, or psychological stresses, without overreacting
- Capable of adapting to circumstances in ways that allow them to function in society and to carry on as a person, meeting their basic needs and interacting with others without discomfort

The hypothetical definition of health, however, is an abstract idea. Many people have a much different idea of health when it comes to their own body or that of their family members. They will describe themselves as healthy even when they have a disability or a chronic problem, as long as the condition is not serious and will not likely affect their life expectancy. Reviewing health questionnaires, one often reads wording like "I am in excellent health except for . . ." Clearly injuries resulting in chronic symptoms are thought of differently than chronic diseases, unless impairment from an injury is severe enough to be incapacitating.

On the other hand, most people would agree that if a person is at risk for a serious disease or injury, has adverse health habits, and an unhealthy lifestyle, then they are not healthy. If they are aware that they are taking health risks, have the knowledge and ability to change their behavior, and are doing nothing about it, they are even less healthy. The idea of "health" therefore includes the reduction, not acceptance, of controllable risks to health and the risk of disability, which are not part of the conventional understanding of health. Dysfunctional health behavior is not only "unhealthy" because a person's risk is increased and therefore their future health status and opportunities are limited, but also because they are engaged in behavior against their own interest, possibly either expressing addictive behavior or operating in denial, which in itself suggests a problem dealing with reality.

Definitions of health must also operate on two levels: individual and population. The literal meaning of health implies wholeness and completeness, both in the integrated individual and in the individual's place in the community. In the community, each individual may be sick, well, or functioning at various levels; however, it is the collective, average, or the distribution of health-status levels that can be monitored, that has social manifestations, and that is of concern to governments and society as a whole.

Anthropologists and medical sociologists tend to perceive health as a social role. A person is healthy when they are perceived as healthy in the community and act accordingly, and they are ill when they assume the "sick role," which is a stereotyped pattern of behavior that features dependency, loss of autonomy, passivity with regard to making health-related decisions, and reduced expectations for contributing to family and the society. Much of the literature of medical anthropology and health behavior is devoted to defining the sick role and identifying those elements of it that are cultural or adopted for secondary gain (i.e., because acting that way brings some benefit, such as relief of anxiety or freedom from responsibility).

Health thus is morphing into a multidimensional idea that takes into account present status, early or subclinical disorders (such as developing diabetes), risk factors for disease, support networks and social factors that provide resilience against disease and disability, and cultural and family perceptions of what it means to be healthy. Many thoughtful people would answer that health also has a spiritual dimension and a mental dimension apart from mental illness.

Obviously, the same definition of health that applies to human beings applies to other species, absent the self-awareness of health-related behaviors. Veterinarians and animal and plant scientists use the word in the same way and in the same sense as physicians and scientists studying human health but do not have the advantage of being able to ask their subjects about symptoms or for an opinion. Community and conservation biologists (ecologists and others studying the relationships of species in a natural or disturbed environment) use the word "healthy" to describe a productive and self-maintaining ecosystem, in much the same way that epidemiologists speak of healthy human populations.

The WHO Definition of Health

The most important formal definition of "health," and the point at which all discussions of a definition must begin, comes from the World Health Organization, in the preamble to its constitution (1946): "A state of complete physical, mental, and social well-being, and not just the absence of disease or infirmity." This definition is quite inclusive and implies a broad mandate for society, and not just for health professionals, to achieve health at this level and implies a "whole" person, intact and high-functioning, living in circumstances that support the person's social role and mental health. This definition is mostly ignored in medicine; however, it is a topic of endless discussion in public health, which is where the ambiguity chiefly lies, mainly because of the social dimension.

The inclusion of social welfare in the WHO definition has attracted much criticism over the years for being idealistic, excessively broad, and unworkable. Nevertheless, it expresses a legitimate aspirational concept of health. The roots of the WHO definition are deep and well supported by research, however. It is well established that social connections and interpersonal interactions play a role in physical as well as mental health and that personal satisfaction and happiness correlates with good health, both as a prior association (causal or otherwise) to health status and as a result. One criticism of the WHO definition has been that it converts all social and many political problems into health problems, which both strains credulity and places expectations for social welfare on the health sector which cannot possibly deal with such a broad burden.

The WHO definition, broad as it is, is still incomplete, however. Health clearly has physical, mental, and spiritual dimensions, yet it is not clear what "well-being" is intended to mean. Does it mean social relationships that are supportive?

Does it mean satisfaction or happiness? Does it mean that a person will not face a high risk of disease or injury in the future? The idea of "well-being" is more relative than the notion of health and implies adaptation to circumstances. If a person is well adjusted to the community's circumstances and has personal resilience, then that person can be "healthy" despite a sick society and dysfunctional family. If a person has come to terms with a bad situation and is at least accepting and forgiving (although hardly satisfied or happy about it) then that person is probably healthier than the people around them who are creating the situation.

In 1984, the WHO introduced a refinement of the definition of health by asserting that health also reflects "the extent to which an individual or group is able . . . to realize aspirations and satisfy needs and . . . to change or cope with the environment. Health is therefore seen as a resource for everyday life—not the objective of living." This addition was codified in the same words in a statement in the 1986 Ottawa Charter for Health Promotion, the foundational document for the field of health promotion, which describes health as: "A resource for everyday life, not the objective of living. Health is a positive concept emphasizing social and personal resources, as well as physical capacities." The implication of this elaboration is that health itself could have varying degrees in a positive direction, just as illness and disability can have varying degrees in a negative direction. One could enhance one's degree of health in a positive direction. The health promotion movement strongly reinforces this third definition of health by emphasizing positive states of well-being and improving the health status of "healthy" people. The popular movements for self-help psychology, "recovery," and the effects of growing up in dysfunctional families and personal growth represent, among other things, expressions of a drive for ever-increasing levels of functioning within "good health." So are fitness and participation in sports and weight control, both of which are also elements of health promotion programs.

"Health" in all of these modern definitions is not framed as a *characteristic* of the person but as a *description* of how the person functions within society to play their social role and meet their personal objectives. If there is an obstacle to doing so, it could be because the person lacks health (in other words, is injured or ill) or lacks the capacity to act (which would be disability), or because the environment they live in presents impediments, is not supportive, or requires abilities that do not match their capabilities (which would be a handicap).

A nuanced definition of health that accommodates these ideas is also emerging from modern biology. It is now known that most people have risk factors for serious disease and that every person has on average 200 potentially serious mutations in their genome (five of which would be lethal if the single-allele gene were to be expressed alone), of which 150 are relatively recent to humanity, having arisen in the last 10,000 years. In other words, none of us are entirely whole or free from mortal error.

Dimensions and Levels of Health

The concept of health, like sustainability, has many dimensions and levels. Some of the dimensions relevant to the questions posed above and that recur in this book are commonly taken in many sources to include physical, mental, emotional, social (how the individual fits into society), spiritual, and environmental (reflecting environmental hazards). Popular culture, when it is not preoccupied with health and beauty, tends to put much more emphasis on individual growth and recovery than the health-care system does.

As with sustainability, these dimensions of health also operate on different levels relevant to the individual, family, community, and population as a whole. These levels include

- personal health, the state of the individual, which is normally the domain of medicine
- function and personal needs, which is normally the domain of nursing for sick people
- public health (defined later), the management of health risks on a community level
- risk, the probability of an adverse health outcome in the future, and how it can be protected by preventive medicine or by public health
- population health (defined later), with an emphasis on policies to keep people healthy
- health risks and behaviors, the determinants of future health that may not have caused ill-health at the a given point but may do so over time; and those determinants often firmly rooted in culture
- capacity and social role, including what the person can do in their social and material environment given their physical health and any impairment they may have; for example, the exclusion of the disabled from the workplace and participation in society can be viewed as a socially constructed form of ill-health
- culture and social role of sickness and health, including how people respond as individuals and on a group basis within their families and society to varying degrees of physical ill-health, impairment, pain, and labels of health status; a common example is perception of disability and the ability to work: a degree of pain that does not reflect serious injury or warn of a complication something to be avoided, with time off work, or something to learn to live

with and to work through. Some families and some cultures place emphasis on the pain, and some place emphasis on not letting it get in the way

- healing and recovery, including self-help psychology, and personal growth, among other things, which express an aspiration for ever-increasing levels of functioning within "good health"

Health for People Alone and Together

One can deduce from the previous discussion that health is not a "thing" or a single characteristic or even a single outcome, but a quality attached to life, a feature of human (and animal) life with many dimensions, all of them relevant to sustainability.

Health can be an attribute of an individual, in which case its domain is medicine, to take care of problems and preventive medicine to protect well-being and reduce future risk. Health can also be an attribute of populations and communities, in which case its domain is public health and its more policy-oriented sibling population health: these are two closely related fields, both of which deal with the health of people on a population level. Communities can be considered to be people living together in a structured order. Obviously sustainability issues can preserve, threaten, or enhance that order.

The Medical Model

The most basic level of health is the health of the individual, which is the basis for what has been called "the medical model." For most of history, the health-care system was seen principally through the eyes of physicians. This point of view later expanded as other health-care practitioners (particularly nurses) became empowered within the system that provides medical care and treatment, but not necessarily from the point of view of preserving or protecting health. For this and other reasons, health has been seen mainly through the lens of health *care*. The problem of health has traditionally been medicalized (seen as a medical problem, to the exclusion of other points of view), and the emphasis has been on provision of services and restoring people to health. In the context of sustainability, however, the emphasis must logically be on preserving, protecting, and (when possible) enhancing health for everyone.

The health of individuals, one person at a time, has traditionally been the domain of medicine—providing individualized diagnosis, treatment, and individualized prevention—and of nursing, which provides care in the form of evaluating individual personal needs, alleviating suffering and discomfort, and administering treatment. (These are broad generalizations.) The "medical model" in figure 1.3 provides a summary of how health care on the individual level is supposed to work. Still, it barely hints at just how complicated individual-level health care can be in practice.

Most of the vocabulary and common understanding of health, what it is and how it is restored or protected, comes from experience with the health-care

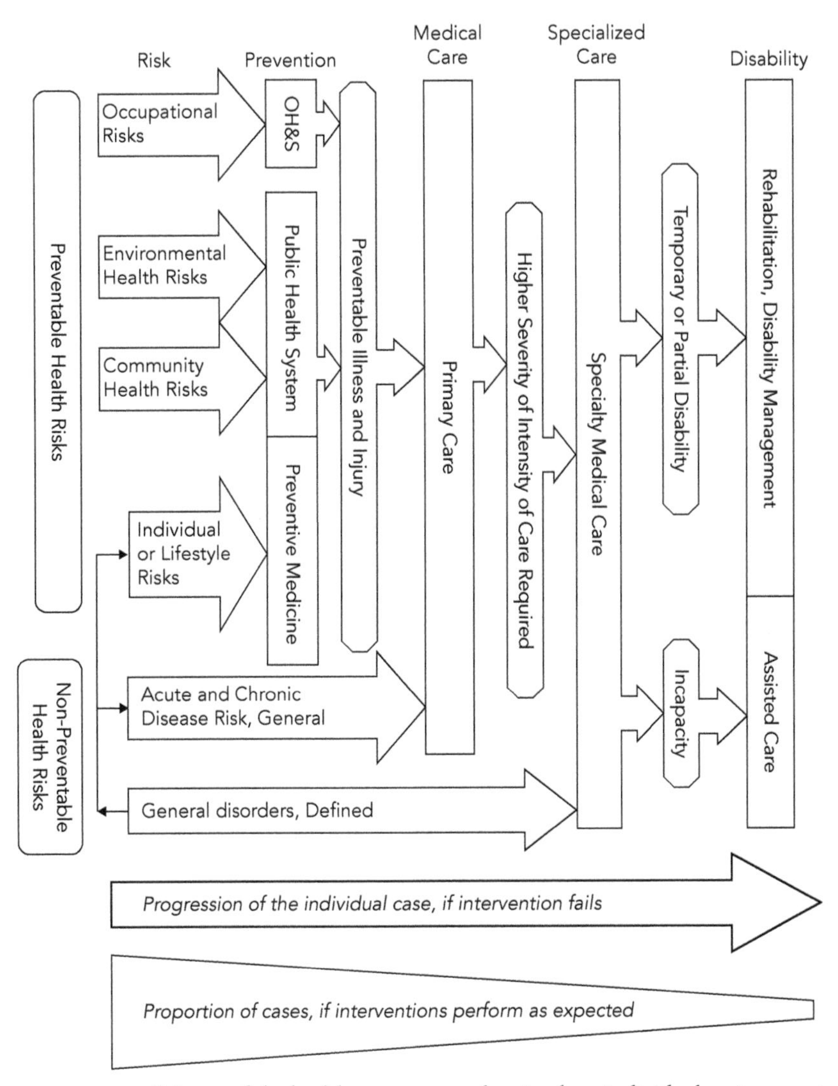

FIGURE 1.3 Schema of the health-care system, showing how individual patients typically move through the system and the absence of close interaction between public health and medical care.

system. Since the mid-twentieth century, thinking about health has expanded considerably to explore other points of view: psychological (not psychiatric), social, cultural, and spiritual aspects of health, illness, and disability. Scholars of health have come to view health as a system rather than an interaction between a patient and a health-care provider. Sustainability has not figured so much into this new way of thinking yet, although that will no doubt change.

Physicians and qualified clinical practitioners (the term of art is "licensed health professionals") think of health a little differently than do people without medical training. For most people, a person is either sick or well. For a health professional, health is a baseline state but a person's state of ill-health lies a position on a spectrum. Perfect good health is the exception rather than the rule as most people have some health problems, even if they are minor ones. Because physicians and health professionals see mostly sick people, they naturally tend to perceive gradations of ill health. There are numerous research instruments, scoring systems, rating systems, and questionnaires that have been formulated over the years to measure health-related quality of life, impairment, social functioning, anxiety, and pain. There are no such categories of wellness, in common discussion.

If a person is in excellent physical condition, with no disease and no impairment, but has a health-related habit such as smoking that is likely to impair health in the future, can he or she truly be said to be healthy? What about a behavior, such as poor sleep habits, that is not a cause of an obvious disease but that reduces a person's potential productivity, satisfaction, and stamina, while still falling within the range of what would be considered normal?

The health status of individuals is an attribute, critical to that individual, important to that person's family, and often the direct result of medical care. The health status of groups of people (i.e., populations) is not exactly the same as the sum total or average of the health status of all the individuals, however. The health of populations has to do with the distribution of health or sickness or disability, is little affected by the status of an individual, and is not affected much by medical care because few people in a population are so sick at any one time that medical care can make a substantial difference.

Health of Populations

The ancient Greeks recognized the distinction between medicine and public health by naming their god of medicine and healing Asclepius and naming their goddess of health and cleanliness Hygeia (one of the daughters of Asclepius). The patterns (distribution) of health in a community (who gets sick, where they are, what might have caused it in the group experience) and how the community can best be protected, is the domain of "public health," which emphasizes prevention and is a field closely related to but fundamentally distinct from medicine.

Figuratively speaking, public health looks at the forest, while medicine looks at the trees. As important as medicine is, "public health" is the health science that is most involved with sustainability.

Health status (sick, well, disabled) is never distributed evenly, and so the average health of a population (for example, the average weight as a measure of obesity) means nothing. Other measures are required that take into account the distribution of health conditions (for example, the proportion of the population that qualifies as underweight, overweight, obese, and morbidly obese). "Population health" is a field closely related to public health that emphasizes the determinants of health in a population, social and physical, with a particular concern for public policy (for example, the cost of obesity to society and how obesity rates might be reduced through public policy).

Understanding health as a characteristic of populations and communities, and not of individuals alone, has advanced greatly in the last fifty years or so, and there is a somewhat bewildering array of academic disciplines, practice specializations, schools of thought, and philosophies to contend with. Box 1.2 provides definitions and descriptions of some of these schools of thought and interest areas.

Public health practice is about clean water, safe food, tracking down outbreaks of disease to determine the cause, intervening to stop an epidemic (for

BOX 1.2

Schools of Thought in Health

Health seems like a straightforward subject, yet it can be very confusing to readers unfamiliar with health-related jargon. In an impressive outpouring of writings and theorizing on the subject over the last fifty years, the health of populations has spawned many schools of thought and different ways of approaching the issues.

The glossary below gives the names of some of these schools of thought and defines some of the more commonly used terms in the literature. Readers should be aware that there is considerable overlap in definition and that it is not only perfectly possible for a particular writer or idea to fall under more than one category but that this is usually the case.

Allopathic medicine. Another name for conventional medicine, rarely used except when contrasting mainstream medicine with "alternative" medicine. Throughout most of the history of medicine there was no proven scientific theory of what caused disease and no standardized approach to evaluation of treatments. In the nineteenth century, rapid advances in science, especially the germ theory of disease, provided the scientific basis; and in the twentieth century advances in epidemiology and biomedical science allowed the rigorous evaluation of treatments. This is far from complete. However, the emphasis on scientific objectivity has led many to criticize modern allopathic medicine as impersonal and placing inadequate emphasis on care and compassion.

Alternative medicine. Approaches to healing other than conventional "allopathic" medicine, especially those that are scientifically unproven or questionable. Examples include homeopathy, naturopathy, faith healing, acupuncture, reflexology, and ayurvedic medicine. The name implies that patients have an "alternative" to allopathic medicine. Osteopathic medicine, which began as a system of joint manipulation (as did chiropractic), has essentially merged with allopathic medicine in the United States, although not in Europe. Some alternative medical systems, such as acupuncture and chiropractic, have become accepted as mainstream for symptom treatment for purposes of insurance reimbursement. Herbal medicine was headed in that direction until research on the efficacy of herbal preparations (which are not standardized by dose) showed little effect for almost all treatment uses and the dangers of some ethnic remedies called attention to their risks. The principal risk of alternative treatment, however, is that patients with serious disease will opt for ineffective alternative treatment instead of effective allopathic treatment and achieve poor results or even, as in the case of Steve Jobs, die unnecessarily. Many allopathic physicians believe that alternative treatments should be evaluated by the same rigorous methods as treatments in conventional medicine and a special branch of the National Institutes of Health was set up to do so. Widespread belief in alternative medicine and public tolerance has made this position politically unpopular. In addition, alternative medicine practitioners often do better at expressing compassion and concern. Most favorable outcomes from using alternative medicine are due to the placebo effect and therefore are very dependent on the setting in which they are given or taken and the patient's attitude.

Community medicine. An approach to delivering health care that emphasizes primary care, the systematic assessment of community needs, and planning local community health services to be responsive to need and for prevention. Now out of favor, this term originally referred to a school of thought that originated with Kurt Deuschle (b. 1923–d. 2003) which had a strong emphasis on "community-oriented primary care" (see below) but later became generalized to medicine with a community orientation, with loss of meaning.

Community-oriented primary care (COPC). A strategy for designing primary care in communities to be responsive to the pattern of diagnoses and medical problems seen in medical practices. For example, health-care providers might organize their practices to deal efficiently with diabetes and obesity in a community with a high prevalence of these closely-related problems problems, or streamline services for geriatrics in a retirement community. COPC is an outgrowth of the community medicine philosophy and can be adapted to an individual office.

Complementary medicine. When alternative medicine is used together with allopathic medicine, it is called "complementary medicine." Most physicians will (often reluctantly) accept but not encourage their patient's desire for complementary treatments, as long as the treatments do not cause harm and have some possibility of benefit: acupuncture is the most common. They are less accepting of herbal remedies because of the possibility of drug interactions when combined with prescribed medication. For diseases that have poor prognoses and that are associated with severe pain, such as advanced cancer, physicians rarely make an objection because they wish to support hope and lift morale. Some complementary medicines are useful for symptom management, even though they have limited effectiveness for cure or healing.

Epidemiology. The science that deals with the distribution and determinants (root causes, underlying conditions, and triggers) of diseases and health-related problems in a population. Epidemiology is heavily reliant on statistics and is considered to be the "basic science" underlying all areas of health related to people in groups, from the assessment of how well a treatment works to identifying, through field studies, the cause of a disease so that it can be prevented. Epidemiology is often described as "the science of epidemics" but that is too simple. (An "epidemic" refers to any disease that occurs at a higher than expected rate.) Virtually every public health professional has had training in epidemiology, regardless of discipline; it therefore provides the field's common vocabulary.

Health promotion. A relatively new approach (since the 1970s) to preventive medicine and public health that emphasizes motivating change in both health-related behavior and public policy to support a healthy lifestyle and healthy choices in life. Health promotion can be practiced on an individual or population level. (Health promotion replaced the older field of health education.)

Health services management. Management applied to the administration of systems for the provision of health care (evaluation, treatment, and rehabilitation), sometimes including provision of public health services. Historically, heavy emphasis was placed on hospital administration, the management of facilities, and supervision of health-care providers, particularly nurses in hospitals, but the field has broadened considerably.

Health services research. A cluster of disciplines focused on understanding the behavior of the health-care system: health economics, health information systems, health planning, ethics, global health, health law, patient behavior, and health policy.

Population health. A particular school of thought in public health that emphasizes the determinants of disease and health risk at the population level in communities and countries and the public policies that improve health on a group level. It is closely associated with the thinking of Kerr L. White (b. 1917–d. 2014) and later with the Canadian Institute for Advanced Research, which elaborated the model used later in this chapter, but the term has become widespread, and is sometimes used just as a synonym for public health.

Prevention science. A collection of scientific disciplines or approaches that contribute to the prevention of disease, either in preventive medicine (where the term is more commonly used) or in public health. The chief prevention science is epidemiology; the term also includes health behavior (mostly drawn from cognitive and behavioral psychology), immunology (especially of vaccines), injury prevention, occupational health, ethics, and many other related fields.

Preventive medicine. That part of practice in medicine, and also the name of a recognized medical specialty, that emphasizes the prevention of illness and disability. In practice, preventive medicine concerns itself with preventing disease or injury (primary prevention), screening for early signs of disease so that it can be treated before it progresses (secondary prevention), and measures to avoid disability when illness or injury occur (tertiary prevention). There are three recognized preventive medicine specialties in the United States: general preventive medicine (which absorbed an earlier specialization called public health medicine), occupational medicine (the largest, focused on work-related disorders and maintaining health), and aerospace medicine. Also importantly, it is the name of one of two fundamental approaches in prevention science identified by Geoffrey Rose (b. 1926–d. 1993): one that involves screening people to identify those at high risk, and then treating them in some way to prevent the disease. (An example would be to perform a routine lab test on a patient who is seen, perhaps for another reason, to determine whether he or she had an elevated level of "bad" cholesterol—and if so to prescribe statin drugs in order to prevent future heart disease and stroke.) The name comes from the observation that this type of prevention is usually done in the context of medical care, not through public health programs.

Primary care. The most basic level of health care, involving provision of basic health care at first contact, definitive care for common disorders, delivery of preventive services on a routine basis, and management of disorders and problems that do not require specialized care. There are many definitions of primary care, many of them shaded with meaning by the various organizations that sponsor them. Most definitions agree that primary care should be coordinated and that medical and nursing care should be integrated with care given by nonphysicians, such as nutritionists and physical therapists. Some definitions, such as those put forwarded by organizations of family physicians, emphasize that primary care is a specialty in its own right, but others emphasize its role as an entry point into the medical care system and gatekeeper for specialized care. Examples of primary care include "general practice" (a term no longer used in the United States but a recognized designation in the United Kingdom) and usual (not specialized) practice in the specialties of family medicine, general internal medicine (internal medicine being a medical specialty for adults, with a variety of subspecialties), geriatrics, and pediatrics. Other examples of primary care include urgent care (walk-in services) and (in the United Kingdom) general practice in the National Health Service. The World Health Organization considers primary care to be essential and a human right.

Public health. An organized approach to protecting the health of individuals and populations emphasizing collective action at the community or population level. This may involve public policy, provision of services (such as mass immunization), investigating outbreaks of disease, control of hazards (such as clean water), educational programs, surveillance and monitoring (of disease rates and health-related trends), scientific research, and many other modalities. Public health has an ancient and venerable history, like but distinct from medicine, but in its current form, it dates from the early and mid- nineteenth century. Also importantly, it is the name of one of two fundamental approaches in prevention science identified by Geoffrey Rose (b. 1926–d. 1993) that involves taking steps to protect everyone, whether they are individually at high risk or not, through interventions such as safe food, immunization against influenza, standards for pollutants and for hazards in the workplace. "Public health" is sometimes casually used by people outside the field, especially journalists, to mean the public funding of health care (as by universal health insurance) but this is *not* proper usage.

Public health medicine. Many public health measures require physicians and nurses to provide or supervise a treatment, such as immunizations, prophylactic drugs (drugs, such as antibiotics or antimalarials, given to prevent an infection rather than to treat it), and well-baby checks. The term is more often used in the United Kingdom than in the United States, and there is considerable overlap with preventive medicine.

Public health sciences. Similar to prevention science, this term refers to the individual disciplines that support public health, of which the most important are usually considered to be epidemiology, biostatistics, environmental and environmental health, microbiology, toxicology, health behavior, health communication, health services management, public health law, and maternal and child health. Schools of public health tend to have additional areas of emphasis or specialization, depending on their history, their location, and the interests of faculty.

example, meningitis), immunizing children against childhood diseases, educating people about nutrition, getting people to stop smoking, encouraging fitness, helping obese people lose weight through participation in organized programs, assessing differences in level of health in different communities, making sure that adults traveling to high-risk places take steps to prevent disease, ensuring that tattoo and piercing parlors use sterile needles (so they do not spread diseases such as hepatitis), and much more. Fundamentally, public health practice addresses things that can be done to protect people in groups, or "populations."

A healthy population is one that has a lower (but not negligible) frequency of disease, less disability arising from diseases and injury, fewer risk factors (such as smoking rates) associated with future disease, and in which disability translates less often into incapacity to perform social roles, such as working at a job. A population's health is much more complicated that the sum of every individual's personal health status in the population. This is where the "population health" model comes in.

The essential elements of the population health model, which builds on public health, are summarized in figure 1.4, adapted from the Canadian Institute for Advanced Research. The model is not perfect, and in its original form placed too much emphasis on economic factors, but with modification it provides a reasonable point of departure. The model postulates five determinants of health of the population as a whole (genetic endowment, physical environment, social environment, health care, prosperity, and well-being). The population health model also goes into greater depth than traditional public health or medicine in considering the social context inherent in the "social environment" determinant, recognizing such factors as place in the social hierarchy, empowerment (akin to the concept of social capacity in economic development theory), social connections, affluence, and nurturing and early childrearing.

It is a tenet of the population health model that social factors—in particular affluence, hierarchy, and equity in the distribution of wealth—are almost as important as physical or biological disease risk factors as determinants of health for populations. A population is more than a collection of individuals. People in

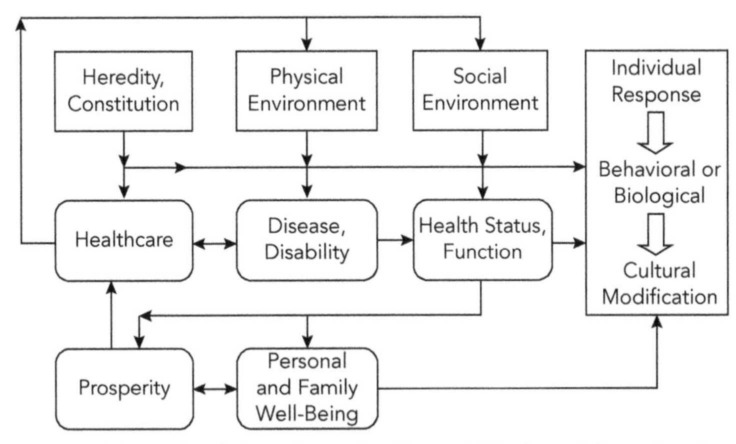

FIGURE 1.4 The updated "population health model," adapted from the Canadian Institute for Advanced Research. The model identifies factors that determine health status for the population as a whole, rather than individually.

groups have structured relationships with one another, whether they live in cities or in relative isolation. They have families, trade and share the necessities of life, interact socially, and share a common culture in ways that may transmit communicable (contagious) disease or puts more than one of them at a time at risk for a chronic health problem. Public health professionals and epidemiologists tend to think of health as an attribute that is distributed throughout a population. Population health theorists tend to think of health as an output, which can be measured by various indicators resulting from a set of social inputs, which represent social and civic investments. Population health is about disparities, gains, trends, and distributions. Population health is distinguished by its emphasis on what can be done through high-level policy to change social and behavioral factors to promote health. More sophisticated analyses are required, which makes this a statistics- and model-based approach.

The links between medicine and individual health have been increasingly influenced by public health and population health thinking. The general narrative of health today in the United States, in public discussion and professional dialogue, draws from thinking about health care and from prevention—and not from treatment-oriented medicine alone. Of particular importance has been the use of public health and population health approaches in clinical medicine, such as epidemiology (the study of the distribution in populations of disease and risk factors) and applications of what may be called "prevention science" in planning health care delivery for the screening and treatment of individuals.

Prevention science is fundamentally about maintaining health and avoiding adverse health outcomes in the future and so is particularly germane to sustainability. It recognizes two different approaches to preventing illness, injury, and disability:

1. The "public health" strategy relies on a universal attempt to change the determinants of risk for the entire population by collective action, such as disinfecting drinking water in a water treatment plant.
2. The "preventive medicine" (or "high risk") strategy relies on a selective attempt to identify and control determinants conferring unusually high risk in a subset of individuals by treating each individual, such as immunizing a child against diphtheria.

Prevention science goes on to recognize three different levels of prevention that at least in theory apply to each approach:

- Primary prevention is the strategy of preventing the occurrence of disease in the first place, by reducing exposure to risk or modifying the person so he or she will not be susceptible (for example, by vaccination).
- Secondary prevention is the strategy of detecting early disease or a marker of risk early enough for successful intervention so as to prevent the disease from developing further.
- Tertiary prevention is the prevention of disease progression (and especially disability) by effective treatment and by measures such as disability management, rehabilitation, and avoiding side effects and complications.

Table 1.3 gives examples of the three levels of prevention.

Environmental Health

To health professionals, "environmental health" means "environmental determinants of health" understood in the traditional public health context as referring to the many sources of contamination, potentially toxic exposures, biological hazards, and physical influences that may directly affect human health. Traditionally, environmental health is a cornerstone of public health, not medicine, and indeed is where public health began in the nineteenth century as an organized field of health protection. Contemporary population health advocates might add that the social, economic, and cultural context in which people live is part of their environment, and in recent years stress (and other psychological hazards imposed on the individual) have been much discussed specifically in occupational health, but these more expansive views of the field have not met with much acceptance. The social dimension of "environmental health" is not much observed in practice. Until recently, then, environmental health was generally understood by public health professionals to encompass the traditional physical, chemical, and

Table 1.3 Strategies and levels of prevention, with examples of each (arbitrary selections)

Levels	Public Health Strategy	Preventive Medicine Strategy	Health Promotion (combining elements of both)
Definition	Intervention at a group level	Intervention at an individual level	Promoting healthy decision making and public policy
Primary prevention	Mass immunization (e.g., against influenza) Environmental health (virtually all environmental health measures are primary prevention) Smoking avoidance Safe sex Treating newborns for diseases Availability of seat belts in all cars; education to use seat belts* Tuberculosis, syphilis, other disease control	Immunization for high-risk patients (e.g., against pneumococcal pneumonia) Smoking cessation programs, individualized Working out with a fitness coach or personal trainer Health app for monitoring blood pressure	Educational and behavioral management programs for a healthier lifestyle* Advocacy for seat-belt or motorcycle-helmet laws Social marketing (campaigns to educate consumers with incentives for safe behavior) Changes in land use and parks to promote walking and exercise Fitness programs Healthful foods available in restaurants and food services; nutritional education*
Secondary prevention	Screening programs for cancer Screening programs for cardiovascular risk Health education for personal health monitoring and self-care Employee assistance programs (mental health)	Blood pressure screening and treatment Colonoscopy Screening newborns for hip disease Tracking the growth of a child through first years of life to identify problems HIV/AIDS screening	Participation in secondary prevention programs Laws requiring that insurance to pay for screening programs Programs that teach self-examination methods (e.g., breast, testicular cancer)
Tertiary prevention	Promote exercise and stretching Low-level sports activities School sports and physical education Access for the disabled in public places (changes environment to diminish disability despite impairment of the person)	Case management Measures to avoid complications of surgery Monitoring recovery Instituting rehabilitation as needed Accommodation (workplace modification)	Intensive case-management programs for diseases such as diabetes (to prevent complications) Dental care Controlling chronic pain

Notes: * The delivery of an educational program widely to the public is considered to fall under the public health strategy even though adoptions are personal decisions: for example, the decision to use seatbelts and individual food choices.

biological environmental determinants of human health: water quality, air quality, food contamination, toxic hazards, and vector control (see chapter 6). The major issues had to do with what caused which effect, how bad it was, and how exposure could best be controlled. This paradigm was recognizably oriented toward public health practice, not social theory or population health.

However, the public often perceives the term "environmental health" very differently, as dealing with "the health of the environment." Environmentalists may see environmental health as missing the point, because the primary concern should be the health of the ecosystems sustaining all life, including humans. Once consequence is that the public health community, which should rightly be in the forefront of environmental issues, has often been marginalized because the environmental movement is using similar words to talk about nontraditional issues related indirectly to health. Because of this, public health professionals and environmentalists often find themselves speaking past each other.

This meant that practicing environmental health professionals can often be found talking about such nasty topics as preventing diarrhea (safe drinking water), mosquito control (which environmentalists often read as either applying DDT or draining natural wetlands), inspecting restaurants, rats, and controlling workplace hazards, all of which are essential to protecting the health of the public. At the same time and sometimes on the same stage, their colleagues in the environmental movement get to talk about sexy issues such as mitigating climate change, low-level (sometimes very low-level) cancer risks, endocrine disruptors, environmental justice, and the connectedness of all things. Since most public health professionals work for local governments, they have to stick to the script and deal with the most urgent problem at hand. Environmental advocates are generally associated with non-governmental organizations with an agenda, such as community groups, they can pick and choose their topics, and they only rarely have to take responsibility for final decisions.

The public health system, although itself highly stressed and almost never adequately funded, has resources to offer in working toward sustainable development. However, the priority for many (perhaps most) environmental health units in local government is to provide and monitor essential public health services, such as food safety, water quality, and pest control. Environmentalists are often but incorrectly persuaded that these traditional public health problems are or should already be under control; they tend to see only the ecosystem issues that threaten to disrupt social and biological support mechanisms. Environmental advocates are aware that the high priority that the public places on health—reflecting its own welfare—makes health issues the most attractive advocacy vehicles for promoting environmental action and awareness. The difference in understanding between advocates and public health professionals often becomes contentious when health issues are put forward as evidence of environmental degradation. Environmentalists may perceive themselves as sounding the alarm on important health issues that illustrate the link between the environment and human health and therefore motivate people to political or social action. Public health advocates may see the same message as alarmist or distracting attention away from environmental problems that are much more likely to cause harm in the short term.

Health Promotion

Health promotion is an applied field of health that draws heavily from population health thinking and from behavioral sciences. In addition to providing a natural interface between health and sustainability on social and behavioral issues, its history represents an interesting model and precedent for the development of sustainability as a movement.

Health promotion is an approach that brought together public health and preventive medicine with behavioral science and fitness, with the objective of encouraging individuals to make healthy lifestyle choices as easily as possible in a social and material environment that supports those choices and allows everyone not only to protect their health and reduce future risk but to improve their health and capacity to function. Health promotion can be considered a "mode" of public health practice, in the same sense that there are modes of science, described in Chapter 2.

Health promotion builds on more sophisticated approaches to health education, on behavioral change, on peer pressure within communities and identity groups (such as large companies), and on rewards systems to create a social climate in which healthful behavior is not only considered responsible but constitutes a new social norm. The essential principles of health promotion were articulated in a document written at a conference in 1986 that has come to be known as "the Ottawa Charter." The Ottawa Charter calls for health promotion on many levels, among them the development of healthy public policy to support individual lifestyle choices. Health promotion is eclectic, in the sense that it uses a variety of health-related approaches, from fitness programs to consumer pricing for healthy foods, and to marking restaurant selections as healthy or otherwise, to urban; it is inclusive and therefore defines strategy more than individual interventions.

Health promotion moves beyond health maintenance into health enhancement, seeking actual improvement in functional and health status rather than prevention of disease alone. As well, the behavior-based approach of health promotion presents sufficiently different characteristics to be considered a distinct "mode" of public health practice. The strategy proposes that individual interventions can be achieved by motivating health-conscious behavior in subjects as groups and by institutional interventions that change the options available for individual behavior. These interventions may result in an enhanced state of health and well-being and therefore go beyond disease prevention, which seeks to maintain the current status.

More sophisticated approaches to health education, peer pressure within identity groups such as large companies, and a rewards system create a social climate in which healthful behavior is not only considered responsible but constitutes a social norm.

The health promotion enterprise sets into motion powerful social forces. The result of these changes can be an exceedingly powerful and self-perpetuating movement. Without peer support, the more healthful behavior may have required extra effort and may have been contrary to social norms: seeking out healthy food choices in a world dominated by fast food, for example. With peer support, the healthy behavior is the accepted standard, and the unhealthful behavior is socially "deviant" (used here in the sociological sense of going against society's norms): bringing fried chicken into a juice bar, for example. The decision of the individual is facilitated and supported through the creation of persuasive forces that make compliance a social expectation and norm. The traditional context of prevention-oriented services is reversed because healthy practices are expected and unhealthy practices are unusual and deviant. Concepts about health, physical fitness, and personable responsibility are continually reinforced by peer pressure.

Health promotion both contributes to sustainability and provides a model for how a second-generation sustainability movement might develop. Health promotion took the established fields devoted to health protection that seek to prevent something bad (such as cancer) and preserve what is good (avoiding disability) for the future and turned them into a drive to improve life and make living better for everyone. In doing so it turned a dutiful but fundamentally "negative" message based on fear (preventing something bad from happening—"avoid this terrible fate"!) into a "positive message" based on gain ("we can all do better and live better lives"!). Such a positive message is much more motivating to most people and achieves the same goal.

This is a viable model for sustainability. Environmental protection, like public health and preventive medicine, is about preventing something bad from happening, repairing damage, and not losing something precious. It is based on the fear of loss and future pain, and the concerns expressed are emotionally negative. Sustainability is about moving together into a place where people feel secure and respected in an environment that supports their aspirations and is pleasing (as well as more natural). The sustainability message is positive and therefore much more motivating than the message of environmental protection but moves society toward the same place. It also provokes fewer conflicts and has elements that everyone can agree upon, regardless of politics or faith.

In a sense, sustainability has many things to teach health, but health, or at least health promotion, has one big thing to teach sustainability: how to change behavior by turning a "negative" message of concern and worry into a positive message of opportunity and even fun, and in so doing convert an aspiration into a movement.

2

Ways of Knowing

THIS CHAPTER EMPHASIZES critical appraisal of evidence. It offers the method of science as the best (but not the only) way of achieving an understanding of the world when predictions are required and decisions must be made. Science and values must be considered together in health and sustainability. Every problem in health is at root a statement of knowledge (science) and of how people should be treated (values). Every problem in sustainability is at root a statement of knowledge (science) and of how resources are shared and rights are protected (values). Thus, although science is a way of knowing objective reality (or at least approximating it by degrees), values are a way of knowing what is right between people and what is important in the world. Science does a poor job articulating values (other than utilitarianism), and value-laden systems (such as religion and culture) do an arguably poor job of describing objective reality (unless one truly believes that faith is the whole reality). These are complementary ways of knowing.

Box 2.1 describes the findings of a study that attempted to tease out the various tributaries of knowledge flowing into sustainability. The study had obvious drawbacks: for example, it fairly obviously understates the contribution from the arts, philosophy, and management science. But in other ways it is highly revealing. When one considers that each of these disciplines has its own way of searching for truth (or beauty or insight) and evaluating evidence, it becomes clear that there is no one way to look at sustainability. One of its main findings, although not emphasized in the paper, is that there is a disconnect between health and sustainability. The discussion in Box 2.1 is also intended to provoke thought as an exercise in skepticism and questioning, or what is often called "critical appraisal."

Ideas in health have an ancient history; they have also experienced many dead ends and false beliefs and have gone through many transformations before modern, evidence-based medicine (which deals with the health of individuals) and contemporary evidence-based public health (which deals with the health

BOX 2.1

A Study of Studies

How might one recognize the constituent disciplines of sustainability, the fields that it embraces and that have fed into its development? One way is to look at who writes about sustainability and where they are coming from. In 2012 a pair of investigators (Quental and Lourenço) in the field of "bibliometrics" (the use of publishing and citation data to evaluate the flow and influence of written communications) looked into this. Using Web of Science®, they searched for papers published between 1981, when "sustainable development" started to be used in the literature, and 2008. Usable papers had a significant number of references that had in their title, abstract, or designated key words "sustainable development" or "sustainability science." The investigators picked those specific terms because previous studies using "sustainability" and "sustainab*" were considered overly inclusive. From this source literature they found 3334 publications, 73 percent of which were original research reports, 16 percent papers in conference proceedings, 8 percent reviews, and 3 percent editorials. The investigators then determined which disciplines the papers came from, as determined by the scope of the journal.

Their findings are summarized in Table 2.1. By far the majority of papers came from the environmental sciences literature and the second largest source was biology. This suggests that writings on sustainable development are grounded deeply in formal environmental sciences such as ecology and the life sciences. Urban studies was third but showed a declining trend that was not offset by combining it with sociology (eighth on the list) and even including anthropology (eighteenth on the list, but it is a relatively small field); this despite the obvious importance of urban issues to sustainability. Among the remaining disciplines that contributed more than 5 percent, economics led the social sciences, and physics and engineering were neck and neck. Not surprisingly, chemistry, agriculture (as "agronomy") and political sciences were represented at more than 5 percent but were by no means dominant. Perhaps surprisingly, earth sciences contributed less than 5 percent, as did management studies. Health sciences contributed a tiny 3.4%, even though this heading would include medicine, biomedical sciences, epidemiology, public health, health services research, and all the other health disciplines that are funded by the National Institutes of Health and all the medical and health research councils that exist in virtually all developed and many poor countries around the world. Philosophy (a field in which a small number of scholars produce prodigious amounts of writing), education (where interest in sustainability runs deep), and mathematics (little of which is unique to sustainability as such) were clustered close together. There was almost no contribution from the arts. Overall, it would appear that sustainability is driven heavily by the sciences and particularly environmental sciences and much less so by environmental studies, and hardly at all by the humanities. Is this correct?

Please see Table 2.1.

It is worth looking at those results more closely. First of all, what about face validity and methodological limitations of the study? Is a study conducted over a twenty-seven-year time period but ending in 2008 representative of the field in 2015? Given that the treatment of "sustainable development" in most of the source literatures massively increased in the later years of this period, what are the proportions really telling us about the development of the literature? Could it be that some small fields, such as forestry (hidden, because it is embedded in "agriculture"), or fields that seem to have contributed little overall, such as agriculture and earth science, were foundational, even seminal in the early years and then were overwhelmed by the volume of contributions from larger fields as their ideas took hold? It is worth considering the following questions:

- Is it a drawback that the authors did not apply inferential statistical significance testing to their list? How would they do this? What would be the point of it? Do the assumptions of the study warrant significance testing or would it be an empty exercise considering the overall uncertainties?
- Is the term "sustainable development" appropriate? Could it be more likely to be used in environmental sciences and agriculture, where the concept is central and the idea began, than in chemistry or mathematics, even for topics that apply?
- Is it reasonable that physics contributed relatively little when the most vital topic of the day, and probably the century, is climate change? Could it be that there is little need for more work on the physics of climate change specifically as it relates to sustainability, because the principles are well known and need little further elaboration? Is it not likely that the physical factors are settled and that physicists now have other things to work on and so have turned the mission of explaining climate change over to the environmental sciences and education?
- Do people who write in the academic literature of the arts (art history, criticism, technique, curatorship, and so forth) address themselves to sustainability advocates or to other artists and scholars in the arts? How many arts journals are registered in the Web of Science® and what topics do they

cover? Does the arts literature really tell us anything about the influence of the arts in sustainability thinking and how the arts help communicate difficult and tacit ideas?

- Likewise, is it credible that management has so little input into discussion of sustainability when the private sector is heavily invested in "sustainable operations" and the "triple bottom line" (ecological, economic, and social sustainability) and sustainability is one of the hottest topics in the management literature? One answer to these questions is that the term "sustainable development" is almost never used in contemporary management science. The term of art in business is "sustainability."
- This study did not state its hypothesis explicitly, but then most studies do not. The implicit hypothesis can be stated formally as an assertion to be disproved, as follows: scientific disciplines contribute equally to the literature of "sustainable development." Was the assertion "falsified" or proven not to be true by the evidence? (Is statistical significance testing always necessary to do this?) Was the process successful in generating useful data?
- Is this junk science, then? Did the authors use sloppy or questionable methods and promote their findings as being more significant than they were because they had an agenda? Are they trying to sell something?

The investigators are not engaged in junk science, and they are not trying to sell something. Not at all. They are engineers on the faculty of a European university. The scrutiny of their work by their peers in this competitive field would have given them a strong incentive to be rigorous and to demonstrate the validity of their method. The investigators performed a perfectly reasonable analysis using standard methods on an inclusive database with explicit descriptions of the study's methods, reasons, and limitations. Their paper is ingenious and well done. They should be commended.

Any effort to look at the dynamics of the scientific literature, however, is just a mobile phone photo from one angle of what is really a crowd scene that is tumultuous and changing. "Sustainable development" may not be the ideal key word, because the term started out in the 1980s as a reference to managing within limits to growth but today is more often used in development economics as it applies to middle-income and poor countries. It has been replaced almost completely in the management literature by "sustainability."

Should the study be replicated to check the validity of the findings? Of course, but how? There is no database in the arts comparable to Web of Science®, which is both complete in registering source journals and comprehensive for the sciences. Other ways of approaching the question would have to be based on different methods, such as citation analysis, to see which fields had more influence on scholars writing in the field.

Finally, the study shows clearly that everything in science begins with the question, or hypothesis, that is put forward. There might have been a very different result if the term "sustainability" were used after 2000.

The study made some very important observations. One is that compared to the other disciplinary areas, the field of urban studies as applied to sustainability has not developed as robustly. The topic will be discussed at length in chapter 9. Another is that there is an obvious disconnect between the health sciences and sustainability, despite their apparent shared interests. That is the topic of this book.

of populations and communities) became the two foundations for practice and public decision making in health. Ideas about sustainability, likewise, are often ancient. The domain of sustainability is also derived from twin taproots, those of ecology and economics. (See figure 2.1, which represents both health and sustainability as trees, with deep roots.) Not all of these ideas (dead branches on the metaphorical trees) were or are correct.

The passion for good health, for a sustainable world, and for protecting the environment often get ahead of what is known. This is one of the hardest realities that professionals and activists face and one of the major reasons they misunderstand one another. There are so many elements to knowing about complicated issues in either health or sustainability (or, for that matter, any complicated topic): facts (which are the basic unit of observation but can be misleading in isolation), known relationships among the facts, a framework for understanding the facts and their relationships together (a theory if it fits them together, a hypothesis if predictive), assessment of the confidence in the accuracy of each (which includes an assessment of measurement error in each or an appreciation of statistical "error" from random variation), assessment of possible bias (nonrandom, systematic error), and (knowing the contingent probabilities and understanding the complexity of the whole) an assessment of confidence in the overall conclusion as a basis for taking action. Separating ideas out and weighing the evidence (i.e., critical appraisal) is central to the inquiry.

This is much more thinking than most people care or are prepared to do about most issues. The public, therefore, depends on experts to sort it out; however, experts are both trained and inclined (because they would not be experts if they were not analytical) to look closely at the details and to look for anomalies and inconsistencies, not verification of preconceived ideas. In fact, the integrity of science is all about resisting the temptation to simply validate a preconceived

Table 2.1 Constituent disciplines of sustainability (as "sustainable development" and "sustainable science") represented in the journal literature, 1981 to 2008 and summarized in Box 2.1. Adapted from Quental and Lourenço (2012). Reanalysis omits multidisciplinary studies 2.8 percent, ninety-two papers

Discipline	Percentage of Papers	Number of Papers	Comment
"Environmental sciences"	70	2318	Obvious. Also likely to be the portal through which highly technical concepts in physics and chemistry are interpreted, digested, and made accessible to sustainability community.
Biology	20	722	Obvious. Ecology is a discipline of biology but also the basic science of environmental sciences.
Urban studies	12	415	Declining proportion over time, despite great interest in urban sustainability.
Economics	11	415	Variability but overall similar proportion 2000 to 2008.
Geography and spatial science	12	370	Variability but overall similar proportion 2000 to 2008, despite heavy use of GIS.
Physics	10	349	Critical in issues such as climate change. Probably overlaps with earth science.
Engineering	10	320	Plausible—"green engineering" is a cornerstone of sustainability.
Sociology	6.0	22	Surprising, may reflect low use of key word in discipline.
Agriculture	5.5	182	Sustainable agriculture is a huge field and forestry (usually counted with agriculture) gave rise to modern "sustainability," so this is surprising.
Chemistry	5.2	172	Chemistry, and increasingly "green chemistry," are foundational but highly technical. Most chemistry probably comes in through "environmental sciences."
Political science	5.0	168	Surprising, may reflect infrequent use of key word in discipline.

(continued)

Table 2.1 Continued

Discipline	Percentage of Papers	Number of Papers	Comment
Earth science	4.6	155	Surprisingly low for a field central to environmental sciences. May be offset in physics for climate change.
Management	3.7	124	Implausibly low influence, probably reflects sample bias since most management journals are not "scientific." Commercialization of sustainability consulting is discussed in chapter 17.
Health science	3.4	114	Very plausible. Environmental scientists write about health but health professionals rarely write about sustainability.
Computer science	1.7	56	Plausible. Sustainable management through remote monitoring and control is a big topic but not as a computer science issue.
Philosophy	1.3	44	Huge interest in sustainability and deep ecology in the philosophy community: likely that papers just did not use key word.
Education	1.0	32	"Environmental education" is huge but may not spill over into scientific literature or use the keyword much.
Anthropology	0.8	26	Plausible. Anthropological concepts infuse sustainability literature (e.g., Jared Diamond), but anthropology is a small field.
Psychology and cognitive sciences	0.8	26	Interesting. May be because psychology and cognitive sciences are about process, not application.
Mathematics and statistics	0.7	25	Plausible. Math is not unique to any given field of application, so why put the keyword in title?
Arts	0.0	1	How many arts journals are likely to be registered in the Web of Science®?

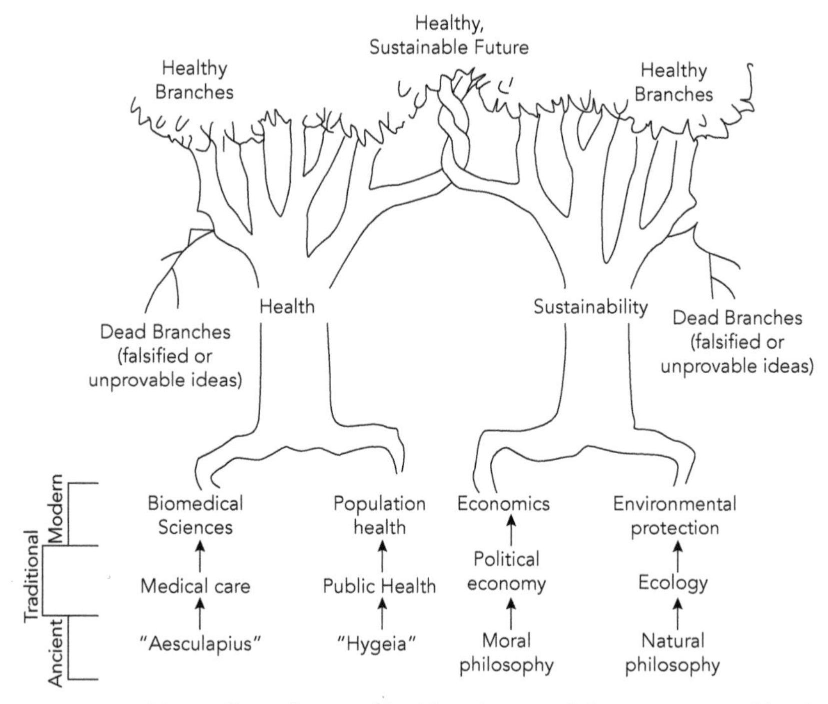

FIGURE 2.1 The intellectual roots of health and sustainability, as interpreted by the author. Both have two taproots, each of them very old and very deep. Health is based on the twin traditions of individualized medicine, and of health protection, with ancient roots in the tradition of public health and protecting populations and communities. Sustainability draws on the taproots of economics which arose, mostly in the eighteenth century, out of moral philosophy applied to politics and public affairs, and from environmental protection, the idea of which is itself ancient but in modern times is grounded in a scientific understanding of nature that came out of the conversion of "natural philosophy" into a modern science of ecology. As drawn, the metaphorical trees are simple but in reality they have many living and dead branches, twigs, and leaves, and are subject to metaphorical wood-boring beetles (frauds), termites and other pests that live on dead wood (false or obsolete ideas), and root rot (bad science). Keeping the trees metaphorically healthy requires skepticism and critical thinking, and sometimes pruning to get rid of obsolete ideas.

idea, however attractive it may be. The preconceived idea may even be moral and "good," but objective analysis, and certainly the principles of science, require that it be questioned. This is not to be contrarian or perverse. Although this skeptical approach does not guarantee that objective truth will be found, examining problems in any other way more often leads to serious mistakes.

Knowing what is true about health and what is valid about sustainability is therefore not so simple. Critical appraisal depends on the accuracy of the "input" (facts and their relationships) and some appreciation for how they fit together (the "theory" or framework). However, people do not come to problems they care about with a clean slate, free of opinions or expectations. They come with beliefs, imperfect knowledge, hunches, hang-ups, emotions, and obsessions. The more important the topic to a person, the more likely he or she is to have preconceived ideas about it. Health and sustainability, including sustainability's origins in environmental protection, are two domains of knowledge that many people care deeply about. As a result, both health and sustainability (and environmental protection), each in their own way, attract considerable speculation, subjective belief, and self-serving rationalization. Both are also haunted by old and obsolete ideas. For example, it used to be believed that flowing water regenerates and purifies itself; after fouling the world's rivers, it is obvious that it does not—oxygenation has a limited but important role for mitigating microbial contamination but does nothing for chemical pollution. Some bad ideas have caused immense harm, as will be seen in the case of eugenics at the end of this chapter.

Standards of Certainty

Activists sometimes get impatient with scientists, whose natural inclination is and should be to be skeptical and to search for the facts. It is also why health professionals and public health practitioners sometimes (well, often) get uncomfortable with both activists and scientists. Lawyers have trouble with all three, because the standards of certainty used to decide something in law are different from science and often different from public policy.

Standards of certainty vary with the situation and the "decider," who is making the decision. Scientists want at least 95 percent certainty, rigorously proven by the process of deduction or by scientific research. Health professionals want to be sure of the facts before they act but are willing to accept that not everything can be certain: they appreciate that sometimes in medicine and public health it is necessary to act before all the facts are in. Lawyers are satisfied with a "balance of probabilities" in civil litigation (what is more likely than not), because that is what the civil courts require (for criminal conviction, it is "beyond reasonable doubt"), and the conclusion can be based on fallible sources, such as the testimony of witnesses. Activists (usually) want to act (most often to ban or regulate) when any credible evidence suggests that there is a possibility of harm or a threat to the environment, invoking one version of the "precautionary principal," that action should be taken proactively if there is any possibility of harm. The harsh reality is that which standard of certainty on which to base an action will be the one that applies in the particular context of decision making: the precautionary principle (which applies to blocking actions and applies primarily in Europe), adjudication in systems where "substantial contribution" to risk is considered to contribute to liability, the courts in civil litigation, regulatory agencies (which are bound by their legislation), and acceptance by the scientific community—in increasing order of rigor and certainty. Congress or Parliament (and other legislatures) are not bound by consistent standards of certainty, and so they make their decision based on however many of their members can be persuaded, by whatever standard they individually accept.

Ways of Knowing

In the domain of health, science rules with some discomfort, because healing and being well are profoundly social, emotional, and intimate. In the domain of sustainability, science provides a foundation, and engineering and management principles rule the application. But this is not without discomfort, since sustainability is fundamentally about values: how people choose to live, the kind of world they want, their obligation to other species and to the planet, and what decisions to make many times a day in the many details of life and commerce. Both domains, health and sustainability, exist to make the world better, to learn more for the purpose of doing so, and to act on what is known. In other words, both domains are really about values, but in everyday practice these values are often hidden in the details and technical issues.

To study any problem, there must be a way of knowing what is right, true, or correct. The ways of knowing what is true are different for different times, cultures, problems, and purposes. For moral values, there are many ways of knowing what is true. For personal enlightenment, there may be many paths, or there may only be one true path. Regardless of the many ways that one can approach truth, and without denying that people may find a common truth to share as a belief, the ways of knowing that we use to solve public problems and to make decisions in a complicated, pluralistic society require a minimum of personal interpretation, which is called "subjectivity," and a maximum of agreement on what is true, which is called "objectivity." For most purposes, that is as far as the discussion on epistemology goes. However, there is a world of nuance and depth beyond the superficial "subjectivity, bad; objectivity, good" rhetoric of science and law and the more realistic "objectivity, cold and pointy headed; subjectivity, good—learn to trust your gut" message of popular culture.

Objectivity requires broad but not necessarily unanimous agreement on the facts and an approach to interpreting them and putting them into a framework that is logical and can be followed by anyone, step by step (if they know enough). Where there is a disagreement on the facts, those who disagree are expected to have valid reasons for their disagreement. Objectivity is the expectation and the ground rule for public discourse and for decision making in this society. This is because a demonstrable fact can, in theory, be agreed upon by all people with an understanding, and a logical framework can be deduced by rules that can, in theory, be followed by all people with an understanding who take it step by step, without repudiation of the whole process or conflict with other rules that do not follow their logic. This is not why science was invented, but it is why science has embedded itself into contemporary society as the preferred way of exploring complicated issues grounded in facts and not opinion, such as problems in health and sustainability.

Objectivity does not mean rejecting values, which are known or felt by other means. Values are no less knowledge than facts and frameworks, but they help solve problems in different ways by helping society choose among options or set their priorities. Objectivity without values leads to a society that rules the people and to a society that fails to function in the real world of competing priorities and preferences. There are no two areas of human endeavor in which it is more important for objectivity and values to be linked than health and sustainability. Sustainability is about knowing objectively what consequences there are for the planet and for society and holding values of stewardship and caring that guide and motivate action. Health is about knowing objectively what causes injury, disease, and pain, so that maladies can be prevented or alleviated; health is also about holding values of compassion and caring that guide and motivate healing and protection. Taken together in each domain, objectivity and values have the potential (because people have to work at it—it does not just automatically happen) to make a society functional and respectful at the same time. This, in itself, is not a bad definition of sustainability.

How does one know facts, and how does one know values? In particular, how does one come to know the facts and values that are used in making decisions that affect other people?

One way of knowing is to accept a revealed doctrine on faith, with or without questioning it. This is to subscribe to a faith larger than oneself. This way of knowing works well for making decisions if everyone involved is of the same faith or holds the same social philosophy, as in the Middle Ages in Europe, China and other Asian cultures when Confucianism dominated, and in much of the Islamic world today. Regardless of the commonality of most beliefs, religious faiths and social philosophies each have their sects or schools of thought and dispute important premises and conclusions. Disputation, as in rabbinical debates and theological arguments over dogma, may illuminate ideas and result in great insights. The essential characteristic of knowing by faith, however, is that what is known is revealed and not proven. Whether there is evidence for the assertion or not, the main approach to knowledge is belief. This is wonderful for the person who comes to a belief or insight, and for the spiritual community to which one adheres, but in a sustainable society in which human rights are respected people would have a right to their own beliefs; pluralism would be celebrated and differences in faith at least tolerated. This means that faith cannot be the only way of knowing if decisions are to be made based on a common understanding, because people will have their own ideas and argue from different premises.

There are other ways of knowing. Science is a dominant form of "public knowledge" (to be defined below) in modern culture and the single most important basis for making decisions about both health and sustainability. However, science is not the sole basis for making decisions, nor should it be. Values, aspirations, priorities for limited resources, trade-offs against other benefits, and the legitimate role of the political system as the mechanism for resolving disputes are all involved and may play an even more important role in the final decision.

Private Knowledge

"Private knowledge" is what one knows, what one thinks one knows, and what one believes to be true. The origins of private knowledge are simple: life and thought. Life is experience, and thought is inference, deduction, and "self-knowledge" (not only knowing ones "inner" self but knowing one's own biases and contradictions and correcting for them). It forms the basis for our actions and decisions in the real world. It may be based on firsthand observation, experience, factual knowledge, evidence in various degrees of abstraction, inference, persuasion, intuition, surmise, belief, or faith. It may be grounded in religion or in what feels right, but it guides a person through his or her own affairs. Private knowledge may contain elements that are widely shared in a culture, but it always includes other elements, such as degree of trust or weight put on consequences of an action, that are highly individualistic. It is simultaneously the worldview one brings to an understanding of the environment, a filter that impedes ideas that conflict with deeply held beliefs, and a prism that shapes and sometimes distorts perceptions of reality.

Issues of health and sustainability are laden with private knowledge: beliefs, assumptions, personal experience, theories, observations, interpretations, and a hierarchy of acceptance from various sources and teachers. Lumped together, indiscriminately, this is called "subjective thinking," a term that obscures as much as it explains. Among other fallacies, subjective knowledge is almost always assumed to be the opposite of objective knowledge: but in fact, they cannot be separated, logically, cognitively, or socially. Ultimately, if one accepts that there is an objective reality, one must believe, subjectively, that a representation of reality is true.

Every individual adult makes decisions on the basis of private knowledge. (This is a tautology, because private knowledge is everything a person knows and believes.) Private knowledge has consequences for others when a person is in a position of responsibility: for example, a parent whose private knowledge may include a true or false belief about health or risk-taking behavior. A senior executive who "goes with his gut" (male gender assumed) or "follows her intuition" (female gender assumed) makes decisions that have consequences for people's livelihoods, including stakeholders, and for consumer products or environmental impact, the public.

On the other hand, an institution or representative imbued with public trust in making decisions, such as courts of law or regulatory agencies, are bound by a responsibility to consider only knowledge (or "evidence") that is admissible in terms of what society agrees is likely to be accurate and acceptable in terms of values. It is not reasonable, in a secular society, for a judge to render a decision based on his or her personal religious beliefs, or to reject forensic evidence because it contradicts some personal belief (e.g., DNA evidence with respect to individuality and evolution).

The responsibility to rise above personal knowledge and to look at some standard of to what is admissible applies only loosely in other important decision-making situations. A legislator, such as a senator or representative in Congress or a member of Parliament, brings their personal knowledge (including religious belief) to the decision on how to vote and what position to espouse. Sometimes personal knowledge conflicts with knowledge that is admissible in the public forum, as will be discussed shortly. Then, a politician must determine whether objective reality is best reflected in what they believe as personal knowledge or by the admissible evidence that society deems to have been derived from methods that are validated and appropriate to the problem (such as science or historical research or bookkeeping). At the moment, one of the most salient examples of this pertains to climate change.

When a group of people share the same (or largely the same) belief system, then personal knowledge becomes the majority thought within that group. Religious communities, for example, adopt the doctrine of their faith, and it is used to resolve problems and to make decisions within the community. This is a form of collective personal knowledge.

Traditional Knowledge

"Traditional knowledge" is a term used mostly in the context of empirical knowledge collected over generations by communities close to the land and sensitive enough to its subtle changes to survive. Aboriginal tribes have accumulated a huge amount of empirical knowledge over centuries of observation and trial and error: this is called "traditional knowledge" in anthropology. Much traditional knowledge has to do with perceptions of change in the environment where the band historically used to hunt, fish, or gather plants of importance as food, coloring, or medicine. This knowledge extends back as legend far beyond history and is much older than science. Embedded in the narrative of their experience and history is a generations-long panorama of change in the environment.

Traditional knowledge has always presented a challenge in law and public policy. At present it is treated as a form of collective private knowledge that is valid for making decisions and dealing with life only within the context of the community that values it. Some people feel that traditional knowledge should be treated like scientific knowledge for purposes of decision making and settling disputes such as legal actions. Others believe that scientific knowledge is the only admissible truth, or public knowledge, in our society. The issue is determining when traditional knowledge is acceptable as the basis for making a decision in the broader society (e.g., in the very contentious and high-stakes cases of land claims and rights of way for pipelines across traditional lands) and when it is only admissible within the culture of the group for which it has deep meaning. In Canada, traditional knowledge is often admissible in aboriginal justice (the territories and certain communities have their own tribal systems, particularly for juvenile offenses), but its acceptance as an environmental tool has come only recently.

The public knowledge of science and history is challenged by aboriginal peoples with vast experience in subsistence on the land, but there is little documentation of that experience beyond tradition. Since the perception of the majority population is that memory cannot be fully trusted, and that expectation triumphs observation, a way of knowing based on long collective experience seemed hopelessly inadequate compared to science. The body of traditional knowledge acquired by aboriginal peoples seemed more worthy of study as folklore by anthropologists than as a way of studying the world.

One historic event changed all that. The first serious effort in Canada to incorporate traditional knowledge into a public process was the environmental impact assessment called the Mackenzie Pipeline Inquiry of 1977. This was an evaluation of the risk of a pipeline project that was never built, but it created an opportunity for the Canadian government and the oil companies involved to contract with aboriginal bands to record traditional knowledge systematically for application in land use decisions and environmental impact assessment. Since the era of the original MacKenzie Valley pipeline commission, the aboriginal experience is no longer patronized or deemed too subjective and emotional for science, at least in Canada.

On one level, traditional knowledge is a compendium of detailed observations about the natural environment, integrated by a structure that combines myth, history, aspiration, and values. This may not be the way that the majority culture has learned to integrate information—although it once did likewise—but it works. The information gathered by successive generations on where to find country foods, their relative abundance and changes over time, and the seasons of their ripening and availability has been key to the group's survival.

The insight to the majority is that traditional knowledge breaks through limits of imagination imposed by the majority culture. One of these insights is a long story about how human beings have been a part of the natural environment and how human culture, at least since the last Ice Age, has been a shaping force for the natural environment. Much of this traditional knowledge has been transformed into the art of creation myths and stories. Traditional knowledge therefore is a rich source of knowledge for sustainability in particular, although often covered by layers of metaphor.

When a cultural group exists as a minority embedded in a country or surrounded by a dominating majority culture, traditional knowledge can be seen as a collective form of personal knowledge. However, where the cultural group is in the majority, that same traditional knowledge is the "public knowledge" of that group or nation, the basis for deciding what is admissible to consider in making decisions.

Public Knowledge

Public knowledge consists of the facts and frameworks that society agrees to accept as valid for purposes of making decisions, solving disputes, planning for the future, and explaining the world. For knowledge to be useful in making decisions, it must be shared widely and agreed upon by a conclusive majority of people in a community. This means that if something important is being decided in a pluralistic as opposed to a tribal or highly homogeneous society, those who make the decisions must accept the validity of facts as a general proposition (that is, that there are facts and that objective reality exists), determine the relevant facts, and agree to the most reasonable way to interpret them. They may disagree on which facts are valid, but the system requires that certain types of knowledge be recognized as factual. Likewise, they may disagree on the interpretation but must agree on basics of the framework for interpretation (e.g., that cause must precede effect, that the absence of free will in a deterministic universe does not absolve a person of guilt for their actions, and that magic is not an admissible explanation).

Examples of public knowledge include testimony that is admitted in a court case, the body of scientific evidence that a regulatory agency requires before a new pharmaceutical product or pesticide can be registered, the costs and benefits calculated before a major construction project is proposed, the accounting of cash flow and balance that indicates the financial health of a company, and the general understanding of history that is taught in school and that forms the basis for national identity, holidays, and ideas about the national interest. (The scholar's deep knowledge of history, based on original sources, is rarely considered public knowledge, however; perhaps it is viewed as esoteric or because scholars are prone to argue.) These shared bodies of common or nominally "verified" or at least accepted knowledge are then used to render a courtroom verdict, make the decision to approve the drug or pesticide, approve and obtain financing for the project, invest in the company, decide how to vote, and whether to go to war.

The final approach is to base understanding on evidence, on demonstrable fact. Leaving aside the thorny issue of validating the fact, this is the approach of history (consulting different sources and converging on what appears to be true), philology (consulting different texts and symbolic representations and determining what is meant), law (deducing from the legal text what should be done based on supplied facts), accounting (tracing the inflow and outflow of money over time), and many other technical fields. In the modern era, one way of looking at evidence and evaluating demonstrable fact has emerged as dominant: science.

Science as Public Knowledge

Science is the socially dominant form of public knowledge, at least in the spheres of health and sustainability. The scientific method is better at elaborating on what is already known than pointing in the direction of where to look for new insight. Scientists are exquisitely sophisticated in using quantitative methods to investigate serial hypotheses to figure out what is probably true. Empirical science has placed so much value on quantification in the scientific method that it is only really now (at least in the health science) rediscovering qualitative methods (structured forms of observation) and different ways of getting the big picture.

Science is the way of knowing that environmental sciences and health sciences use to study a problem and to determine what is true. The distinguishing characteristic of science, as will be shown, is that the scientist is always seeking to disprove his or her conjecture rather than prove it. Other ways of knowing work hard to accumulate evidence in support of an idea, but science is distinguished by insistence on reflex skepticism that demands that an assertion be tested before it can be considered true. Only then is it considered provisionally true. This method, which is fundamental to science, is called "falsification." Thus when a true scientist questions an assertion or idea, it does not necessarily mean that the scientist believes that it is not true. It means that the scientist is suspending final judgment until the idea has been tested against evidence.

Too often in sustainability and environmental protection, skepticism is misinterpreted as invalidation and denial masquerades as skepticism. The method of science is to carefully and thoughtfully construct a provisional concept of reality and then to attack it ruthlessly. Sound science is based on a process of probative testing (called "falsification") that achieves much greater certainty than simply compiling facts. That process has to be respected, even if it looks from the outside as if the scientist is denying the weight of evidence. A genuine scientific approach to a problem means to look at data skeptically and test ideas of how the world works against solid evidence. To the extent that sustainability rests on accommodating facts, therefore skepticism, not uncritical advocacy, is essential to progress in achieving sustainability, as it has already been for advancing health.

Good policy and good decision making in sustainability and health must be based on sound science. However, the level of certainty that can be attained with sound science is rarely available or sufficiently complete for advocating or making complicated decisions. They require different levels of certainty. Therefore it is perfectly possible for a scientist to be persuaded that a given chemical is dangerous or that a particular hazard needs to be regulated while accepting and admitting that the scientific evidence is not conclusive. The two positions are not mutually contradictory. They simply reflect different standards of certainty: what needs to be done based on what is known with uncertainty and what is known to be true, however incomplete.

In both sustainability and health, good policy and responsible actions depend on sound science. If the science behind it is sound, there is no guarantee that the policy will be good and effective, but it is more likely to be well considered. If the science behind it is poor, then the policy will not work even if by dumb luck the right option is chosen, because it cannot be justified, and so in the face of opposition the decision will probably not stand.

Modes of Science

Scientists make a critical distinction between two "modes" of sciences: basic and applied. Basic, in this sense, does not mean "simple"; it means fundamental. Basic science is often thought of as science for its own sake. Applied science is the application of science to achieve an intended goal. It builds on basic science and contributes to technology and engineering by finding ways to solve problems. Technology, taking up where applied science leaves off, develops the science into a specific application or product. The difference between technology and applied science is somewhat arbitrary; however, in general, applied science achieves a foundation for developing a new technology, and engineering builds on the applied science to make the technology work.

These two modes are well recognized by scientists. However, there are other possible modes of science. One of them is a science that would document abuses or problems arising out of technology and social actions and monitor the effectiveness of solutions. This has been called "critical science" and it is directly relevant to sustainability and to health, particularly public health.

Science can also be a check on the misapplication of science and technology. The most important examples of the conscious use of science as a check on itself and on technology have come from sustainability, health, and the environmental sciences. In one famous example, physicians Eric and Louise Reiss (b. 1920–d. 2011), assisted by the then-unknown environmental scientist Barry Commoner (b. 1917–d. 2012) and other colleagues, collected baby teeth from American children in St. Louis in 1961 to demonstrate that levels of strontium-90 bound in the teeth had risen steadily as a result of nuclear weapons testing, and then abruptly and massively after 1963 as a result of atmospheric testing in the Pacific. The results persuaded US President John F. Kennedy and Soviet Premier Nikita Khrushchchev to conclude the first Nuclear Test Ban Treaty in that year. The "Baby Tooth Project" became the nucleus for organizing several early nonprofit environmental groups concerned with sustainability, most notably the Scientists' Institute for Public Information, which together with the Union of Concerned Scientists (founded 1969) pioneered science in the public interest and science-based political advocacy.

A prominent philosopher of science, Jerome R. Ravetz (b. 1929, American working in the United Kingdom, see figure 2.2), described a new and then-emerging mode of science in the 1960s that he called "critical science." Critical science is the mode of science that acts for the public interest and documents problems of technology and society. Critical science is more like basic science than applied science because it is fundamentally about discovery and solving a puzzle rather than achieving an objective. What defines critical science is that the problem under study is of human creation and is usually relevant to sustainability and health.

But accurate, responsible science is not enough by itself to guarantee responsible and responsive public policy. Good science does not guarantee a good decision

FIGURE 2.2 Jerome R. Ravetz (b. 1929), philosopher and critic of science and originator of the concept of "critical science." Dr. Ravetz was born in Philadelphia and was educated in the United States but did his graduate work at Cambridge, naturalized as a citizen of the United Kingdom, and now lives in Oxford.

Source: Photo courtesy of Jerome R. Ravetz, all rights reserved.

or good public policy, because there are many other considerations to be made; however, bad science virtually guarantees bad public policy. As observed before, even if a decision made on bad evidence actually turns out to be valid under the circumstances, it cannot be defended and so is likely to be reversed sooner or later.

Junk Science

In both health and sustainability, falsehoods, popular ideas, and science coexist and get confused. A particularly unfortunate situation occurs when bad science or intentionally misleading science gains currency in public discourse and decision making and practice.

"Junk science" is misleading work that is uninformative or plain wrong by design, incompetence, or ignorance. (Those are important distinctions that the label of "junk science" obscures.) The notion of junk science goes beyond isolated facts that are wrong and implies work that is misleading in its direction and distracting to mainstream science. It assumes a scientific framework that is incorrect and an agenda that results in distorted science. It implies that the investigator is deluded, misled, uncritical, or committing a fraud but does not necessarily imply which.

Originally, the term "junk science" had a distinct partisan flavor and was rarely used except to label and invalidate studies that suggested a need for regulation or intervention for environmental protection, or to impugn scientific evidence submitted in lawsuits. The implication was that such studies were manufactured to influence politics, media coverage, support pending legislation, or justify regulation. The term arose from the political right, and for a while it was used mostly to label and marginalize research that might support sustainability measures such as on climate change (called "the mother of all junk science stories" in 2006 by Steven Milloy, a prominent partisan science writer). However, in the modern world real-life junk science was actually pioneered when the tobacco industry systematically distorted science for purposes of countering health advocacy.

Junk science does exist, unfortunately, on "both" sides. (The idea that there are only two sides to any advocacy is naive, but that is how issues play out in media and politics.) Advocates of sustainability and health, to be effective, need to know this and must learn to recognize junk science on whichever side it presents itself.

Junk science comes in the following five general categories:

- *Absence of falsification.* When the investigator seizes selectively on data that support his or her idea and forgets the importance of falsification. (e.g., cold fusion, or over-emphasizing the "flat part" of the global warming curve, which has several reasonable explanations.)
- *Bad facts.* The investigators produce work that is wrong or misinterpreted and then try to shift the paradigm by promoting it as a breakthrough (e.g., studies on chrysotile asbestos that purported to show that it was safe).

- *Self-deception.* The investigator is so persuaded of an outcome that he or she rationalizes any finding and does not see the total argument (e.g., uncritical support for the idea that a substantial amount of human disease is caused by mycotoxins, which are toxic chemicals on the surface of mold spores, at exposure levels commonly encountered in homes).
- *Outright fraud and deception of others.* This form of junk science usually does not last long because science is fundamentally a self-correcting system of knowledge (e.g., a wildlife biologist in Spain published over thirty articles on parasites in various species of migratory birds using apparently fabricated data).
- *Design to fail.* A particularly nefarious form of junk science includes studies that have design flaws that guarantee that they will not show a positive finding or meet their goal (e.g., epidemiological studies with insufficient statistical power to show the effect they are supposedly testing).

Scientists have excellent powers of reasoning but often cross the line into rationalization when they are convinced that something is true but cannot "prove" it. This both predisposes them to error if they seek shortcuts around the discipline of method and the doctrine of falsification and encourages acceptance of a plausible but questionable study without the discipline of skepticism. Understanding that, the very best, world-class scientists almost invariably default to the doctrine of "falsification" by retaining their skepticism until the last moment rather than coming to conclusions on the balance of evidence, which is how the rest of the world works.

Many authors have proposed indicators for the detection of junk science. None of them are foolproof. Box 2.2 provides a list of criteria that suggest when a study merits the label.

The usual corrective for junk science is dispute followed by constructive neglect, when scientists feel that there is nothing more to be gained by arguing. Unfortunately, there is a pronounced tendency, particularly among activists, to consider skepticism and critical review of studies in environmental sciences, including those that support sustainability interventions, as tantamount to obfuscation, denial and rejection of sustainability as a policy. This paradoxically impedes the legitimate scientific argument required to sort through the facts, determine the validity of explanations and theories, and to ground policy on sound science, but it is understandable because abuse has been so prevalent in opposition to environmental protection and health regulation.

When proponents of junk science are challenged or refuted, they often respond with one or more of the following arguments:

- Their new knowledge is threatening to the scientific mainstream/power structure and is therefore being suppressed.

BOX 2.2

Indicators that a study may be "Junk Science."

Many authors have proposed indicators for the detection of junk science. None of them are foolproof, but junk science is much more likely when the following conditions are present:

- The investigator presents the revolutionary findings directly to the media before they are published in a reputable journal, and often accompanied by news promotion.
- The investigator comes from another discipline and has no obvious preparation for work in the field (e.g., an engineer who publishes a study on a complicated topic in environmental epidemiology without the assistance of an epidemiologist).
- The magnitude of the revolutionary effect is small, difficult to resolve, and hard to recognize or it is of borderline statistical significance. (Scientific findings at or close to the limits of detection are subject to a lot of noise and error.)
- The results fail to improve, and statistical significance to increase, with further investigations and replications of the revolutionary finding using better methods.
- The revolutionary finding is based on detection of a pattern or an image after the fact and not one predicted in advance. (The human brain has hardwired pattern recognition algorithms that can be persuasive for seeing patterns that are not there. Optical illusions are a simple example of this.)
- The revolutionary finding results from deduction or observation alone, not from a testable hypothesis using systematic methodology.
- There is no "hard evidence" for the revolutionary finding, only anecdotal reports or indirect indicators.
- Validation of the revolutionary finding rests on an accumulation of evidence, not a test of the hypothesis by falsification. (Also, when the investigator assumes that an assertion or finding is necessarily true because it has not, yet, been shown to be false.)
- The investigator worked alone and other reputable scientists did not visit or observe the work in progress and other investigators in the field knew nothing of it before it was announced. (This scenario is very rare in contemporary science, which is a highly social and collaborative activity.)
- The investigator invokes a new "law" or principle of science to explain the revolutionary finding. (An example is the claim that the exposure-response relationship, discussed in chapter 5, does not apply in a particular situation.)

- Circular citations, in which the paper or report of the revolutionary finding depends heavily on references or documentation that is not explicitly cited and often refers back to itself, to papers that seem irrelevant, or to the investigator's own other papers.
- Theories that claim to represent a "holistic" approach (see Chapter 3) in which the revolutionary finding can only be seen in a total system and cannot be explained by interaction or identified in the component parts. (Reductionism and its limitations are usually blamed for failure to replicate or appreciate the revolutionary finding.)
- Rhetorical flourishes that feature exaggerated claims, obscure language or jargon (that does not make complete sense to others in the field), frequent citation of facts that do not seem directly relevant to the claim or that imply that the burden of proof is on science to disprove the assertion, rather than on the proponent to demonstrate that it withstood testing by the method of falsification.
- Personalization of the claim or debate over the validity of the revolutionary or contrarian finding, often with allegations of discrimination or conspiracy (e.g., articles that label those who accept that climate change is caused by human actions, including legitimate atmospheric scientists and climate change mitigation advocates as "warmist," a usage that began in 1989 and now has 814,000 hits on Google).
- *Ignoratio elenchi*, which is the term in logic for addressing a conclusion that is actually irrelevant to the argument and used as a diversionary tactic (e.g., challenged to validate the effectiveness of a questionable medical treatment, the proponent produced data on how many had been treated and for what conditions, but not the outcomes they experienced).
- Resistance to having data peer reviewed or examined by experts. (However, resistance or at least objection is sometimes warranted. Sometimes investigators are harassed in the name of review, as in the case of calls in Congress for an investigation into the key studies supporting regulation of particulate air pollution by the US EPA, which ignores that the entire data set was reviewed and comprehensively reanalyzed by an equally qualified but disinterested research group, not once but twice, with identical results.)

- Science has not advanced enough to understand their logic.
- The paradigm is shifting/is starting to shift/will shift in X years and scientists critical of their work have not yet caught up.
- Other scientists are not doing the experiments correctly.
- There is plenty of data in support of their theory, but there may be just a few inconsistencies that they can easily explain if given a chance.

- They are a lone voice raging against a monolithic scientific establishment that seeks to silence them. (Variation: they are persecuted because of their independence and refusal to compromise. Second variation: they are persecuted because the biomedical-healthcare-pharma-industrial complex has a vested financial interest in perpetuating the disease. Third variation: they are persecuted because of "political correctness," usually presumed to be imposed by environmentalists.)

Sustainability rests on a foundation of environmental sciences. Health rests on a foundation of biomedical and population sciences. It is critical to be vigilant against junk science and to demand scientific rigor, because although good science does not guarantee good public policy or good decision making, bad science always results in bad policy and bad decisions: even if the final decision is the correct one, it cannot be justified and will soon be reversed for expediency.

Junk Science Can Kill: Eugenics

Among the most terrifying examples of junk science has been eugenics, whose dark past continues to haunt discussion of one of the most important aspects of sustainability: the balance between population and resources. Eugenics needs to be studied here not only as an example of junk science, but also to set the stage for the discussion of population which comes in the next chapter. Knowing about this skeleton in the closet of population studies is important because it explains why many people are uncomfortable with population studies and why the topic tends to be neglected. Understanding the history may allow the principles of population and resource balance to be examined more objectively in the chapter following, without the baggage of a history of abuse.

There have been many examples over the ages of junk science attaining a wholly unwarranted influence in public life and harming society. Most of the worst examples have been related to "race science" (the effort to find an anthropological theory to support the idea of racial superiority) and to "eugenics," a distortion of genetics as it was understood early in the last century that gave rise to terrible abuses. Unfortunately, population issues became inextricably entangled with these bad ideas by the early twentieth century, partly because of bad science and partly because of the misguided enthusiasm of some brilliant people.

The doctrine of population control, combined with the old idea of "survival of the fittest" (which actually preceded Darwin by many years) and a highly distorted understanding of heredity, led to a corrosive but popular pseudoscience which quickly caught the popular imagination. The name "eugenics" comes from "eu," Greek for good, plus genetics. The overenthusiastic embrace of eugenics in the perceived risk of overpopulation by many Malthusians (the overly invested followers of Thomas Malthus, who will be introduced in the next chapter), led to a sense of urgency to do something about a series of interrelated but often falsely perceived issues that included overpopulation, the burden of increasing numbers of people born "unfit" (particularly by mental incapacity), unwanted immigration, the declining numbers of the upper or ruling class, the mixing of races, homosexuality, refractory poverty, crime, colonialism, and public policy regarding education. Eugenics came with a built-in social program, political support (entrenched doctrines of colonialism and racial superiority), the endorsement of some of the greatest scientists of their age, and a veneer of scientific credibility. Serious scientists, political leaders, social reformers, philosophers, business leaders, and artists (whether socially conservative or progressive) at various times and to varying degrees signed on enthusiastically to the idea of eugenics in the hope—the illusion, really—of creating a better world by breeding better human beings.

The central idea of eugenics is that the human race not only had to be saved from overpopulation of the unfit but could be improved by selective breeding of superior individuals and races. This could be done, it was asserted, by encouraging people with favorable qualities (such as higher intelligence) to reproduce and, even more importantly, by discouraging reproduction among people with unfavorable traits. These unfavorable traits included a long list of physical conditions, such as epilepsy and mental "deficiencies" that were thought by the science of the time to be hereditable (e.g., low intelligence, criminality, and personality disorders). The doctrine of eugenics was in some ways a sanitized and more "acceptable" version of nineteenth-century "race science," a thoroughly discredited and debased version of biology that was used to defend the institution of slavery in the American South and, after the US Civil War, to insist on the inferiority of African American people and justify the denial of their rights. Eugenics sounded more sophisticated, was endorsed by prominent scientific leaders of the age, and had a false veneer of then-modern science.

Eugenics started as a simplistic extrapolation from the limited science of the day, particularly naive ideas about genetics, but it quickly turned into a social movement that condoned involuntary institutionalization, forced sterilization, and racial discrimination. It became widely accepted, even popular among the upper classes of the United States and the United Kingdom, including many prominent scientists of the time. Eugenics emerged as part of the agenda of the Progressive movement, a powerful political reform movement in the early 1900s that is normally thought of promoting a period of social advancement. Eugenic ideas about institutionalizing and sterilizing the "unfit" were passed into law and acted upon vigorously, first in Indiana, then especially in California, western Canada (Alberta and British Columbia), and later in North Carolina. The result was widespread abuse and human rights violations, mostly involving poor minority populations, especially African Americans and Native Americans. Forced sterilization continued in the United States and western Canada for many years, long after World War II. Some of these abuses were not corrected until the 1960s and 1970's and aging victims are still coming forward occasionally for compensation.

Eugenics achieved sufficient scientific respectability to be part of the university curriculum in many countries, including Germany, which was the leading country in science of the day. In the 1920s, the ideologists of the National Socialist Party (Nazis) saw in eugenics a way of purifying and improving what they considered to be "the master race," or "Aryan race" and reducing the influence of less desirable inferior people, including, in their view, Jews, Roma, Slavs, and all people of color. When the Nazis came to power in Germany in the 1930s, the California compulsory sterilization law became the model for similar legislation in Germany. Eugenics provided a pseudo-scientific justification for the ideas of racial purity that were central to Nazi ideology and Hitler's teachings.

Combined with indigenous anti-Semitism in German society and religion, ruthless political expediency, and a veneer of respectability due to acceptance of eugenics in the English-speaking world, the Nazis committed appalling abuses in addition to compulsory sterilization, including separation of married partners, segregation of Jews from the mainstream of German society to prevent "contamination," and propaganda designed to dehumanize and objectify people who were considered to be from "inferior" races. This led directly to the so-called euthanasia program, which led to the murder of thousands of people with chronic diseases, mental deficiency, or mental illness. The next step was extermination of undesirable races and subgroups (such as homosexuals) to promote "racial purity," and this was duly attempted in the infamous death camps, which killed millions. Only the defeat of Germany in the Second World War stopped the evil in Europe.

A detailed critique of the scientific theories of eugenics is beyond the scope of this chapter and would be pointless anyway. Every aspect of this pseudo-science is riddled with errors, from the assumption that undesirable traits can be weeded out of a population to the definition of what constitutes an undesirable trait. It is enough to say that eugenics was not just a result of ignorance of basic genetics. It was an aggressive and malignant movement that raised the beliefs of its adherents above human values and refused to admit scientific evidence even long after its errors were revealed.

Today, echoes of eugenics can still be heard, not only in racist rants, but in nuances of public debate on issues such as immigration, supposed plots by the Establishment to sterilize and weaken African Americans, and resistance to vaccination and family planning programs in disadvantaged communities. It persists especially in conspiracy theories, such as the idea that HIV or Ebola were diseases intentionally developed and introduced for the purpose of taking out minority populations, but that got out of control. Once such long shadow is "the

Plan", a conspiracy theory that persists in many African American communities that was to have involved reducing the size and fragmenting black populations; in the 1920's many eugenics activists actually did advocate an agenda, called the "eugenics plan," to do just that.

However, the malign legacy of genetics is perhaps most powerfully reflected in what is not said, in the silence of mainstream advocates, United Nations agencies, and non-governmental organizations on the topic of population control, especially since the 1970's, when population control was last at the center of the debate on resources and the environment. Because population studies had played such a foundational role in the development of eugenics, proponents of controlling population found themselves just one step removed from the justification for atrocities and linked without intent to a disgraceful chapter in scientific history. This unfortunate association compromised the legitimate observations of demographers (population scientists) about population dynamics, survivorship, and risk. Eugenics tainted and marginalized the field of population studies long after eugenics was eventually rejected by mainstream science.

Since eugenics cannot be ignored, it is a responsibility of everyone who is involved in the schools of thought that it touched to recognize it, see beyond it, and refute its errors.

Keeping Track

When the idea of "public knowledge" was introduced earlier in this chapter, the ways of knowing that it embraced were not limited to science. Mention was made of monitoring and, perhaps improbably, accounting. Both are important modes of public knowledge and play a critical role in sustainability and increasingly in health. The reason is that both monitoring data and accounting data, while they may not yield new knowledge, track trends and disclose the underlying condition of the situation or system they are following. They make the performance of complex systems explicit, convert data into information that can be analyzed and interpreted, and allow stakeholders and the public to follow and understand how the system is doing—whether it is a company, an ecosystem, or protection against disease. Monitoring is discussed further in the next section.

Monitoring may take many forms, among them tracking tree cover in an urban space, or energy consumption or the installed capacity of wind turbines to generate electricity or disease incidence or contaminant levels and body burdens of chemicals in the human population. This process allows managers, economists, scientists, public health professionals, and anyone else using this approach to determine the performance of the system without measuring and studying the entire system as a whole. Thus, economists use economic indicators, such as retail sales, the consumer price index, and savings rates to monitor whether the economy is slowing down, approaching inflation, not investing enough, and so forth, without having to measure every aspect of economic activity. Similarly, lake ecologists might track the chemical oxygen demand, biological oxygen demand, pH, temperature, species count and distribution, biological productivity, and selected contaminant concentrations to derive a trend for how a body of water is coping with stress. They then track the indicators that most closely reflect the trend. Public health professionals track indicators such as disease frequency (through physician reports and hospital discharges), positive laboratory tests, measured pollutant levels, proportion of children vaccinated, proportion of adults who smoke or engage in other risk behavior, and so forth.

Financial accounting is a similar and especially elegant process. A common metric (money) is used to measure inflow and outflow, investment and retained earnings, expenditures on materials, tax liability, debt, and many other processes that reflect the performance of the organization. As noted in the last chapter, this has given rise to a way of tracking sustainability performance in business called the "triple bottom line" because it follows performance as measured by three sustainability accounts: business/financial, environmental, and social. The triple bottom line will be revisited after a deeper discussion on indicators demonstrates how the environment and social "accounts" can be tracked, since they cannot be monetized.

Ecological and Health Indicators

Tracking indicators is an effective method for studying and managing ecosystem services and their adequacy in meeting quality and quantity needs for human sustainability. Some characteristic indicator of the system or situation is determined, validated, operationalized through some sort of data gathering or tracking system, recorded, and analyzed and interpreted for trend. Validation is an especially crucial step, because the indicator has to reflect something fundamental in the system and must reliably change in a way that is predictable and interpretable when the underlying conditions of the system change.

Indicators are used to determine the status of a complex system when it is impractical or unnecessary (as when the essential variables are known) to conduct a detailed study on all its parts. A collection of relevant indicators is like a dashboard, providing information on the system in much the same way that the dashboard of an automobile provides information on its operations (speed, revolutions per minute), condition (whether the engine is cold or overheating), consumption (the gas gauge), and risks (warning lights for low oil pressure or low tire pressure). Indicators are used to measure performance of the system while it is operating, to guide corrections, warn of impending problems, and to compare with similar systems and improvement or decline in performance in the same system. Indicators can be "trailing indicators" (also called "lagging," or "status" indicators), indicating the status of the system as it is, or *leading indicators*, identifying trends that will affect its status and performance in the future. (The terms "trailing" or "lagging" and "leading" indicators are more often used in economics, policy, and management than in biology and public health, but the idea remains valid.) For example, tracking emissions of carbon dioxide to the atmosphere defines a leading indicator that predicts future atmospheric carbon dioxide concentration, which is a status indicator to the atmospheric scientist but a leading indicator for the climate expert. Temperature corrected for season and location over time is a status indicator. Atmospheric and climate change are reflected in the trend over time in the status indicator.

Ecosystem indicators are measurements than provide insight into the status of the ecosystem and how well it is responding to stress. For freshwater bodies, for example, reliable indicators may include the diversity of invertebrate species, the chemistry of the water (such as pH or chemical oxygen demand), health of fish in the habitat, and the presence or absence of overgrowth by algae (suggesting "eutrophication," or contamination by excess nutrients). These individual indicators can be combined into a panel of indicators (such as the dashboard mentioned above) that together give a status report on the body of water. Such indicators have been essential to ecosystem conservation efforts because they allow progress to be tracked in a systematic way.

Health indicators are similar measurements for human populations, based on statistics that reflect the health status of most people in the community, such as the overall death rate, perinatal mortality, cancer incidence, smoking prevalence (a leading indicator that predicts future lung and cardiovascular disease), and indicators of cardiovascular disease risk (such as cholesterol levels, although this requires laboratory testing). Clearly, when medical tests are performed on an individual, the results are indicators of that person's health status. Health indicators are measurements that apply to a population, such as a community or a country. On a population basis, health indicators are summaries that, ideally, reflect risk level for serious disease in the population as a whole (for example, the frequency of heart attacks in the population), the distribution of risk in the population (the frequency of obesity in the population), the presence of exceptionally susceptible groups (such as the prevalence of diabetes), and the direction the population is headed in the future (dietary fat consumption or cholesterol levels).

The best single indicators are simple measurements or counts. This is because they are easy to administer and less prone to error than complicated indices or computed indicators with many variables. They are robust, easily collected in the field or obtained from statistical data that are already collected, and are not appreciably affected by small errors. They must be practical to obtain, well documented by scientific studies, and relevant to the public as well as decision makers.

For example, a measurement of peroxyacetyl nitrate (PAN) in air is a valid scientific indicator of the severity of photochemical air pollution but it needs to be interpreted for public policy. An indicator based on PAN that advises the public that eye irritation is to be expected will get attention and have face validity.

Some indicators are derived through calculation and are called "composite indicators." The familiar "air quality index," for example, used by most cities to communicate the level of air pollution, combines measurements of the most significant pollutants (fine particulate matter and ozone, and variably oxides of nitrogen, sulfur dioxide, and carbon monoxide) into a simple, useful numerical scale or a warning, usually color-coded, that advises the public when they should restrict their activity. The air quality index is therefore a composite indicator, because it combines information on several parameters into one scale.

One of the most important health indicators in any population is the death rate, or *mortality*. The crude mortality rate is a composite, consisting of the number of deaths in a given year (a count) divided by the number of people in the population from which the deaths are being counted (the denominator) to yield a rate (so many deaths per 100,000 people per year). However, the crude death rate would be misleading if a population with many old people (such as Japan) were compared to a population dominated by young people (for example, Brazil). So epidemiologists recalculate the rates for each age category, apply them to a single chosen age distribution to get a rate for the entire population that can be compared. This is called *age adjustment*. Another composite indicator is the calculation of how many years of life are lost through premature mortality, compared to an expected life span: *person-years of life lost* (PYLL, pronounced "pill"). PYLL gives a much more nuanced picture of deaths in a population that can be easily translated into economic loss and impact of disease, but it requires more information and some underlying assumptions. (An even more refined indicator is "disability-adjusted years of life lost," or "DALYs.")

However, health indicators are not only specific for environmental health and sustainability issues, even for occupational diseases. Many diseases, such as asthma, have many causes, environmental exposures among them. For such conditions, trend tends to be more important than the absolute number of cases or the rate: because if environmental factors are making the problem better or worse, but other factors are biological or remain the same, at least the direction of change can be determined.

The World Bank has been particularly concerned with setting priorities for environmental management in sustainable development projects. In 1998 the World Bank published a seminal report on the use of available data for the rapid assessment of environmental hazards in the mining sector of Bolivia. The approach was familiar to public health professionals. It employed a matrix of potential hazards and informed judgments of the likelihood of exposure in order to establish priorities for hazard control and remediation. The Bank and other international agencies have become increasingly reliant on indicators because it is imporactical to measure everything all the time.

The "Triple Bottom Line"

Sustainability management goes deeply into the way the business is conducted (for example, dependence on supply chains). Effective sustainability management includes managing reputational risk, monitoring regulatory compliance, energy conservation, pollution prevention, benchmarking against best practices, minimizing waste, and identifying opportunities for environmental gains (for example through reclamation, habitat protection, and reuse) and "green" technological innovation and process substitution.

In order to track progress in sustainability in enterprises, one needs indicators and metrics. The business consultant John Elkington (b. 1949), an expert on corporate responsibility, was engaging the issue of business sustainability in 1994 when he coined the term "the triple bottom line," which refers to tracking progress using appropriate metrics for financial, social, and environmental performance. (See figure 2.3.) The model is based on accounting principles and divides business performance, respectively, into business performance as a "financial account" (drawing on standard financial and business indicators), social performance as a "people account" (using indicators for community investment, ethics, and corporate responsibility), and environmental or ecological performance as a "planet account" (using indicators of resource utilization, compliance, and product stewardship) each of which must be considered separately and each of which must be "balanced" over time and show a return on investment. The overlap between financial and social sustainability represents how "equitable" the distribution of wealth, compensation to workers, job creation, and investment in the community has been. The overlap between financial and environmental sustainability represents how "viable" the business enterprise is, in terms of financial stability, costs, resource utilization, and efficiency. The overlap between the social and environmental accounts represent how "bearable" the burden has been on society, taking into account environmental justice issues, the ecological footprint and resource consumption, and occupational health and safety performance. Achievement in all three represents progress toward sustainability, represented by where all three accounts overlap.

The triple bottom line (often abbreviated "TBL" or "3BL") has become a standard business tool, especially for reporting to investors in the annual report of a company. The triple bottom line has been criticized, however, mostly on the grounds that 1) measurement is imprecise (especially in the social account), 2) the financial account will continue to dominate decision making because of responsibility to shareholders (see chapter 13), 3) the three-account system is not a reflection of reality because of failure to account for externalities, and 4) corporations cannot be trusted to represent their own performance accurately.

The triple bottom line also stimulated the development of "sustainable accounting," a school of thought and movement within the accounting profession that has developed relevant indicators and means of tracking, documenting, confirming, and auditing them, so that performance can be verified. Sustainable accounting provides means and methods for providing reliable information to investors, other stakeholders, and the public. The accounting profession relies heavily on consistency and comparability, and so a major step forward was recently achieved with the creation of the Sustainability Accounting Standards Board (currently co-chaired by Michael Bloomberg [b. 1942], former mayor of New York City and a major force in business information systems, and Mary Schapiro [b. 1955], former chair of the Securities and Exchange Commission).

Clearly, with its roots in sustainability accounting, the emphasis in developing the triple bottom line has been conservative, to adhere to costs that are documented and benefits that are reasonably estimated. It has a particular problem assessing performance in health, which is surprising because methodologies for assessing health status, health risk, and health effects on productivity are validated and relatively easy to apply.

Refinements in the method may come but there must be more attention paid to health before the triple bottom line can be considered a satisfactory profile of sustainability performance. Health is usually tracked in the social account. Despite the high state of development of health statistics and monitoring methods, it is not clear that sustainable accounting is using all the information

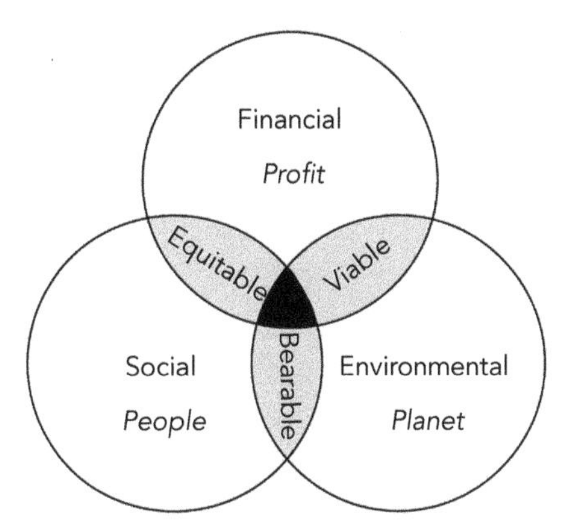

FIGURE 2.3 The "triple bottom line," conceptualized by John Elkington, consists of three "accounts," representing the economic, environmental, and social performance of a company. Where they come together represents sustainability. Health is considered in the social account.

available to track health gains or even that health necessarily belongs in the social sustainability account, at least exclusively. For example, the relationship of state of health to work productivity is measurable and important in a workforce, and the methodology for assessing the effect is well developed and validated in the field of "health and productivity" studies. However, this effect belongs in the business and financial sustainability account, not the social sustainability account and is logically a form of investment. Similarly, health issues associated with environmental performance (improvement as well as hazards) may be better calculated by the methods of risk assessment (as used in regulatory analysis) than by valuation methods, because they usually pertain to future risks and behavior or regulation taken to avoid those risks. Health measures also tend to be driven by insurance costs in business because present costs are obvious and already recorded, and future costs for health care can be determined by actuarial methods. Insurance costs, however, are a business expense, not a particularly informative indicator for social well-being. The triple bottom line therefore does not account for health particularly well. Fortunately, it has the flexibility and conceptual robustness to be improved and refined and to accommodate a more sophisticated accounting of health.

The triple bottom line is a major conceptual breakthrough and has put business, not regulation of environmental management, at the forefront of the modern sustainability movement. This is enormously important in itself, and not only because it lowers resistance to measures that might otherwise be considered antibusiness or a restraint on free enterprise. It is also important beyond the fact that acceptance by business advances adoption of sustainability as a cultural norm, institutionalizing it in society and consumer expectations. The simple truth is that the acceptance and integration of sustainability into business is key to building a sustainable future. Businesses, not communities or governments, run things, make things, and deliver the goods and services that the world needs. One of the great achievements of the sustainability movement has been to recruit the cooperation of business, whatever the motives or degree of engagement.

3

Catastrophic Failure

THIS CHAPTER ADDRESSES abrupt and unwanted transitions, discontinuities in sustainability, usually with extreme consequences for health, that force a change in society and often the global order. Among the topics discussed is population, which has always taken its urgency from perceptions of impending catastrophe.

The term "sustainability" must include the meaning that society and the order of things can continue without special effort, large investment, rescue, or intervention. Sustainable development achieves this by design and by minimizing the requirements of the sustainable economy so that it can be maintained without exhausting resources or compromising the future of the society it supports. However, sudden discontinuities and unanticipated adverse events do occur and can be a destabilizing shock to the sustainable system, especially if they are disproportionately large and unpredictable. Catastrophic failures are arguably the greatest threat to sustainability, whatever their cause. Preparation for disasters and catastrophes is therefore essential for sustainability, and planning for them is a requirement of sustainable development at every level, from the enterprise to the community and to the highest level that social organization has yet to achieve: on the level of cooperation among nations. Table 3.1 describes the relevance of disasters and disaster prevention to the values of sustainability presented in chapter 1.

A "catastrophe" is an event that is unexpected and that ends something of importance. By definition, a catastrophic event or failure compromises sustainability with respect to the existing situation (i.e., the status quo). Catastrophes are events that overturn the existing order; they are also uncommon and can not only cause extensive damage and misery with real suffering but also may destroy the community's ability to respond to and to control the damage—and in so doing, change the conditions for life afterward. The original meaning of the word (in English, since the mid-sixteenth century) meant that the status quo was knocked down by the event ("*kata*" means "down" in Greek, "*strophein*" a

Table 3.1 Sustainability values table: Catastrophic events

	Sustainability Value	Cultural Component
I.	Long-term continuity	Disasters and catastrophic failure limit measures for sustainability in the short term and compromise long-term continuity.
II.	Do no/minimal harm	Disasters force short-term expedients and disruptions that may cause harm to the environment while sustainable measures are suspended.
III.	Conservation of resources	Disasters compromise conservation of resources for the future, both by creating immediate non-negotiable demands and by damaging or destroying infrastructure.
IV.	Preserve social structures	Disasters may do harm to social structures, depending on whether they follow the pattern of "natural" or "technological" disasters, as described below.
V.	Maintain health, quality of life	Disasters compromise health and the quality of life in the short term; the degree of the disruption, and whether it is catastrophic in magnitude, determines whether this will carry over into subsequent generations.
VI.	Performance optimization	Disasters require diversion of resources and so reduce economic, social, and environmental performance; however, because of the requirement to make new investments and the opportunity to replace old structures, they can sometimes stimulate economic performance in the short term and can lead to more sustainable and resilient alternatives.
VII.	Avoid catastrophic disruption	Disasters carry a risk of catastrophic disruption that may set back sustainable economic and infrastructure development; for example, the 2010 earthquake that struck Haiti.
VIII.	Compliant with regulation	Disasters interfere with compliance with regulation and best practices; for example, during the response to the tragedy of the World Trade Center in 2001, the health risk was compounded by failure to provide basic personal protection for the responders, causing avoidable disease later.
IX.	Stewardship	Disasters initiate their own logic of necessity and often sideline stewardship and moral or social responsibility for anyone other than the immediate victims.
X.	Momentum	Disasters may foster stewardship and responsibility in dealing with other people but only if there is an effective and shared response; they rarely extend to include other species or the protection of the environment until the emergency is passed.

turn). A catastrophe is absolute and represents a complete failure and thus a forced discontinuity with what has gone before; it is usually (unless used ironically) the reversal of something big, such as a community, a country, a civilization, or an institution, rather than an individual's fate. In European history, the great example of a true catastrophe was the Lisbon Earthquake of 1755, with a magnitude of 8.5, which killed well over 10,000 people, impoverished one of the richest countries in Europe, ended the building phase of the first global European empire, caused philosophers to question the meaning of life, and moved theologians to ponder on whether God is just. It was the inspiration for Voltaire's famous satire *Candide* (1759), which is a savage, sarcastic parody of optimistic thinkers of the day who, he said, believed that the world they lived in was "the best of all possible worlds" despite the suffering all around. By contrast, in 1960 an even bigger earthquake, with a magnitude of 9.5 (the largest ever recorded) killed 6,000 people in Chile; but the impact was purely local, and the country, which was accustomed to earthquakes, carried on otherwise changed. Although enormous and tragic for the people and communities directly affected, this even larger earthquake resulted in no appreciable social change, and so cannot be considered a catastrophe.

The even older word "calamity" (originating in the fourteenth century in English) has much the same meaning as catastrophe but without the required sense of the established order being threatened. A calamity is a huge disaster, representing a huge misfortune or setback, whether it happens to the community or to an individual; however, it does not necessarily change the direction of society at the time. "The Great Calamity" is a name often given to the Irish Famine of 1846–1851, when the potato blight caused widespread crop failures and starvation, aggravated by British government policies that blocked measures to provide relief. During the tragedy a million people died, and 1.6 million (about 25 percent of the Irish population at the time) emigrated, mostly to the United States and Canada. Politically, almost nothing changed in the aftermath of the tragedy, but socially and in terms of attitude, everything changed. From that time on many historians believe Irish independence was inevitable, although it would take more than a half-century longer to achieve it.

The definition of a "disaster," on the other hand, is more limited. The term refers to an event that exceeds the ability of the affected community to handle the problem. A disaster can be a two-car collision on stretch of highway far from a hospital, or it can be a near-national emergency such as a lethal hurricane.

Catastrophic failure is not necessarily the same thing as a single catastrophic event, since a failure may take place over time and involve a slow, inexorable slide into unsustainability. An event that can properly be described as catastrophic suspends the possibility of achieving sustainability and requires a reach for survival at almost any cost.

The most credible event external to the planet that could lead to a catastrophe of planetary proportions would be a collision with a large asteroid or other high-mass "near-Earth object" (NEO). How high must the probability be before action is justified? Assuming that something could be done, if the end of all life on earth were a real possibility, would one chance in a thousand (meaning 999 to one odds *against* the possibility) be too much to tolerate? Almost certainly, based on past history, the nations of the world would require much greater certainty in the assessment of risk before committing resources that might amount to a significant fraction of the world's wealth. On the other hand, if the cost were relatively low, such that one nation could do it (say on an extended space mission), the task would probably be eagerly taken up by countries as a point of national pride. Somewhere between the two extremes are two lessons for sustainability and application of the precautionary principle: the first lesson is that even in the case of the most devastating consequences imaginable, there will inevitably be a trade-off between uncertainty and commitment. The second lesson is that the most reliable unifying force in any society is an external and existential threat.

Existential risk other than by astronomical mishap has been a realistic possibility only since the development of nuclear weapons, a risk that achieved its own form of sustainability through the morbid doctrine of "mutually assured destruction" (the fear of mutual annihilation if either side acted) and the fortuitous limitation of massive nuclear capability during the Cold War to two superpowers. Even so, it is known that the world came extremely close to (and possibly within days or hours of) a nuclear exchange between the United States and the Soviet Union during the Cuban missile crisis, suggesting that the political status quo was far less stable than the world realized. Notwithstanding the metaphor of brinksmanship, the threat of nuclear annihilation is entirely unsuitable as a model for managing the existential threats posed by ecological disruption such as accelerated climate change.

The possible health consequences of global atmospheric change represent the biggest prospect for global catastrophe at this time, the most serious threat to the global environment, and the catastrophic situation most likely to cause permanent changes in civilization and planetary stability since the prospect of thermonuclear war receded in the 1980s.

Global Atmospheric Change

Catastrophic changes do not have to occur suddenly or just in one place. They can occur over a period of time and in different places. However, when this happens, and the effects are protracted and distributed, the process may not be recognized as catastrophic at the time. The ecological catastrophe popularly

BOX 3.1

Ghostpine: A Cautionary Tale

The community of Ghostpine (not its real name) is a small town in the Canadian province of Alberta, not far from the city of Red Deer, on the north end of the long, narrow Ghostpine Lake. The countryside is rolling prairie, pretty during the summer, and there is little rainfall in that season. Few people live in Ghostpine year round, but the population swells to a few thousand during the summer months.

Ghostpine has become a popular recreational area for the vast rural region surrounding it in southern Alberta. The snow melts quickly in the spring there, opening the lake early for jet skiing and windsailing. In May, families move into their vacation homes, rented or owned, for the season. Schoolchildren and adolescents come by the score to the several summer camps on the lake. Many of the families are affluent and have successful farming operations in the region. There is a large country club for golf and a nice restaurant, a movie theater, and a locally famous ice cream shop. People who have homes there feel that they have a good life.

From the standpoint of environmental sustainability, however, Ghostpine has its issues. Fishing is not allowed in the lake, although it is home to many species of fish, including walleye, burbot, northern pike, and yellow perch. As everywhere on the prairies, the aquifer (the water table in the ground) is relatively close to the surface, and residents depend on well water, which they of course want to protect. The highest point in elevation in the vicinity is a hill 200 meters above the lake. Groundwater movement from the hill is toward the lake and the town. The lake is drained by a creek that meanders to the distant and picturesque Red Deer River. However, the main problem of environmental management facing Pine Lake in the 1990s was what to do with trash.

The town had a "landfill" that was basically just a dump situated in a hole in the prairie outside of town. It was unlined, which meant that chemicals from decomposing material could leach into groundwater, and it was uncontrolled, which meant that anyone could put anything in the dump because there was no supervision. And the dump was reaching its capacity after several decades. The dump presented all of the problems of unsecured landfills, as described in chapter 11, except rats. (There are no rats in Alberta.) For a large landfill, traffic and noise can be problems, but this one was fairly small. In the short term, however, the biggest problem for Ghostpine was that the landfill was running out of space.

So to address the end of the useful life of the Ghostpine landfill, the regional waste management authority that included Ghostpine developed a proposal to replace it with a lined "sanitary landfill." The landfill proposed for Ghostpine was to have had all of these modern features. Landfills tend to be highly contention, as demonstrated in chapter 11 (see boxes 11.2 and 11.3).

The cost of the new landfill was estimated to be Can$2 million, which would be recovered over years from "tipping fees" (charged every time a person brings trash to the landfill). This was a lot of money for a small community. To be economically viable, the new landfill would have to accept trash from the several small communities in the surrounding area, which together had a permanent population of at most a few thousand.

The authority assumed that this would be a big improvement because the potential impact on the environment would be far less than already existed with the unmanaged, nonsanitary landfill. The proposed project was approved by the provincial department of the environment and approved and advocated for by the provincial department of health. Most local government officials were in favor of the landfill, but the local health unit was concerned about the possibility of adverse health effects.

Once the project was announced, however, local residents were strongly opposed and organized to stop development of the new, sanitary landfill, even though it would have seemed to benefit their community by providing a means for trash disposal and would be an improvement over the existing facility. The residents cited concern over groundwater contamination, traffic, and nuisance problems, all of which the authority thought it had addressed satisfactorily. However, it soon became clear that opposition was rooted in fears that such a facility would change the community. Residents thought that it would attract more people to Ghostpine, overcrowding the community, changing its cozy local character, and leading to unwanted development. Above all, they deeply resented being the recipients of other communities' trash and having the town labeled as "the landfill site." The residents' association spent Can$60,000 and considerable time opposing the proposal by petitioning, attending hearings, lobbying politicians, and filing appeals.

Finally the matter was referred to the provincial public health appeals board, where the final decision would be made. The board accepted the new landfill site and plan.

Construction was authorized. Immediately there were problems. During installation, the contractor failed to install the liner to specification, causing doubts about its integrity. (Torn liners are a common problem in landfill construction.) It had to be repaired, at considerable expense. There were numerous technical problems that raised more doubts about the landfill and shook confidence in the authority. In the end, construction costs of the landfill were far over budget with an embarrassing history of problems. Because of the escalating cost, the operating cost of the sanitary landfill rose to Can$70 to 80 per ton, compared to Can$33 per ton at the next-closest landfill, which was a forty-five-minute drive away. To be viable, the landfill had to receive trash from a wider area than expected and to charge higher fees. This was not the deal that residents had been expecting.

The residents were angry. The project became a symbol to them of government mismanagement and the focal point of political anger and resentment.

Then, at 7:05 PM on 14 July 2000, a tornado that had formed on the prairie to the west during a thunderstorm struck Ghostpine Lake. There was not enough time for residents to act on warnings from Environment Canada. Twelve people died in the tornado and much of Ghostpine was destroyed. For kilometers around, crops were destroyed by the storm. Ghostpine was devastated.

In the aftermath of the deadly storm, Ghostpine rebuilt and is again a pleasant vacation village and retreat. The tornado did not affect the landfill itself, but nothing was quite the same in the town again. The new landfill could not comply with its permits and went bankrupt financially and had no prospect of financial viability. The authority that sponsored it went Can$1 million into debt because of the project. The landfill was closed in 2003. Residents had to haul their trash forty-five minutes away, until a new solution could be found. There was concern that people in the area would begin simply to dump trash without any controls, because of the high hauling costs and inconvenience.

There are many lessons to take away from the experience of Ghostpine. The town officials were not necessarily wrong in their decision. The townspeople were not necessarily right in their opposition. People are sensitive to perception of their communities: they do not want something they perceive as nasty in their backyard. (See chapter 11.) Concerns over health effects are often used to oppose an unwanted project. Planning and implementation are two different things. The alternative to the new landfill was still going to be worse. Reality and perception are not always aligned. People often act in ways that do not seem to be in their own interest.

Above all, the lessons of this story are that it is not possible to control everything and that bad things do happen. Sometimes they happen unexpectedly and are beyond human control, and sometimes they happen despite all good intentions. In all things related to the natural environment, the built environment, and sustainability, a little humility is in order.

known as the "Dust Bowl" that began in the 1930s in the United States and in the 1920s in Canada, greatly complicated the severity of and slow recovery from the Depression in the central rural regions of both countries.

Existential threats due to ecological disaster on a global scale are more likely to be distributed than concentrated. At one level, the health implications of global ecological change are easy to predict. If the planet expires, so does humanity. However, the details at levels short of the existential are much more complicated. They involve multiple outcomes, widely distributed sources, regional variability in outcomes, and the actions of many parties with conflicting interests—all competing for advantage and self-interest. Whether human extinction would occur in any but the most dire scenarios of accelerated climate change seems unlikely; but there could be catastrophic changes in society long before survival is threatened. Conceivable futures in a hot, dry, food-poor world are likely to be impoverished, brutal, depopulated, and retrograde, and unlikely to support the level of organization and cultural development consistent with a safe, healthy, and decent life and society. Such cataclysmic effects therefore make the threat presented by climate change an existential one even if human beings were to survive.

This section will not attempt to cover the science behind climate change—for that the reader may seek out the series of authoritative reports of the Intergovernmental Panel on Climate Change (IPCC), perhaps the most effective scientific but least politically persuasive bodies in the history of the United Nations. Rather, this section will provide a brief overview and will emphasize the point that climate change is the biggest of a set of problems related to changes in the earth's atmosphere that have widespread implications. This global atmospheric change has as its roots a burgeoning world population demand dependent on an unsatisfactory technological base for supply. The result is driving pressure caused by dramatically increased participation in the modern economy, both globalized and local. This increased participation is evident in both affluent and emerging economies, both of which rely on polluting technologies that destabilize the earth's atmosphere.

Readers of this book will already know that the enhanced greenhouse effect is likely to result in changes in regional climate and weather patterns and local weather disturbances, resulting in health problems related to heat stress, natural weather disasters, changes in vector distribution and consequently infectious disease distribution, unreliable crop production, and social disruptions leading to violent behavior, civil unrest (historically mediated largely by high food prices), and stress on the health-care system.

Global climate change and ocean acidification are not the only global environmental threats resulting from atmospheric change. However, they represent the most serious threats: those most likely to result in catastrophic change by the definition given earlier and as an existential threat to vulnerable communities. The fundamental problem is atmospheric change, driven by the twin forces of industrialization and modernization by the world's population, and of inadequate technology that is incompatible with natural systems and unsustainable. In this way, the drivers of change are much the same in the pressure they put on limited resources and other serious sustainability problems.

However, other atmospheric changes provide reason for hope. Stratospheric ozone depletion and increased incident ultraviolet irradiation at surface would have ultimately resulted in an increased incidence of skin cancers including melanoma, increased frequency of cataracts, accelerated skin aging, and possibly alterations in immunological response. As it turned out, stratospheric ozone depletion has become a model for international cooperation to mitigate global atmospheric problems, since the Montreal Protocol of 1987 set in motion a sequence of international agreements and actions that halted deterioration and may have reversed the process. Transregional transportation of pollution, as with acid deposition (sometimes called "acid precipitation" or "acid rain") is another example of atmospheric change that has been successfully addressed and mitigated by both effective regulation of emissions and international cooperation. The long-range transport of air toxics has been substantially reduced by international cooperation, particularly the transport of persistent organic pollutions such as obsolete organochlorine pesticides, dioxins, and polychlorinated biphenyl compounds (PCBs) that migrate in the atmosphere and tend to concentrate in northern latitudes. This important reduction has come about specifically through restrictions on production and release of many of the toxic agents under the Stockholm Convention on Persistent Organic Pollutants (known as the "POPS Treaty") and implementation and regulation on the national level. The recent (2014) historic agreement between the United States and China on sequentially phased reduction of greenhouse gas emissions is a breakthrough in practical climate change mitigation.

Global climate change, stratospheric ozone depletion, transregional air pollution transport, atmospheric migration of toxics, and local air pollution should therefore not be considered as separate problems. They are much better understood as different manifestations, spatially segregated, of a common problem, which is an inefficient, dirty, and inadequate technological base providing for too many people given the finite capacity of the environment.

Figure 3.1 shows how atmospheric changes are driven by modernization (economic and social), demand, modern technology, and energy policy, resulting in a wide range of air quality issues, ranging on a vast spatial scale from climate change to the most local atmospheric conditions, such as indoor air quality and airborne occupational hazards. Atmospheric change is used in chapter 7 as an example of an important analytical model called "DPSEEA" (adapted as "DPSEEA + C").

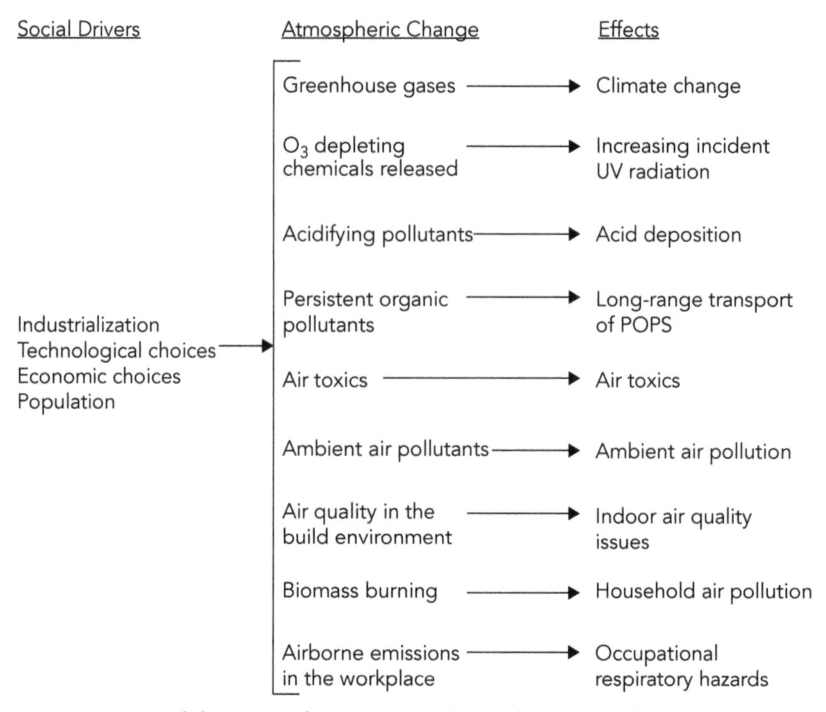

FIGURE 3.1 Social drivers resulting in atmospheric change, and the many issues they create, ranging in scale from global to "microenvironments" in the home and workplace. The top four issues are discussed in the text; two issues on the right, climate change and stratospheric ozone depletion, are potentially catastrophic on a global scale.

Of the various atmospheric changes described, two qualify without question as global and catastrophic in their implications: climate change, and stratospheric ozone depletion.

Global Climate Change

The single most likely cause of global catastrophic failure in the current era is global climate change. The primary issue for global climate change is the exaggeration or enhancement of the so-called (and misnamed) greenhouse effect, a critical physical feature of the earth's atmosphere that makes life in its present form possible.

The natural greenhouse effect contributes to the stability of the world's temperature by preventing the escape of radiant heat into space. This maintains the "biosphere" (the space on the planet capable of supporting life) within a temperature range conducive to existing forms of carbon-based biology. In effect, electromagnetic radiant energy infrared passes through the atmosphere to warm the earth, but the heat so created is reflected back by molecules of "greenhouse gases," principally water vapor and the natural background level of carbon dioxide (CO_2), with some naturally occurring methane (CH_4), together with methane released from human activity, primarily from rice paddies and other agricultural sources but also fugitive emissions of natural gas. These gases permit the passage of visible and ultraviolet (UV) radiation but reduce the escape of infrared (IR) radiation from earth by absorbing at the IR energy wavelength and then re-radiating the energy as heat. This re-radiation goes out in all directions, but a large part goes back toward the earth, where it came from. The net effect is to prevent loss of heat energy from the planet by restoring much of the infrared energy that was lost, not by acting as insulation. Greenhouse gases therefore play a role in maintaining surface temperature far out of proportion to their fraction in the atmospheric (for carbon dioxide, a mere 0.03 percent). Without the global greenhouse effect and the carbon dioxide concentration in air that causes it, virtually all radiant heat from the sun would be reflected back into space, and the surface temperature of the earth would be many degrees below freezing.

The comparison with a greenhouse is imperfect; in a real greenhouse, heat is retained by an insulating blanket of warm air that is retained in the space, not by return radiation. The earth is, in effect, receiving back by re-radiation energy that would otherwise radiate out into space as heat loss, like a potato wrapped in aluminum foil. The aluminum foil reflects heat back into the potato; it does not prevent heat from leaving the potato by insulation, as anyone knows who has picked one up after baking.

The greenhouse effect has been known for over a century, and its essential role in enabling life on earth was obvious to the first person who recognized it, the eminent Swedish scientist Svante August Arrhenius (b. 1859–d. 1927). The significance of Arrhenius to modern science can scarcely be overstated. After an inauspicious beginning to his career (not unlike Einstein), he made fundamental contributions to chemistry in the theory of acids and of salts, electrical conductivity in solutions, and physical chemistry (the kinetics of chemical reactions under physical conditions). His recognition of the greenhouse effect (which he apparently named) came directly from his interdisciplinary training as a physicist and chemist. Arrhenius correctly understood that an increase in carbon dioxide in the earth's atmosphere would lead to overall warming.

The surface temperature does fall if incoming radiation is blocked by stratospheric dust particles, as has sometimes happened after massive volcanic eruptions. There is some evidence that particulate emissions to the lower atmosphere are doing just that, but not enough to counter the enhanced greenhouse effect. The implications of an artificially enhanced greenhouse effect have been pondered for decades, but in the early 1980s global cooling even became a serious concern in the context of nuclear war or volcanic explosions, which would raise up clouds of dust into the upper atmosphere. But by the late 1980s attention had shifted back to the less dramatic but equally ominous implications of an enhanced greenhouse effect. The idea of reflecting heat back by adding sulfate aerosol to the atmosphere has been proposed as a possible intervention if and when climate change reaches a critical point threatening human sustainability. This, however, is full of short-term risk if the target were over- or undershot; and, because the sulfate loading only persists for about a year, this intervention would have to be repeated at least every year for decades, until greenhouse gases decay to sustainable levels. "Geoengineering," which is discussed further in chapter 10 because of its close relationship to energy technology, should be understood to be a last resort in the event of catastrophic failure, not a desirable solution or an acceptable alternative to prevention.

Root Causes and Drivers

The planetary greenhouse effect is natural but limited and always changing in the long term. The earth has never been completely in equilibrium with respect to atmospheric carbon dioxide and climate. The atmosphere has changed constantly over geological time and the evolution of plants and when it did the level of carbon dioxide has changed, often profoundly. Water vapor, which in the natural atmosphere plays an even greater role in radiative forcing (the effect of re-radiating heat back to earth), increases with surface temperature through evaporation, and so there is already, at the simplest level, a complicated interplay between carbon dioxide, other atmospheric processes, and temperature. Geochemical and paleobiological evidence demonstrate conclusively that atmospheric carbon dioxide is a major controlling influence on climate and has been since at least Paleozoic times (hundreds of millions of years ago, when life was recovering from the first great extinction). A close association between carbon dioxide levels and contemporaneous estimated temperature is discernible in all these studies. However, because these changes take place over many thousands of years, life on the planet has usually (not always) had time to adapt. This time is different.

Superimposed on the natural greenhouse effect is an additional, "enhanced greenhouse effect" caused by release of "anthropogenic" (generated by people) "greenhouse gases." These greenhouse gases released by human activity are mostly carbon dioxide (at about 70 percent of the radiative forcing effect), methane (at about 20 percent, because although methane is thirty times more effective as a greenhouse gas, its effective lifetime in the upper atmosphere is only twelve years), and nitrous oxide from agriculture (less than 10 percent, which is low in actual concentration but with hundreds the effect of carbon dioxide and lasting more than a century).

At a planetary level, the physics is even more complicated than this, involving absorption of electromagnetic radiation by carbon dioxide and re-radiation to the surface of the earth, and cooling of earth's surface by radiant heat loss, and planetary redistribution by convective air currents, convective ocean currents, and evaporation of water. On a regional level at the surface of the earth, local weather conditions are driven by complicated meteorological factors that make prediction difficult. Although the long-term and planetary or continental trends are clear, the short-term and local trends remain difficult to discern.

The enhanced greenhouse effect added to the existing greenhouse effect drives the average global climate to be progressively warmer, just as carbon dioxide levels in the past have driven climate. This warming then releases methane from natural biological sources, such as wetlands, and also reduces the ocean's capacity to hold carbon dioxide. The result is that average warming due to the enhanced greenhouse effect drives a "positive feedback loop" that makes the total greenhouse effect that much worse. There is a demonstrably and rapidly rising concentration of these so-called greenhouse gases in the troposphere (the layer of the atmosphere closest to the surface of the earth), primarily carbon dioxide, methane, oxides of nitrogen, and the less abundant but much more potent chlorofluorocarbons, ozone, and even water vapor (which is more abundant in the atmosphere due to rising temperature and an example of a positive feedback). Some of this diffuses into the upper atmosphere and contributes to the enhanced greenhouse effect. Methane, especially, may reach a tipping point at which atmospheric concentrations dramatically increase due to release from wetlands and permafrost due to rising temperature.

The primary driver of this atmospheric change is energy use, discussed in chapter 10. Greenhouse gases produced by human activity are mostly derived from the combustion of fossil fuels, especially coal, as a reflection of power generation required for economic development, consumer demand, and transportation. Methane mostly comes from agriculture and other biological sources such as decomposition of rotting vegetation but some leakage of methane from the natural gas distribution system is contributing. Since accurate recording began, the rate at which these gases have accumulated in the troposphere has accelerated sharply, largely because of the magnitude of the increase in combustion of fossil fuels.

Temperature

Although global warming trends are usually expressed as average temperature, this way of looking at the problem is misleading and purely an artifact of the poor spatial resolution of computer models. Global warming is not expected to result in a uniform warming punctuated by excessive heat waves. Current concern is much more often focused on extreme weather conditions and the resulting climate-associated disasters, flooding, and extreme variability that may result. Regional climate changes will dominate the new climate regime.

Oscillation in the periodic strong upwelling off the coast of Peru (El Niño) is a major factor in determining planetary weather patterns. The effects of a global warming trend may be largely mediated by dynamic ocean process. Ocean warming would disrupt ocean currents and establish anomalous flows of air, moving the current path of the trade winds and jet streams. As well as causing more prolonged droughts, global warming could also increase the frequency of severe precipitation, especially in the tropics. The result may be to increase the frequency and severity of violent weather disturbances such as hurricanes, tornadoes, severe rainstorms, floods, and blizzards. Such natural disasters have many public health consequences. Prolonged drought would certainly be a serious risk, especially in inland areas closer to the equator.

Regional predictions are much more difficult than global predictions and are likely to be confounded by local factors. A major source of uncertainty is the lack of knowledge of the cloud cover, which could be increased by ocean evaporation but may also cause paradoxical ground cooling. According to most models of future climate, rise in average temperature is generally likely to be less at the equator and in high latitudes and greatest in mid-latitudes. Significant warming has already occurred in polar oceans. Although average temperature increase by itself may impose considerable stress, an even more serious concern is exaggerated temperature extremes. The range from low to high will probably become wider because of disruption of stabilizing forces in winds and ocean currents. Winters may be colder and the summers considerably warmer than at present on average, with temperature excursions far in excess of those now seen and pushing limits of habitability in the tropics.

Global temperature rise will profoundly affect agriculture, forests, and fisheries and other essential economic activities. The magnitude of this impact is now barely discernible but will grow. One major unknown is the effect on coastal ocean upwelling patterns and the consequences for fisheries, coastal climate, and the integrity of the ocean environment.

Mortality in urban populations is affected more strongly by weather systems (so-called synoptic weather conditions) than by heat alone. High heat and humidity are a particularly lethal combination. Individuals who are susceptible because of health conditions to heat-related illness, including the elderly, persons with cardiovascular and respiratory disease, and those on medication or abusing alcohol, are at greatest risk of dying during a heat event (often called a "heatwave"). Likewise, those who cannot afford or have restricted access to air conditioning and safe drinking water and who live alone and in solation are at increased risk. Heat events impose a strain on many essential health and social services.

Urban residents will not necessarily acclimatize to the heat over time. The human body undergoes physiological adaptations to tolerate increased temperature over time. Mortality from heat stress is therefore conditioned by acclimatization of the population to sustained increased temperature which occurs over relatively short periods of time: that is, the degree of physiological adaptation to heat that has already occurred in the days and weeks before the heat stress. In large populations, this is reflected in the lower threshold for heat stress-related illness and higher mortality during heatwaves in northern cities such as Toronto (33°), where average temperatures are lower and high temperature extremes are less frequent, than in (relatively) southern cities such as Shanghai (36°), where average temperatures are higher. If there is a general, even summer warming in a given year, the effects of a heatwave will be less than if the background temperature were lower. However, that is not likely to happen with global warming. Episodic heatwaves will result from increasingly chaotic weather driven by the energy input from climate change and so are likely to occur more suddenly, often before people have time to adapt physiologically.

Extreme Weather Events

Another major outcome driven by climate change is extreme weather, in the forms of storms such as hurricanes, tornadoes, and monsoons. The dramatic increase in the severity of storms worldwide during the last ten years (although not necessarily their frequency in the United States), is a harbinger of what to expect and, arguably, the first wave of events as the new patterns emerge.

The principal effect of the enhanced greenhouse effect is on the energy budget of the planet. The consequences of all the retained heat needs to be understood as a phenomenon of physics. Heat is energy and an increased energy input means that dynamic systems controlling weather will become more extreme and variable, not more consistent and regular, as the popular imagination originally seemed to assume. Global warming will certainly not be uniform. It is resulting in more frequent and severe, even violent, weather systems and occasional anomalies of short-term and regional cooling.

Climate is changing worldwide because energy is being continually added to the global weather system (wind, oceans, land surface, ice, open fresh water) through progressively enhanced entrapment of radiant energy from the sun because of the accumulation of greenhouse gases. The nature of any complex, unstable system is that adding energy to it increases its instability, in the case of weather for wind and precipitation levels. In the case of weather, the system is constantly active so that every "event" has an immediate consequence elsewhere in the system, and changes in underlying conditions act on a system already in motion that never achieves equilibrium.

Of even greater concern is that the system will shift from one quasi-equilibrium, in which storms were played out within a range of energy levels, to another quasi-equilibrium at a higher energy level, in which storms are much more destructive.

Oceans

The ocean is the sink with the largest capacity to absorb carbon dioxide and heat, therefore to mitigate the effect on the climate; however, this capacity is not unlimited or without consequence.

The capacity of the oceans to absorb carbon dioxide is finite in two important and closely related respects. The first is the amount of carbon dioxide that can be absorbed and the second is the amount that can be buffered to prevent chemical changes in seawater. The oceans are the most important carbon sink on the planet, by far, so diminution in the capacity to absorb more, which is manifested in a slower rate of absorption, is a serious and alarming danger signal of accelerated climate change ahead. The accumulation of carbon dioxide in the atmosphere that promotes the enhanced greenhouse effect diffuses into sea water and causes a gradual acidification due to the formation of carbonic acid. The rise in pH (measuring acidity) of seawater and indicators of a biological effect from ocean acidification are a second alarming danger signal. The result may be a second global atmosphere-related catastrophe as the productivity of the oceans suddenly falls, food derived from the sea becomes scarce, and environmental services provided by oceans that mitigate climate change weaken.

Ocean warming, melting of large deposits of frozen water, such as the Greenland Ice Cap and the Ross Ice Shelf in the Antarctic (which at the time of this writing was occurring rapidly), together with thermal water expansion, is beginning an inexorable rise of sea levels that because the earth's rotation tends to be higher in the west than the east. And like water in a basin being carried in a car, the water piles up against one side through inertia. A worldwide rise in sea level is beginning to destroy coastal wetlands both by flooding and by saltwater intrusion, jeopardizing the complex ecosystems that include most of the world's richest fisheries, threatening already-endangered and highly productive mangrove swamps and causing many estuarine freshwater supplies to become undrinkably saline. Salt water intrusion into coastal aquifers may compromise the fresh water supply of coastal cities.

An important consequence of global warming will be the rise in sea level, due both to expansion of the sea water mass and to melting of polar and alpine icecaps. This will submerge coastal wetlands and disrupt their ecosystems. Low-lying coastal cities, particularly at the mouths of river deltas, will be at risk of tidal flooding. The problem is most likely immediately to threaten major commercial and cultural centers such as Shanghai, Venice, New York, and New Orleans, tidal wetlands such as southern Louisiana, south Flordia, the salt marshes of the Baltic, and mangrove swamps in Asia, and low-lying coastal plains such as Bangladesh and the Netherlands or the coastal regions and barrier islands of the Carolinas. Unfortunately, some flood control measures may inadvertently cause conditions that increase the risk of disease, as in the case of flood control channels that may increase the risk in Bangladesh of *kala-azar* (a parasite infection medically known as cutaneous leishmaniasis).

Some extremely low-lying islands, such as Tuvalu, Kiribati, and Vanuatu in the South Pacific, may be inundated and lose habitable land mass. One mitigating factor could be the growth of coral reefs and sand deposition, which causes atoll-based islands to elevate, at least for some at a rate similar to the rate of sea level rise. However, for coral regeneration and growth to play such a role assumes that local conditions do not disturb sand deposition and that coral remain healthy, which cannot be assumed because of ocean acidification. It also assumes that atoll elevation will keep pace with sea level rise, which is also unlikely.

Exaggerated tidal action will present a significant problem. Tide control is an expensive and uncertain undertaking that may not be sustainable forever (in civilization terms). The Thames Barrier, the world's most ambitious tidal surge protection system, cost 1.6 billion pounds sterling to construct and has already saved London many times as much through prevented flooding. Evacuation of some low-lying areas may be a more sensible approach in many places but would still cost a great deal. In Canada, to name just one country, the Bay of Fundy and the St. Lawrence River valley may face significant exaggeration of tidal bores affecting even upstream cities such as Québec and Montréal.

Coastal communities with heavy development, such as New York, New Jersey, Florida, and North Carolina (not just the Outer Banks), also face exaggerated storm surges during hurricanes, making it certain that destructive events will be worse along coastal communities during storms, especially during high tide.

Ocean Acidification

Occurring in parallel with climate change, and also driven by carbon dioxide in the atmosphere but through a different mechanism, is acidification of ocean water. Dissolved carbon dioxide forms carbonic acid in water, the same reaction associated with carbonated soft drinks and the reason fizzy mineral water tastes slightly acidic. Because life forms in the ocean such as coral polyps have evolved in a relatively constant pH (the main measure of acidity), there is limited capacity to adapt to rapid changes. This, together with warmer temperatures, causes "bleaching," which is depopulation and die-off of coral, so called because the coral reefs lose color when the polyps die. This effect threatens reefs around the world's oceans, especially combined with overfishing of species such as parrotfish and the loss, for unexplained reasons, of sea urchins. Such fish species and sea urchins graze on seaweed, freeing space on reefs that coral can compete for attachment sites. Coral are considered among the species most sensitive to acidification; therefore the depletion of coral and bleaching of coral reefs is regarded as a warning sign of ocean health and an indicator that other species will soon be at risk.

Also, increasing ocean temperature, may result in phytoplankton blooms and "red tides," creating a toxic hazard and depleting nutrient species in littoral ecosystems worldwide, not unlike what is presently observed in the Red Sea.

Terrestrial Ecological Change

Carbon dioxide is absorbed by the oceans and by biota and is therefore continuously removed from the atmosphere; but not enough carbon dioxide is being absorbed to stop the driving effect of the enhanced greenhouse effect. The oceans absorb large quantities of carbon dioxide but already have reached a level resulting in increased acidification (because carbon dioxide in water forms carbonic acid). Since the oceans are the major "sink" for atmospheric carbon dioxide on the planet, this is alarming. The most significant terrestrial sinks for carbon dioxide appear to be the Amazonian rain forest and the northern temperate (sub-Arctic) boreal zones of the northern hemisphere. The integrity of the Canadian and Siberian boreal forests are therefore of global importance in mitigating global changes in climate.

Increased carbon dioxide availability does produce a "fertilizer effect" that causes more rapid growth of plant life and especially forests, although at some cost in terms of desirable species and nutritional value of the vegetation. Reduced nutritional value is not only a concern for crops but also for animals at higher trophic levels. In experimental studies of tropical ecosystems, more vigorous growth of plants did not necessarily result in increased biomass and therefore carbon storage but was associated with more rapid loss of soil carbon and leaching of mineral nutrients. For crops, at least, this suggests a detrimental effect on long-term yields. The fertilizer effect may also promote the proliferation of undesirable species. For example, ragweed, a major cause of allergies in North America, thrives with elevated carbon dioxide levels.

The distribution of vegetation is already changing drastically, in a short period of time relative to the past rate of change on earth. Forest canopies may disappear from many temperate zones and replaced by grassland and savannah. The distribution of many plants, including many such as ragweed that produce common allergens, will shift correspondingly, as would the many insect species dependent on specific plant varieties. Changes in the composition of vegetation on large land masses may cause even wider temperature swings, particularly in temperate zones, since forests moderate temperature. However, in the short term, tropical areas that have been recently cleared for cultivation may be most heavily affected. When formerly homogenous ground cover is broken up, forces are set in motion that lead to more rapid desertification and loss of soil fertility.

Agriculture

Wildly swinging but on average rising temperatures and unpredictable rainfall patterns will make agriculture highly uncertain, despite longer growing seasons because of the higher average temperatures. (Agriculture is discussed more broadly in chapter 9.)

Drought, which at the time of this writing is exceeding recorded history in California and other parts of the western United States, presents a serious risk to food security worldwide, not only in terms of local production and global food supply but in pricing, since poor yields and scarcity drive up prices, and may do so to levels that the poor cannot pay. The predictable result could be a major reversal of the very real progress recently made on food security. Worsening malnutrition, aggravated poverty (with reallocation of family income to food and away from education and other expenses), corruption and hoarding of food supplies, and violence over food distribution are all possibilities.

A combination of many effects could lead to adverse crop yields and even local food shortages, even in North America. Areas now devoted to grain crops may become drier. Topsoil could be lost in dust storms as in the terrible Dust Bowl years of the 1920s (in Canada) and 1930s (extending to the United States) which devastated the prairies; soil conservation techniques are much more advanced today, of course. The consequence of temperate zone warming would also include changing growing conditions, including a decline in soil moisture, which would seriously impair grain production. For example, as the prairies get hotter during warm weather, they will become more arid, lose much of their surface vegetation and tree cover, and become even hotter. The effects of increased temperature and aridity may offset the increased growing yields predicted for many crops as a result of increased availability of carbon dioxide and the possible expansion northward of regions suitable for growing grain and canola. Rich countries able to pay for food security may find themselves competing with a more populous hungry world.

Another major threat to agriculture would be the expansion of the range of crop pests. This is likely to result in more intensive pesticide use. Fortunately contemporary pesticides are less toxic than in the past and biopersistent than before, but greater dependence on pesticides carries its own risks. Some of the new pesticides, such as the neonicontinoids, may threaten ecosystem stability in other ways (specifically for their purported effect on honeybees).

Health Consequences

An obvious and highly visible consequence of climate change has been the redistribution of insect vectors of disease. The ranges of some vectors are likely to change, most probably expanding into temperate zones now free of these diseases.

On a global scale, this is already beginning to cause public health problems, with vector-borne infectious diseases expanding their range. Increased local temperatures also promote survival over winter, faster maturation, and increased biting frequency of mosquitoes. (This is explored in greater detail in chapter 6.)

The number of hot days and the average daily temperature may increase as a result of global warming, but it is unlikely that there will be a relatively even rise in average temperature. There may instead be a marked increase in the frequency and duration of heat waves. This is likely to result in human mortality, as in the past concentrated among populations with less access to high-quality housing and technology such as air conditioning. The effect of sustained excessive heat on mortality is diffuse, affecting all causes of death and not just cardiovascular causes, where the effect is most obvious.

Violence

It has long been known that civil unrest and violent antisocial behavior tend to increase in frequency in very hot weather, leading to the possibility of social destabilization during such periods.

Three scholars at the University of California at Berkeley conducted an examination and meta-analysis of sixty studies based on primary data sets on climate and risk of conflict, from ancient times to recent years. Violent behavior and conflict, including local violence, civil unrest, and civil wars demonstrated a correlation with temperature change in tropical countries. Conflict was determined by many measures (intergroup violence, warfare, civil conflict, violent crime such as rape). The studies, which individually already showed wide agreement, in the aggregate showed a strong correlation between local temperature or loss of rainfall and conflict. The pattern is consistent across studies, despite the heterogeneity of populations, demographics, cultures, and even time periods. The relationships were nonlinear and usually showed a wide plateau of relative stability close to the average temperature or rainfall. However, the curves showed a more or less simple and steep linear relationship with increasing change in temperature. (The relationship is more complicated for cooling change in temperature and some of the curves are J-shaped.) Episodic outbreaks of violent and antisocial behavior are associated with heat waves in more temperate climates. The conditions underlying such unrest is likely to become more frequent anyway in an era of economic chaos, food shortages, and sociodemographic transformation associated with urbanization. This could aggravate the detrimental effects of climate variability in both rich and poor, especially but by no means exclusively in the tropics.

This does not mean that temperature itself, irritability from discomfort in heat, water scarcity, or other consequences of climate change, acting alone, directly cause conflict or are the primary driver for it. However, it is clear that weather and climate events predictably contribute to the risk of violent conflict.

A warming trend associated with increasing risk of violent behavior would reverse a historical trend toward less violence. Overall, the contributing factors to war and mass conflict appear to be in long-term decline, as has been extensively documented by psychologist Steven Pinker. In Pinker's analysis, human behavior is less violent than it has ever been known to be, although it does not seem that way because of media attention and subjective insecurities.

Inadequate Response

Extreme weather conditions are discussed later in this chapter in the section on disasters and can, of course, occur with or without a direct connection to climate change. A robust response to climate change would also have the beneficial effect of increasing resilience to natural disasters and reducing loss from extreme weather events.

However, steps to mitigate or accommodate climate change have been extraordinarily weak. One of the few exceptions has been urban planning and housing in the Netherlands, where, among other innovations, houses have been designed to sit on floatable foundations so that they will rise with changes in sea level or with storm surges. That urban planning measures are already skipping the mitigation phase and proceeding to accommodation suggests that the most farsighted observers have no confidence that mitigation will occur fast enough to prevent a potentially catastrophic outcome. Political partisanship, institutional inertia, denial of the science behind climate change, and a strong reluctance to change makes climate change the leading candidate for future global catastrophe and an illustration, documented in every day's headlines, of the failure of society to respond to an existential issue of ever-increasing urgency.

To date, management of climate change has been quite ineffectual. The "Kyoto Agreement" (the popular name for the Kyoto Protocol of the Framework Convention on Climate Change) was an agreement negotiated under the auspices of the United Nations among 192 countries of the 195 that signed a treaty in 1997 calling for action on climate change (the "Framework Convention"). These countries committed themselves to reducing greenhouse gas emissions, particularly carbon dioxide, by means of regulation to achieve country-specific targets for overall reduction based on their GDP and history of industrialization, with poorer and developing countries committing to less reduction than rich and already industrialized countries. The agreement was acknowledged at the time to be a first step only, because even the reductions stipulated would not be enough to achieve carbon neutrality or substantially mitigate climate change. However, in the event performance was even worse, in large part because the United States did not sign or observe the Kyoto Agreement, Canada later withdrew (in 2011) and because other significant high-carbon-emitting countries (such as China and Russia) were protecting their industrial base and economic growth. In practice, therefore, the Kyoto Agreement, while ineffective itself, became a "proof of concept" that a better agreement could actually work, as the relative success of the European Union in reducing carbon emissions demonstrated the feasibility of the approach. In the end, carbon emissions did fall but as a result of the global economic recession rather than a concerted effort under the agreement. Subsequent efforts to negotiate a second, more effective treaty have failed. (At the time of this writing, another effort is planned for Paris in 2015.) However, the likelihood of future success is not negligible because of a number of trends unanticipated when the Kyoto Agreement was first negotiated: turnover of industrial facilities and the conversion of technology to higher efficiency, resulting in less emissions from traditional sources, the increasing energy efficiency and alternative energy sources of the European Union, the decline of the coal industry in North America (but not in China or India), and the heavy commitment of China to renewable energy sources. These factors, and the proliferation of "green technology," lend hope that when the world gets serious about climate change, the means to respond will be there to do it. The recent (2014) agreement that commits first the United States and then China to carbon emissions reduction, based on the history and recent development of their economies, may represent a new and more promising point of departure.

Ozone Depletion and Ultraviolet Irradiation

In contrast to global warming, the health and sustainability issues associated with stratospheric ozone depletion appear more manageable, and considerable progress has already been made under the Montreal Protocol. However, this has been an "easier" problem to manage than climate change, which is not to say that it has been easy at all. The public health consequences of depleted stratospheric ozone are narrower and more easily projected. The sustainability issues are more easily managed, since removal of ozone-depleting chemicals form commerce, especially the chlorofluorocarbons, was not exceptionally difficult once there was international agreement. However, had international agreement not been achieved, the effects of unrestrained stratospheric ozone depletion would certainly have been literally catastrophic.

The lessons for climate change are clear because many greenhouse gases, including water vapor, methane, nitrous oxide, and various low-level gases (such as cholroflurocarbons, hydroflurocarbons, and sulfur hexafluoride) are both greenhouse gases and potent ozone-depleting gases. The problems are closely related and simply represent different effects of similar atmospheric changes.

The consequences of stratospheric ozone depletion are due to increased exposure to ultraviolet light. Ozone in the stratosphere, particularly, absorbs ultraviolet light completely in the UV-C range (200–290 nm) and a large proportion in the UV-B range (290–320 nm). Inside the cell, UV-A (320–400 nm) is absorbed by proteins and DNA, UV-B by all nucleic and by aromatic acids, and UV-C by all cellular constituents; absorption may lend to breakage of covalent bonds in critical macromolecules and to carcinogenesis, accelerated aging, and cataracts. Those at greatest risk for direct effects of UV exposure on skin are people with fair skin who sunburn easily and the few with rare skin conditions that predispose to UV-induced injury, including albinism and diseases involving deficiencies in the ability to repair broken DNA.

Because low temperatures inhibit the relevant chemical reactions, depletion of ozone is most severe over the poles. The stratospheric ozone layer was observed to be thinning in depth over Antarctica in the 1980s. Repeated observations have confirmed the attenuation and have charted its progress. The "ozone hole" (once reported to be as large in area as North America) placed southern regions of Australia, New Zealand, Chile, and Argentina at risk. In the northern hemisphere, the process never advanced as far.

Before the compounds were withdrawn from commerce by the international treaty, release of chlorofluorocarbons into the atmosphere occurred through industrial activity, leaks, or the decommissioning of old refrigeration and air conditioning units, as well as by use of aerosol cans using the compounds as propellants. Chlorofluorocarbons release chlorine by photolysis in the atmosphere; this free chlorine scavenges ozone and destroys it. One chlorofluorocarbon molecule may destroy as many as 10,000 ozone molecules. CFCs also are potent greenhouse gases, so their removal from global use also made a significant contribution to climate change and demonstrated that atmospheric change is a complicated issue with multiple effects.

Stratospheric ozone depletion is not to be confused with tropospheric ozone accumulation; ozone is an air pollutant in the lower troposphere and a greenhouse gas throughout the troposphere but a vital protective shield against potentially lethal UV-B irradiation in the stratosphere. Stratospheric ozone is regenerated by photolysis of oxygen and is minimally affected by migration of tropospheric ozone upward into the stratosphere.

The human health effects of increased ultraviolet irradiation that are due to ozone depletion include higher risks of cancer. Most concern centers on skin cancer but intense sunlight exposure may also possibly impair immunological responses such that there is an increased risks of systemic malignancies and other conditions. The effects of ultraviolet light on the immune system are thought to be selective rather than generally suppressive, but the phenomenon is too recently discovered for sure extrapolation. At the same time, there is evidence that vitamin D in the form (D_3) produced by exposure to sunlight may be protective against cancer. Clearly insufficient sunlight is detrimental, by reducing circulating levels of vitamin D_3. Thus, some ultraviolet radiation exposure through sunlight is important to good health but too much is dangerous. This is a difficult public health message to get across to people, so the emphasis usually, and probably appropriately, falls on avoiding excessive exposure.

Relatively minor but cosmetically significant effects may include accelerated aging of skin and perhaps increased frequency of pterygia, small wedge shaped tissue webs on the whites of the eye (this is a minor problem). Although unlikely to result in serious health problems, given the vanity of human nature these cosmetic effects may be just as significant in motivating action as the major health effects, if more people were not only informed but became persuaded that sunlight ages them, personally, rapidly and unattractively.

Skin cancer was and remains the biggest concern, however. Increing exposure to ultraviolet radiation in sunlight increases the risk of nonmelanoma skin cancer (particularly squamous cell carcinoma), malignant melanoma (an aggressive, often lethal skin cancer), actinic keratitis (a premalignant condition), cataract, and retinal degeneration. Even at higher latitudes, as in Canada, exposure to sunlight is the principal risk factor for nonmelanoma skin cancer. The biological amplification factor for sunlight-induced nonmelanoma skin cancer is approximately 2 for basal cell carcinoma and 1.5 for squamous cell carcinomas. This translates to an increase in incidence of approximately 2 percent for basal cell carcinomas and 1.5 percent for squamous cell carcinomas for every 1 percent reduction in ozone concentration. However, the amplification factor decreases with increasing latitude, markedly above 30°; this implies that residents of the far North may, paradoxically, be at less increased risk for the carcinogenic effect of increase ultraviolet penetration than those at lower latitudes, despite thinning of the ozone layer predominantly over the poles. The effect at tropical latitudes would be proportionately greater.

Substantial progress on curbing chlorofluorocarbon generation and release on a national level was first with the Vienna Convention for Protection of the Ozone Layer in 1985 and later with the much more stringent Montreal Protocol on Substances that Deplete the Ozone Layer in 1987, both negotiated under the auspices of the United Nations. The Montreal Protocol and its successor agreements have represented a singular success in international environmental diplomacy. Despite progress, substantial new problems emerged, among them the emergence of a black market and smuggling trade in CFCs. Even so, production was curbed (there were initially exceptions for poor countries) release virtually halted, and the ozone layer began to recover. By 2013 it was reported to be nearly normal in thickness.

This issue demonstrates that international cooperation can be effective in managing even enormous global health risks and threats to sustainability.

Catastrophic Failure of the Commons

A particular kind of catastrophic failure that virtually defines sustainability is the problem, or "tragedy," of the commons. This concept, in its modern form, arose out of an appreciation for the potential for mismatch between the number of people in the world and the availability of resources such as cultivatable land and mineral resources. This concern is central to issues of sustainability, has been a major theme in discussions of population for many years (see chapter 2 on the baggage this field carries), and has deep implications for health, so it will be presented in some detail.

Malthusian Crisis

The most traditional concern related to environmental sustainability is the predicted mismatch between population and environmental resources first articulated as a social theory in 1798 by the Reverend Thomas Malthus (b. 1766–d. 1834). His book *An Essay on the Principle of Population* (1798) became, and still is, one of the most influential texts in social science, and indeed in human history.

Malthus observed that population growth, if unrestrained, is exponential but that agricultural growth, by the technology of his era, was mostly limited by land area and so mainly grew incrementally—although he acknowledged that innovation in his day was already expanding productivity. Over time, he reasoned, population growth had to outstrip the growth in food supply and so result in famine and starvation or widespread malnutrition and disease. This Malthusian crisis would limit population by "positive means" (his peculiar term, in the sense of forced correction by disease or famine) and most efforts to relieve the plight of the poor or increase food availability would only delay and increase the magnitude of the inevitable catastrophe. The only solution he saw was to limit population voluntarily by measures that emphasized "virtue" (to the conventional way of thinking of the times): moral restraint such as celibacy, delayed marriage, and not marrying until a couple's income was secure. (Safe and effective contraception was not available at the time to poor people, although rich people had a form of condom. Dangerous or ineffective abortion methods and infanticide were the means of last resort for the desperate.)

As natural history became science and began to study populations systematically, the exponential increase in population turned out to be a fundamental principle of biology, whether applied to human populations or to other species, large or small. The basic assumption that resources limit population proved to be correct at the limits, where population growth is unconstrained and critical resources (such as nutrients) are fixed. The maximum population that a resource base or ecosystem can support is now known as the "carrying capacity" (abbreviated as

"K" in the literature of ecology). Malthus anticipated (but could not estimate by how much) that the carrying capacity for human beings would increase with technology and improvements in food production. He also had no way of knowing the degree to which his ideas would be oversimplified and distorted over the next 200 years, although he was criticized during his lifetime for his bleak outlook.

Malthus was a Cambridge-educated scholar and clergyman from Surrey, England, who became a teacher and professor at the East India Company College, an institution dedicated to the education of civil servants and functionaries in the British colonies. Malthus's thinking was therefore steeped in religion, political theory, economics, and imperialism, with particular reference to the British colonial experience in India. He was also a highly contrarian and pessimistic thinker who in other ways went against the intellectual currents of his time. His theories still, often in subtle ways, pervade contemporary thinking about sustainability and resource management.

Malthus's opinions exerted considerable influence in the intellectual life of his time in and beyond England, where it informed thinking that eventually led to such important intellectual developments as the theory of evolution and the school of thought known as pragmatism. He is considered to be both the founder of scientific demography (the study of populations) and a pioneering theoretical economist (he "discovered" the theory of rents, which is fundamental to modern economics). His startlingly brutal theories and pessimistic attitude were shaped by his vocation as a clergyman and his worldview that human nature could not change and that social improvement could only come from grace as a gift from God.

Just as Voltaire's *Candide* was a response to optimism in the aftermath of the Lisbon Earthquake, Malthus wrote his epic work at least in part as a response to what he saw as the overly optimistic attitudes of influential thinkers of the day, such as the Marquis de Condorcet (b. 1743–d. 1794) in France and Jean Jacques Rousseau (b. 1712–d. 1778) in Geneva. These philosophers believed in unrestrained social progress and the possibility of improving not only the state and material well-being of peoples but the betterment of humankind generally. Malthus was reactionary in that he did not believe that people could create a better world, only God. Malthus lived at a time when the Industrial Revolution was just getting underway and the changes that it was introducing were justifiably feared. The merit and dangers of social reforms as embodied in the French Revolution were also widely discussed and when the Revolution turned into chaos and then persecution and then dictatorship, people were again disillusioned. Where other people saw the difficult birth of a new age and trampled but ineradicable seeds of progress, Malthus saw little hope.

Although not an unkind man in his personal life, Malthus argued against relief to the poor and excessive charity on the grounds that they delayed the inevitable correction, encouraged dependence instead of virtuous hard work, and undermined moral restraint. Instead, he argued for measures that placed a considerable moral and behavioral burden on people, advocating late marriage and sexual restraint as a means of population control to prevent catastrophic overpopulation. There was a deep class-based root to this argument in that Malthus (himself upper class and with a private income) and others of his social stratum assumed that poor people shared a lack of ambition, a weakness for vice (such as drink and prostitution), overindulgence in sex, and an inclination to use easy or expedient methods (such as abortion, which at that time was usually ineffective or very dangerous) instead of confronting moral dilemmas through religious faith. Malthus also thought of overpopulation and the suffering it caused as a form of natural justice that punished people who lacked self-restraint.

After Malthus, it was assumed that a more general global crisis of population mismatched to resources, initially assumed to be food, later and variably water, food, and economic capacity, would result in massive death and disease, which would reduce the earth's population substantially. Predictions were often highly specific, as in the best-selling, highly influential 1967 book *Famine 1975! America's Decision: Who will Survive*, written collaboratively by the Paddock brothers, William C. Paddock (b. 1922–d. 2008, American plant pathologist) and Paul Paddock (n.d., American diplomat). Fortunately, despite predictions that a global food shortage would happen imminently, it obviously did not. Instead, famines have been local and regional and often tied to civil unrest. The most apparent conflict involving agricultural production and sustainability in recent years has been the upward pressure on global food prices, including that in 2011 and 2012 resulting from the diversion of corn production to biofuels, a sustainability objective (although in the case of corn ethanol, an imperfect one).

Since Malthus's famous prediction, famine has indeed been a sporadic condition but more often linked to ineffective distribution, corruption, and policy failure than to global imbalance of supply to a population consuming food at the limit of productive capacity.

What happened, in retrospect, is that although the Malthusian crisis may be said to have operated as expected on a local level from time to time, most of the great famines of history have resulted from: 1) catastrophic crop failure (most of the famines of Europe and more recently in Africa and famously in northern China in 1928 and 1929), 2) political mismanagement (such as the forced collectivization of farms in the Soviet Union and more recent famines in North Korea), 3) deliberate policy (such as Stalin's created famine-genocide in Ukraine from 1932 to 1933, and in Cambodia under the Khmer Rouge in the 1970s), or 4) acts of war (almost constantly during the Thirty Years' War in Europe in the seventeenth century, frequently during both world wars).

Likewise, conflicts over limited resources (mostly water) seem to have been more of a secondary factor shaping the conditions of conflict than a direct cause of outright warfare. Examples of water-related conflicts include tension over water supplies between Israel and neighboring Arab states and the contribution of the massive famine in Syria to unrest and rebellion in the country which led to civil war and then a wider proxy war that is raging at the time of this writing (2014).

On the other hand, food production increased almost exponentially as a result of greater agricultural productivity, agrarian reform, the "Green Revolution" (a systematic scientific approach to increase crop yield led by the great agronomist Norman Borlaug, [b. 1914–d. 2009]), and methods to prevent spoilage during storage, preserve food, and control pests. A major factor has been the globalization or at least regionalization of food supply and other critical resources, so that food imports that sustain a population are drawn from well beyond its home territory, although this places a stress on more distant and usually much larger areas (this is called the "ecological footprint", a concept developed by William Rees, b. 1943). Although many of the food production methods over the last century have undesirable effects, such as a reliance on monoculture (lack of diversity in planting) that results in ecological simplification and instability, on limited genetic stocks of food crops (creating stands that may be uniformly vulnerable to pests or ecological stress such as drought), and on heavy use of pesticides and fertilizers (with unwanted effects on land and run-off into waterways). Future technologies, including but by no means limited to the highly controversial issue of genetically modified organisms as food, suggest that improvement in crop yields will continue and even raise the possibility of other viable, if difficult to visualize, means of providing protein-rich food, such as recent projects using tissue culture to produce consumer-acceptable meat without the heavy energy input required to raise livestock. (See chapter 9 for more discussion.)

The result has been a repeated postponement of the predicted consequences of overpopulation until population growth began to decline as a result of economic development, lower poverty rates, and voluntary birth control, and involuntary limitations of family size (mostly in China as a result of a "one child" policy, which is now being relaxed). In other words, over the twentieth century the carrying capacity for the global population increased more than expected while the population burden slowed its growth, forestalling the crisis. Although global population levels are still rising and may still exceed carrying capacity extrapolating from the projected rates, the Malthusian crisis has been averted and the prospects for it in the future seem remote. Even so, the challenges of climate change and resulting drought in many parts of the world (acutely in California at the time of writing), depletion of "fossil aquifers" (groundwater that does not become recharged from surface percolation fast enough to replenish itself as a water source) used for irrigation, are now presenting another profound set of challenges. However, in a globalized market economy these factors are more likely to affect price than critical availability. Local food shortages, political and social inequity, hoarding behavior, reactive food security policies, and food preference changes and substitutions are therefore more likely than widespread starvation.

It is now clear that global catastrophe, if it occurs, is more likely to be incremental, regional, and insidious or episodic, with a cumulative effect on health rather than a rapid killing off, as envisioned by Malthus and his successors (called "neo-Malthusians"). Catastrophic failures on a global or widespread multiregional scale are certainly possible through climate change (with parts of the world becoming essentially uninhabitable), failure of the oceans, provocation of global cooling through interference with ocean transport of warm water, and more frequent and severe weather-related disasters. Humankind evolved the capacity to protect the species from environmental stress through physiology, behavior, aggregation into protected settlements (which in other species might be described as colonies), the institution of the extended family or clan, cultural adaptations for protection, and technology. Health effects of environmental change are therefore always mitigated to some extent except in the direst circumstances. Health consequences, when they occur, are therefore more likely to be expressed on a local level and to represent failures of resilience.

The Concept of the "Commons"

People manage their private property in their own interest, which normally means maximum gain for the owner over a relatively short period, regardless of the consequences for sustainability. People manage property that they do not own but for which they are responsible, as managers and senior executives, in keeping with performance incentives and the level of accountability to which they are held by their job or position. People have a tendency to neglect or abuse community-held property because the benefit to their personal interest tends to be much larger than the cost to them as their share of the collective value of the resource and because they rationalize away or do not accept the cultural expectations of responsibility to the community. How to manage shared property, which encompasses assets held in common for and by the community, has been a fundamental question in economics and public administration.

As it applies to in sustainability and environmental protection this problem has come to be known as the "tragedy of the commons." This fundamental concept is related conceptually to the idea of the Malthusian crisis and imbalance between demands imposed by an increasing population and a limited resource base. The "commons" has broader applications and other dimensions, but it has become inextricably linked with population studies because of its intellectual history.

Although not new, the idea was powerfully articulated in a landmark essay, "The Tragedy of the Commons," published in *Science* in 1968 by Garrett Hardin (b. 1915–d. 2003), a professor at the University of California at Santa Barbara and a staunch modern proponent of Malthus's theories. Immediately, the essay became an essential part of the literature of ecological theory and, later, sustainability.

The theme of the essay was introduced as a parable. Members of an agrarian community (such as a village in colonial New England, where this institution did exist) share a common field known as the "commons," where all residents could freely allow their cattle to graze without restriction or fee. This tradition imposes no penalty or unit cost for access on the owners of the livestock: they may add cattle freely. On the other hand, the livestock owners gain benefits by turning more cattle out to graze in the commons: they gain more cattle, milk, and meat than they would have otherwise. There is every incentive for each villager who owns livestock to add more and more cattle. However, the commons is a limited resource. It may have been able to take this use in the beginning, when the small number of cattle allows grass to re-grow and the resource is renewable. Soon the point is reached at which the commons cannot support more cattle (i.e., it has reached carrying capacity). However, there is no feedback mechanism through effective social control to discourage villagers from adding yet more cattle because, relatively speaking, they still benefit. The damage to the commons results in diminishing yield for everybody; but for any one livestock owner, their share of the decrease in total value of production is small relative to the benefit of adding one more head of cattle. As more cattle are added, the commons then becomes over-grazed, and the grass cannot grow back fast enough. A tipping point is then reached in which the commons itself is destroyed and the resource fails for everybody. The time between when the tipping point is reached and when the resource fails catastrophically is brief and allows little time to adapt or mitigate the collapse. Restoring the resource involves severe disruptions and deprivation to everyone. In the metaphor, taking the commons out of grazing entirely to let the grass grow back takes years: during this time, the commons would be unavailable to anyone.

Hardin clarified that the tragedy he described was the failure of the *unmanaged* commons, which had not been subject to socially determined restrictions on use. The "commons" is a "free good" in economic theory, a resource readily available with little or no initial user cost. The "tragedy" is that past a certain level of use the resource deteriorates by overly intensive exploitation or becomes inadequate to meet all the demands placed on it. What began as a "free good" becomes a "scarce resource," and real costs are incurred for additional use; the "commons," however, are freely available for unrestricted exploitation, so each individual attempts to maximize his or her own gain at the expense of his or her fellow exploiter. Everybody puts one more cow in the common pasture until there is no more pasture left. From the individual's point of view he or she benefits directly by a unit gain (of one more cow), but the cost is distributed among all the users (less food for all the other cows). Because the system's total performance fails, everybody loses in the end as everyone tries to shift his or her cost onto everyone else, which in economics is called "externalization" of costs. The concept applies equally well to any shared resource.

Examples that are literal applications of the tragedy of the commons are commonplace in ecology and ecosystem sustainability issues: the burning and deforestation of the Gran Chaco wilderness in Paraguay to expand agriculture (principally cattle ranching), the overharvesting of ocean fisheries as well as local fouling and pollution from poorly managed aquaculture, resistance to curbing greenhouse gas emissions, and the externalization of health costs to the public from sources of air pollution. The broader implication of the commons is that everyone belongs to a community that represents a shared legacy, a commons of society, and lives in an environment which is shared not only with other people but with other life—a commons of the material world. The idea of the commons is now one of the fundamental principles of sustainability, having entered the discussion through ecology and environmental protection.

The argument advanced particularly by Hardin was that overpopulation was the most important application of the principle, with each new person drawing from the commons resource. Its single-sentence abstract ("The population problem has no technical solution; it requires a fundamental extension in morality.") carried two unambiguous messages: first, the fundamental problem underlying resource supply, pollution, and equity is population; and second, the only solution lies deeper than science and beyond ethics and law, in the realm of morality and the culture- and religion-grounded terrain of right and wrong. This was a frightening prospect at the time, and those who came of age in the environmental movement at about that time were suitably terrified: this was because the solution could only lie in a goodness that humankind had conspicuously failed to exercise up to that time and seemed unlikely to muster in the ensuing future.

Deconstructing the tragedy of the commons has consumed a lot of intellectual energy since the idea was first put forward. Hardin clarified that the tragedy he described was really social and represented the failure of society to agree on institutions to manage the commons, not a natural law. The fundamental questions were: Why is the resource not managed? Was the commons ungoverned out of principle or negligence? Does the management vacuum exist because of a failure to foresee that market forces fail in a public space, or because of an ideological blind spot experienced by advocates of collective action? If the central problem has as much to do with management as with resource constraints, is it really a valid metaphor for unrestrained population growth or is it a better metaphor for the inevitability of social controls against a romantic but improbable vision of disciplined self-restraint?

In practice, of course, not all users draw from the commons equally. There may be a small number of heavy exploiters and a large number of light exploiters, as in the case of polluting industry. There may also be a large number of exploiters each with a small individual effect, such as automobile drivers. The attempt to distribute costs among other users creates what are called "diseconomies" or "externalities," where an "external cost" is imposed on another party, in effect, to subsidize one user's gain. For example, a polluting power plant saves the cost of cleaning its stacks by letting the entire community, which shares the same resource of air, suffer injury to health, amenity (comfort and aesthetics), and property, and exposing distant populations to mercury from its emissions. These costs are externalized to the residents of the community, who individually may or may not suffer dramatically but in the totality, which is often invisible and always hard to document, have a poorer quality of life and poorer health. Recent US EPA regulations on coal-fired power plants, together with the phasing out of many plants as they reach the end of their cycle, have imposed much tighter restrictions on these sources.

The most serious problem with externalizing essential costs is that the enterprise (company, country, society) has no pricing or cost signal by which to monitor the scarcity or value of the resource and so goes on wasting or abusing it. There is a strong trend in environmental policy to internalize currently external costs by demanding an accounting that is factored into the cost of production and has to be weighed in management decisions, for example, by carbon taxes or tradable permits or severance taxes and user fees. (This is in part the motive behind reporting the "triple bottom line", as outlined in the previous chapter.)

The cost of foregone or future use is also significant. Polluted water must be purified downstream before it can be used again, and that involves a cost, which is usually greater than treating it before it is discharged but which is paid by someone else. Nonrenewable resources that are depleted become unavailable for future use and must be replaced by other materials or technology. The depletion of resources and the replacement of function by new technology is a major theme of applied science and engineering and has been a driving force for technological advancement. Historically, this applied particularly to scarce mineral energy, nonrenewable resources, and wood. It has always been assumed that resource depletion will impose limits to future growth in the face of expanding population.

In the past, it was always assumed that technology and economic development would be limited by resource availability. However, another such limit is the capacity of the environment to absorb and remove contaminants and to withstand pollution. "Loading capacity," is the capacity of a medium, such as the atmosphere, to accept pollution without irreversible adverse consequences. In later years, it became the functional equivalent of the concept of carrying capacity, in managing ecosystem sustainability as it applied to the capacity of the commons to take pollution loads. Loading capacity turned out to be the pivotal concept for the great successes in mitigation of acid-forming emissions resulting in "acid rain" pollution in the 1980s and 1990s. Loading capacity emerged as the metric driving policy choices for long-range pollution transport on a regional macro scale, such as stratospheric ozone depletion and especially acid precipitation. Unfortunately, constraints on growth by limits on loading were largely ignored until just a few years ago in the case of greenhouse gas emission. Somehow, for climate change, both loading capacity and carrying capacity seem to be marginalized concepts.

Conventional calculations of cost are oriented toward human use, not ecological sustainability. A deep ecologist (see chapter 12) finds this offensive, since human use cannot morally or effectively value essential natural functions such as resilience and evolution. Natural systems incur cost in the form of instability, as ecosystems are threatened, resources are extracted, and pollutants are injected faster than and in quantities above the capacity of the environment to adjust.

One way of expressing future instability and the likelihood of damage is by estimating risk. Risk is difficult to predict and is even more difficult to interpret as a cost in economic terms. Risk assessment methodology, which derived from anticipation of human risks such as the health risk of pollution, is emerging as a valuable analytical tool for ecological damage and the loss of valuable environmental services. However, some in the field of sustainability feel that it has been oversold because ecosystem damage is cumulative, and every episode of destruction or ecosystem simplification leads to effectively irreversible loss. (This argument is also related to theories of ecosystem restoration and the idea that the cost of restoring an ecosystem is so great that it is only practical on the smallest scale and that a restored ecosystem is never identical to the natural one that was lost.)

In 2009 the Nobel Prize in economics was awarded to Elinor Ostrom (b. 1933–d. 1912, American), for her work on this problem, which laid out specific rules for governance of "common-pool" resources. They were intended to address this long-standing problem fundamental to sustainability of ensuring fairness, equity in access, and accountability. The following is a paraphrase (especially the "user" terminology) and adaptation of her rules as she interpreted them for the public:

1. Define clear boundaries (what are limits to the resource, who is a "user" entitled to access).
2. Adapt rules for managing the common resource to local conditions and requirements (so that the user community can recognize the rules as appropriate and applying to them).

3. Involve stakeholders (those affected by the rules, including but beyond the users) in formulating and, if necessary, modifying the rules.
4. Affirm and validate the right of the stakeholders to make their own rules and to have them respected by the larger community and relevant authority (usually government).
5. Monitor and document users' behavior.
6. Develop a set of sanctions that are appropriate to violations of the rules (one penalty, such as exclusion, will not work for every infraction).
7. Resolve disputes in a way that is open to all users at low cost (implying that disputes should be kept out of the legal system whenever possible).
8. Empower stakeholders at all levels, with responsibilities and accountability commensurate with their role, their influence on the system, and in layers of interrelated institutions and enterprises (creating a balance of powers and constraints).

Ostrom's solutions were an answer to Hardin's bleak vision and conviction that the tragedy of the commons was inevitable and refractory. Ostrom showed that the commons could be managed after all and that they could be managed by ordinary people once they were equipped with the appropriate organization. Hardin did not live to see the culmination of Ostrom's work. They both agreed, however, that the commons cannot be effectively managed by a market-based solution.

The Commoner-Ehrlich Debate

An important debate that occurred in the 1970s has shaped attitudes and theories regarding sustainability today and should be understood as a foundation of contemporary sustainability theory. At the time, there was a tendency for those concerned about environmental quality, sustainable resources, and about population to break into two opposing camps, based as much on personalities as on philosophies. One camp followed Hardin and Paul Ehrlich (b. 1932), a professor of biology at Stanford University, in emphasizing population as the root cause of environmental crisis; the other followed a school of thought led by Barry Commoner (b. 1917–d. 2012, see chapters 1 and 2), a highly influential biologist and social activist who spent much of his career at Washington University in St. Louis, and the British physicist and ecologist S. P. R. Charter (n.d., active 1962) in emphasizing that technological change was primarily responsible for environmental degradation and caused pressures that overpopulation mostly amplified. Notwithstanding their many shared views, Commoner called much of Hardin's essay on the commons into question, while at the same time tacitly supporting many of Hardin's premises.

Commoner demonstrated in *The Closing Circle* (1971) and other works that the rise in individual consumption and in population could not, even in combination, account for the dramatic rise in environmental degradation in the United States since the Second World War. His conclusion, supported by an extensive analysis of postwar American production and marketing practices, was that the basic technologies of industry, agriculture, and transportation developed in such a way that they are drastically incompatible with natural environmental cycles and processes. He asserted that the search for more profitable production and marketing has led to lopsided technology, successful at meeting narrowly defined goals but miserably failing to suit natural systems, which are closed cycles (see chapter 4). In his writings, Commoner cited many pages of examples of then-recent changes in automotive design, electric power generation, packaging and bottling, manufacturing processes, and farming practice as being self-defeating in the total picture of the commons. Although recognizing the important role of individual lifestyle and population stabilization, he demonstrated that the single most important factor in achieving environmental reconstruction is control over technology and its rational use.

A comparable study would have suggested, as do comparative policy analyses today, that Japan, and now South Korea—a late convert to ecological concerns—and even China are anticipating ecological pressures by acting now to design, produce, and market products that are lower in impact on the environment. (China is a world leader in sustainable energy, as well as a world leader in nonsustainable consumption.) They have learned from history. At the same time, the United States, the early international leader in ecological reconstruction, is economically incentivized to preserve its less efficient infrastructure, slowing adoption of disruptive new technologies with less ecological impact. The process is playing out now with the slow, ponderous movement toward alternative energy technologies, which has been confused in the United States by numerous dead ends (such as corn-derived ethanol) and backtracking (the persistence of coal and the resurgence of oil production).

This analysis is entirely consistent with Hardin's proposal that each user of the commons is acting rationally from his or her own limited point of view in a commons in which there are no rules. Whether on a conscious accounting of profit and loss or as a cultural manifestation of the "Protestant ethic" (the important theory by pioneering sociologist Max Weber (b. 1864–d. 1920), that entrepreneurial and managerial practice is influenced by culture, for which he used religion as a proxy), the business owner or manager who exploits the commons in any of the ways Hardin described is behaving no more unethically than anybody else who tries to maximize his own advantage in any other commons, be it the law, politics, or morality. As long as the commons exists unmanaged, over-exploitation is inevitable. When the commons becomes managed and mutual benefit through cooperation becomes the goal, a whole new set of rules (i.e., a new ethics) is introduced. The problem is that the new ethics must be closely aligned with ecological reality and biological sustainability, or the result will be a well-behaving population devouring the commons in yet another, perhaps more efficient and equitable fashion. One might argue that this is a description of the modern market economy.

Commoner emphasized total systems, science, and social responsibility. The more tightly knit group of population control advocates, of which Ehrlich was the undisputed leader and Hardin a leading light, emphasized overpopulation, politics, and government reform. The emotional charge surrounding the schism obscured the reality that their differences were based largely on emphasis and that their philosophies tended to be complementary rather than contradictory.

An example of the emotional tone to this matter was the offense expressed by Commoner at sentiments expressed by Hardin in a 1974 editorial to the effect that Americans are outnumbered and surrounded by rapidly increasing hungry and potentially hostile populations and must prepare to defend themselves and preserve civilization by defense of national territory. Aid to those in other countries who need it was deemed counterproductive and only likely to increase the total of human suffering by increasing the number of the dispossessed, which was a concept taken directly from Malthus. Hardin termed this view "lifeboat ethics" and in so doing caused a convulsion in the environmental movement. This position is clearly antithetical to environmentally sound development because it would create insupportable pockets of overconsumption in a world of irreversible environmental degradation and depletion. In a world in which societies and nations are inextricably linked and environmental concerns are global, it is an anachronism to think of sealing off territory for environmental protection. In 1974 this was not so obvious. Commoner responded melodramatically, if perhaps appropriately, by explicitly labeling Hardin's position "inhumane" and "abhorrent" and proposing that poverty promotes overpopulation, not the reverse. Unfortunately, in the process of attacking Hardin's editorial and "lifeboat ethics," he dismissed without adequate consideration Hardin's concept of the "commons," which was the crux of Hardin's argument.

In the 1980s and later, there was a strong backlash against the emphasis placed on population control in the 1960s and 1970s, fueled in part by the demise of eugenics. Leaders in many countries and minority groups within the United States became suspicious that population control efforts were intended to hold them back for the benefit of economically dominant Western countries or the majority white population in America, respectively. For many years thereafter, the issue of population became so contentious in international development circles that it was seldom addressed directly. (Chapter 2 explains one of the reasons for this—a desire to distance oneself from eugenics.) US foreign assistance also deemphasized birth control and voluntary population stabilization, in part because of domestic political divisions over abortion and whether support for international agencies indirectly supported the practice. The opprobrium attached to coercive policies such as the "one-child" policy in China, a misguided forced sterilization program in India, and the policy of encouraging abortion as a principal means of birth control in countries such as Romania (then under a particularly brutal Communist dictatorship), discredited many more meritorious interventions for individual birth control and for population policy. The result would have been two lost decades in terms of redressing the balance between population and resources had it not been for two unexpected developments.

The first unexpected development was the technological, de facto expansion of the resource base. For example, copper was once a highly valued commodity for electrical wiring in applications such as communications, as well as electrical power transmission. The development of fiber-optic transmission and later wireless technology put an end to much of the demand for copper and created a glut of copper for recycling, which dramatically lowered the price of this commodity (to the detriment of producing countries such as Zambia). In addition, the development of technology such as cellular mobile phones allowed resources to be allocated much more efficiently in developing countries: for example, cell phones provided a means for farmers, fishermen, and small business entrepreneurs to check market prices before delivering their goods to a buyer. Trade also became cheaper and easier to conduct with developments such as containerized shipping and then the Internet. Globalization allowed much greater access to distant resources through established and reliable trading channels. The downside of this resource-seeking trend is the intensive exploitation of renewable resources, such as ocean harvesting or deforestation (for example, to plant palm oil in Southeast Asia), damaging the ecosystem and producing a classical "tragedy of the commons" effect.

The second development that marginalized the importance of population control in the population-resource equation was the dramatic fall in population growth experienced in almost all countries. In middle-income and most poor countries, population growth has declined in step with education, empowerment of women, social opportunity outside the family, and economic security, such that many sons are no longer required to protect against poverty and neglect in old age.

Population growth has also declined precipitously in most rich countries (the exceptions are in the Middle East, where traditional family values coexist with oil wealth and modernity). The declining birth rate in developed countries, particularly Japan, where the population is shrinking and rapidly aging, in Europe, and (until recently) Québec (a province of Canada but a distinct society), has even raised renewed fears that there will be insufficient young workers in these countries to support productivity and provide critical services. Immigration has been the preferred response, together with "pro-natalist" policies of subsidies, parental leave, and tax incentives to encourage larger families, as in France. Historically, the critique of developed countries has been that their relatively modest populations consume such a disproportionate amount of world resources. Underlying the rhetoric that developed countries such as the United States are "overpopulated" is the reality that the currently developed world got to a point of economic privilege first and has the resources and the technological means to obtain resources wherever it can. Other rich countries also use financial means to ensure advantage, such as countries in the Middle East that hedge their food security concerns by buying control of large tracts of land in more fertile and water-rich areas for agricultural production dedicated solely for their consumption. A globalized economy and the purchasing power of strong currencies ensures supply for the populations in these countries.

The population-resource balance situation is somewhat different in underdeveloped nations. Population-resource imbalance has played a much more important role in impeding the achievement of a tolerable quality of life outside the "developed" (or "overdeveloped") societies and slowing economic progress when economic development is stagnant or collectivized; however, as the early experience of China and the more recent example of India have shown, population alone has not stopped economic development. Countries with large and rapidly growing populations, such as Mexico in the 1980s, have been able to break through to middle-income status, and for some the presence of a large internal market may be an advantage, as the later experience of China has shown.

The world perspective on population issues in developing societies has grown considerably more sophisticated since Hardin's essay. Population is now seen more as a primary driver than a proximate cause, expressing its effects indirectly by influence on other factors (such as poverty, competition between ethnic groups, and land issues). Competition for diminishing resources remains a potential and often an actual cause of conflict among and within developing nations but is expressed by aggravating existing divisions. Many of the localized wars that have taken place in the last forty years may fall into this category, even though most are attributed to ideological, religious, or ethnic differences. For example, the tragic conflict resulting in genocide that occurred in Rwanda in 1994 was thought by many observers to be fueled at least as much by competition for scarce farmland as by racial division between Tutsis and Hutus. The causes of other conflicts are easy to attribute to overpopulation superficially but this connection usually proves to be elusive because the effect of population is indirect.

The UN Conference on the Human Environment (now known as the first "Earth Summit"), which was held in Stockholm in 1972, explicitly linked the environment and development, long before the Brundtland Report. It was a landmark international conference, where the importance of poverty in the equation came to be recognized. More recently, the follow-up "Rio+20" conference in 2012, while disappointing in its results for global management, at least established equity and sustainable development on an equal footing with ecological sustainability: this came about not only because its principles were just but also because the cultural dimension is an inextricable part of sustainability through generations, as will be argued later in this essay. Issues of social equity, and particularly the emancipation of women in developing societies, are now coming into focus and will likely provide the next conceptual step forward. In the past there was a tendency to dismiss social and religious issues as stubborn obstacles to putting into place concrete actions that were objectively necessary. Now, there is a growing realization that these social and religious issues stem from the intrinsic mechanisms each society uses to solve its internal problems. They are keys to corrective action, not irrelevant obstacles.

Hardin focused especially on the idea that the family is the natural, fundamental unit of society and targeted the notion that family size must be a decision of the family itself alone. Denying the inalienability of this particular right, which is among those enumerated in the UN Universal Declaration of Human Rights, he suggested that reproduction be subject to a mechanism of "mutual coercion, mutually agreed upon" by all affected parties. (By 1992, under fire, he had to clarify that this did not mean rigid or enforced social control, but that "coercion" could take the form of incentives such as allowing income tax deductions only for the first child.) Hardin's fellow advocates of population control, such as Ehrlich, were considerably more reluctant to give legitimacy to any form of state-sponsored population control, except as a last resort for survival. Commoner denounced what he saw as repressive implications of compulsory population control on the basis of human freedom. However, he did not leave the argument there. Instead, Commoner proposed an alternative mechanism for population control, in effect taking the argument of Malthusian population biologists into their own camp.

It is well known that as nations become more affluent the birth rate drops and population growth declines. This "demographic transition" was considered in the 1970s to be inadequate to meet the coming crisis and not necessarily applicable to rapidly changing developing societies of today by Ehrlich. Commoner, on the other hand, concluded that the magnitude of the demographic transition had been underestimated and that it was already dramatically under way in developing nations. Since the demographic transition is based on the decisions of individuals in the population who voluntarily limit their reproduction as the death rate declines and personal security increases, this would represent one instance in which collective individual judgment results in a rational and productive direction for the society as a whole, contrary to Hardin's virtually blanket denial that this ever happens. By the 1990s it was obvious that the demographic transition was much more powerful than was anticipated in the 1970s. However, it was also clear that national populations in countries such as China, India, and Mexico had grown so large that the "momentum" of population growth, even with lowered birth rates, is still adding huge numbers of people in absolute terms.

With the benefit of hindsight, it appears that Commoner had the best of the argument, because population growth has indeed slowed with affluence, education, and material security. Commoner concluded that one of the most promising avenues for population stabilization, and a necessary concomitant for any successful program, would be national development (carefully planned), security, and, especially, reduction of infant mortality so that larger families would become unnecessary for a family support system and would be discouraged by the expense of extra mouths to feed. In this he anticipated the Brundtland report's call for sustainable development, the polarization of the world diplomatic response along "North-South" lines (a then-popular shorthand term meaning the relatively affluent northern hemisphere and the relatively impoverished southern hemisphere, a distinction which is now breaking down), and the recognition that poverty is the single most critical factor driving resource depletion, inadequate pollution control, and the effective (but not relative) failure of population control in the developing world. What he did not anticipate was that a linkage would be made with the status of women. The provision of an education for women and conferring societal status on them, thus empowering them to make their own decisions, are now recognized to be the keys to population control and one of the most important steps in the transition from a traditional culture to a society capable of preserving its values (new and old) in the face of global competition and encroachment.

"Natural" and "Technological" Disasters

Catastrophes, calamities, and disasters have certain characteristics that condition the response of societies and their resilience after the event. In the literature of emergency management and the sociology of responses to disasters, an important distinction is made between "natural" and "technological" disasters.

A "natural" disaster is an event that occurs without perceived human agency, such as the natural disasters described above. In the past, this definition would also extend to sporadic unanticipated epidemics or even to partially socially mediated adverse events, such as famines and setbacks in warfare, in which it was believed that arbitrary supernatural forces were at work (such as antics of the Greek gods). The critical element in the definition is that regardless of the true cause, the public or community does not perceive the cause as being the responsibility of a person or of other people. Such disasters are often referred to as "acts of God." The perception of a natural disaster is therefore that it was unavoidable, was probably unpredictable (at least insofar as timing was concerned), and that no one person or group bears responsibility or can be "blamed" for what happened. As a consequence, communities who experience natural disasters tend not to assign blame or consume emotional energy in debating who was at fault. Affected residents tend to accept their loss or a more limited liability coverage (if insured) or blame themselves if they went uninsured. Such communities tend to recover relatively rapidly, whether the event is a hurricane, flood, earthquake, or wildfire.

A "technological" disaster, on the other hand, is one that is perceived as an "act of man," for which human agency is responsible and blame can be assigned. Although the term is frequently misused to apply to disasters involving machinery or technology, the term as originally coined referred to any event of human agency, involving human design, intent, error, negligence, or failure. Such events tend to be heavily freighted with approbation (sometimes misdirected), moral outrage and anger, blaming, and the pursuit of litigation. The community may be united at first but often fragments into factions because of competing interests, especially if the responsible party is an employer perceived as a financial support for the community and some residents fear for their jobs. Fragmentation also occurs when some members of the community stand to gain from legal judgments or insurance indemnification and others do not. Such communities often find it difficult to "get past" such events because of constant reminders, new revelations, and the need to rehash details continually in preparation for lawsuits.

Prominent technological disasters in recent years have included the collapse of substandard factory buildings in Bangladesh in 2014, killing more than 1000 low-paid workers in the garment industry; a near-fatal collision between a badly driven ferry in New York in 2010 and again at the same dock in 2013; the dust explosion at the Imperial Sugar refinery in Georgia in 2008; explosion, fire and tragic loss of life and the ensuing oil spill among workers on the Deepwater Horizon, a British Petroleum oil-drilling platform in the Gulf Coast in 2010 that also resulted in the world's largest offshore oil spill to date. The world's worst disaster involving toxic exposures occurred in Bhopal, Madhya Pradesh (India) in 1984, when a leak from a pesticide-producing chemical plant using methyl isocyanate caused a cloud of gas to envelop the sleeping city, immediately killing about 3,800 residents and ultimately killing a conservatively estimated 10,000 people and causing disabling illness in thousands more; the low levels of compensation paid to survivors of the incident was widely criticized and demonstrate how contentious these issues become.

Some disasters cross the boundary, beginning as natural disasters and becoming technological disasters, with even more serious sociological consequences. The Fukushima crisis began with the Tōhoku earthquake (9 on the Richter scale), which caused a historic 40.5-meter high tsunami that flooded and destroyed safety systems in the Fukushima Daiichi nuclear reactors, revealing critical gaps in Japan's preparedness, nuclear engineering safety, and national energy policy, with its overreliance on nuclear power. This turned it into an unequivocal technological crisis. Had it been restricted to the earthquake and tsunami alone, the event probably may still have been perceived as partly technological because development in the area was unrestricted, although it would have been understood that the previous tsunami had occurred centuries before and was not well known to residents. Even so, the Tōhoku earthquake and tsunami alone would still have caused an exceptionally long and drawn-out response for a natural disaster because of the loss of life and community dislocation and the degree to which it exposed the vulnerability of Japan's aging and increasingly dependent population.

Reasonable decisions may look foolhardy in retrospect. Box 3.1 tells the story of Ghostpine, a community that faced an environmental problem that should have been manageable, followed by an economic reversal out of proportion to its financial resources (a "technological disaster"), followed by a completely unexpected and tragic natural catastrophe (a "natural disaster").

A case can also be made, of course, that the accelerated frequency and increasing magnitude of extreme natural weather events may in the future turn them from natural disasters to technological disasters, in the sense that they will have been aggravated by climate change of human agency and that people who knew better let conditions develop that allowed them to happen. It would follow that people in a position of authority who failed to act or impeded measures to reduce or mitigate climate change are responsible. This would be an insurance and legal nightmare, but it appears to be playing out in some US state policies and, in North Carolina, even a law prohibiting government agencies and employees from planning for climate change by including predictable effects, such as sea-level rise, in their planning, zoning, and emergency management decisions.

4

Pollution and Contamination

THE DIFFERENCE BETWEEN "contamination" and "pollution" is at the heart of any serious discussion of toxicity and health effects. The words are often used interchangeably and carelessly, but the differences in meaning are essential to understanding the relationship of pollution to health. The words also carry more subtle meanings for perception and cultural attitudes toward pollution.

Contamination is the more general term; pollution is a subset of contamination. Pollution implies both that harm has already been done and that the cause is the result of human action. A body of water such as a lake or stream, the water underground (aquifer), the atmosphere, the soil on a tract of land, foods, consumer products, and any other medium may be contaminated without necessary being polluted. When the word "contaminated" is used alone, it does not imply that the source of the contamination may or may not be from human activity and that harm may or may not have been done. Thus, referring to a situation as "contamination" alone, means that the speaker or writer is often reserving judgment or leaving it open whether the situation represents pollution: this is either because it is not clear that harm has been done or that the source is not proven.

Pollution, together with resource depletion and ecosystem damage, is one of the three important threats to environmental sustainability. Not all types of pollution have significant implications for ecosystem sustainability, but they all have implications for sustainable development. Short-term effects from pollution may affect human health and do harm, but their health effects may be short lived and local. The types of pollution that have the greatest significance for ecosystem sustainability are those that persist and that have consequences and effects that span many years and may deny future generations their use of critical resources, such as polluted groundwater. Effects on ecosystems may have the further effect of reducing resilience in an ecosystem and making it even more difficult for a biological community to stabilize at a sufficiently productive level to sustain itself, let alone to provide environmental services to people. More deeply still, pollution is the result of unsustainable practices that compromise the environment and quality of life. Processes that produce pollution or that require toxic chemicals have simply become unacceptable, tolerated only when there is no alternative. Therefore, reducing and eventually eliminating pollution is, along with ecosystem protection and resource conservation, the essence of sustainability.

Pollution compromises sustainability in many ways, as exemplified in table 4.1, which lists some of the associations following the order introduced in chapter 1 for the cluster of values in the definition of sustainability. However, some elaboration is necessary to explain the brief entries in the table. First, the terms of the table are written specifically for "pollution," which can be considered in terms of sustainability as contamination that has an actualized or potential effect, whether on human health or the environment, that results in making the resource unsuitable for other uses, either by people or within the ecosystem.

Pollution that continues unabated obviously may cause serious damage to health, cultural materials, and the environment. In the short term, the effect on the ecosystem may be reversible, and there are many examples of rivers, estuaries, and bays returning to near-natural conditions after pollution is mitigated. However, these are exceptional cases. In general, even reversible pollution has some lasting effect on the environment that persists and after recovery is never quite the same as the natural ecosystem. Likewise, the effect on people may be reversible because the human body has means to adapt, at least in the short term. Yet high short-term exposure can overwhelm mechanisms of adaptation and prolonged exposure to pollution usually leads to effects that are irreversible. Exactly which risks are involved and how serious they may be depends on the specific "exposure regime" (situation, duration, and concentration) and the toxicity profile of the agent.

Pollution may compromise conservation of resources in many ways, including

- rendering them unusable for long periods of time (e.g., pollution of groundwater)
- greatly increasing the cost and availability of resources that could otherwise be put to better uses (e.g., use of contaminated industrial sites, known as "brownfields," for other uses such as housing, schools, and clean industry)
- corroding and degrading materials (ozone attacks rubber and makes it brittle, sulfur dioxide erodes metal surfaces, photochemical air pollution affects plants)
- interference with natural processes that sustain ecosystems and support natural biological communities, often to the point of destroying them

Pollution has social consequences, some related to its direct effects and other related to managing the risk and consequences of pollution; much of the social

Table 4.1 Sustainability values table: Contamination and Pollution

	Sustainability Value	Sustainability Element
I.	Long-term continuity	Pollution, unabated, may cause such serious damage to human health, materials, and to the environment that an otherwise sustainable system cannot withstand the damage indefinitely without degradation.
II.	Do no/ minimal harm	Pollution harms the environment, usually by toxicity to key species and simplification of the biological community.
III.	Conservation of resources	Pollution may compromise conservation of resources in many ways, as described in the text.
IV.	Preserve social structures	Pollution has social consequences, some related to its direct effects and others related to managing the risk and consequences of pollution; much of the social impact comes from divisive issues of who is to blame and who benefited from pollution.
V.	Maintain health, quality of life	Pollution compromises health and degrades the quality of life apart from health (for example, by reducing visibility and enjoyment of sports and recreation).
VI.	Performance optimization	Pollution reduces economic, social, and environmental performance by imposing "diseconomies" on the society, which by definition are costs that are not paid by the enterprise doing the polluting and are therefore imposed on the community.
VII.	Avoid catastrophic disruption	In extreme cases, pollution can cause catastrophic events. (One example was the Minimata Bay incident, in which mercury polluted ocean waters in Japan caused widespread serious illness with permanent disability.)
VIII.	Compliant with regulation	Pollution is obviously contrary to stewardship and, because it imposes harm on others, is irresponsible and unethical. It also opens the polluter to legal liability.
IX.	Stewardship	The act of pollution demonstrates a failure of stewardship on the part of the responsible party and tolerance of pollution represents a failure of stewardship on the part of society.
X.	Momentum	Pollution does not go away by itself unless the source stops, and so it is not self-correcting; mitigation requires design and intervention to control the source or, at even greater expense, to clean pollution up downstream.

impact is mediated by issues of who is to blame and who benefited from pollution. Pollution causes social conflict, for example, by

- creating conflict between those who benefit from producing pollution while avoiding the cost of cleaning it up ("externalizing the costs" in economics terms) and those who must suffer its effects
- causing illness and diminished quality of life that people perceive, understand, and deeply resent
- creating disparities in exposure and health depending on income levels (and the means to escape pollution) and where people live (creating problems of "environmental justice")
- raising costs for managing consequences, including health care, environmental protection, and use of shared resources, such as water

Pollution reduces economic, social, and environmental performance by imposing "diseconomies" on the society, which by definition are "externalized costs": these are costs that are not paid by the enterprise doing the polluting and that instead are imposed on the community in the form of adverse effects due to pollution or costs to clean up pollution, which are almost always many times higher than it would have cost the polluter to do so. When these externalized costs fall on a defined group in a discriminatory way, especially when that group is disadvantaged, then the social injustice this creates becomes an issue of "environmental justice" (discussed in detail in chapter 11).

It may seem obvious that pollution is unsustainable, but it has been tolerated to a greater or lesser extent throughout human history. At its extreme, pollution can cause catastrophic events and force change. One such memorable event was the Great Stink of 1858, in which the odor from untreated human waste dumped into cesspools and cesspits around London and discharged into the Thames River creating an intolerable situation and forcing a reluctant Parliament to build an effective sewerage system. Another was the Great Smog of London in 1952, in which a combination of dense fog, not uncommon in London, combined with high-sulfur air pollution and particles from coal-burning emissions (producing an acidic, chemically reducing form of air pollution) caused an estimated 4,000 deaths, mostly from respiratory disease and heart attacks. A similarly intense air pollution problem is playing out now in many cities in China, where intolerable pollution levels, this time of oxidant air pollution (characterized by ozone and oxidizing chemistry) have grown so severe that the national government is intervening.

On the other hand, and as will be discussed below, "contamination" will be defined more generally, as the presence, in minor proportion, of something that does not "belong" there—whether or not it has an effect, and whether it occurs from release into the environment or is a natural occurrence. The word "contamination" will therefore be used here as a broad category indicating the presence of an impurity or inhomogeneity where a pure or contaminant-free medium would be expected. This definition is not the same as for pollution and requires close attention to the meaning of "impurity" and what truly "belongs there." (See figure 4.1.)

Contamination is usually thought of in terms of human activity and adverse consequences, but in environmental chemistry and geosciences it is well understood that levels of contaminants, such as arsenic and mercury or radionuclides (a form of energy contamination), can arise from natural sources releasing a radioactive substance (such as radon) into water. In human activity, "contamination," here meaning the introduction of impurities, is also sometimes intentional, as in doping semiconductors, where a controlled level of impurity is critical to performance.

Measurement technology has advanced by orders of magnitude in a few short decades, especially when one considers that the field of analytical chemistry is only about a century and a half old. It is easily possible to detect contaminants at levels of parts per trillion or even below, at concentrations highly unlikely to

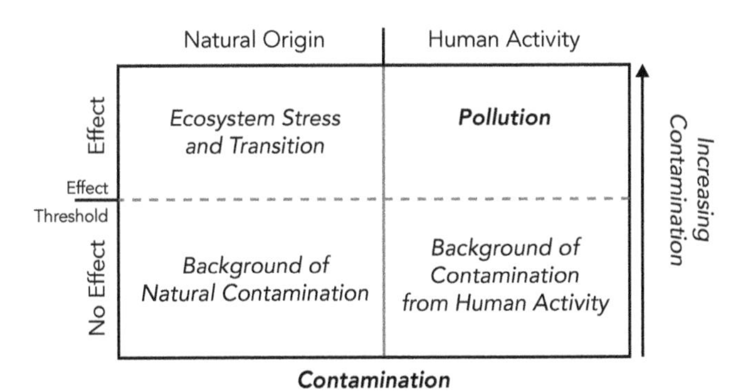

FIGURE 4.1 "Contamination" refers to the presence of an impurity, something that does not "belong" or is unexpected in a medium, which is present in amounts of minor proportion. Contamination may or may not have an adverse effect (the dashed line indicates a threshold for effect); this is often hard to discern and document. Some contamination occurs from natural sources; when it is great enough to have an effect, the contamination may cause naturally occurring transitions and stress biological communities or entire ecosystems. Contamination by human activity that has an effect is called "pollution." Note that although the outer rectangle represents the "universe of all contamination," there is no room inside or outside for "pure" media. That is because in environmental chemistry, contamination is always a matter of degree. Aside from laboratory-grown crystals and other artificial situations, there are always impurities present and no medium is ever totally free from contamination, although it may be present at very low levels.

be associated with health effects. (Some chemicals do produce effects at such low levels, usually because they bear a similarity to hormones and other biological agents.) This does not mean that minimal or relatively minor levels of contamination that do not lead to a detectable effect are never a problem or that they are always tolerable. Minimal evidence of contamination may be a warning that something is happening to cause an unexpected or uncontrolled release and that the environment is under threat. It may mean that standards for environmental protection are being violated and that if the problem continues, significant pollution will result. The effect may also be difficult to detect and document, as has been the case with some chemicals that act as endocrine disruptors. It may also mean that the impurity was not expected to be in the medium but was actually there all along and went undetected.

Thus, recognizing contamination and where it crosses the line into pollution depends a great deal on whether there is an adverse effect and on whether the amount of the contaminant can be measured. Detection and measurement of contaminants is an issue of technology that is fundamental to modern environmental chemistry and therefore sustainability.

From Contamination to Pollution

A fundamental issue in environmental sciences is determining whether the release of a chemical (such as a potentially toxic agent) or a biological agent (such as bacteria or virus) or a form of energy (usually radiation or heat) constitutes a relatively mild degree of contamination, with few or no consequences, or pollution sufficient to require that some action be taken in response. For the purposes of this chapter, contamination will be defined as the general case: something has been released into the environment that does not belong there.

Contamination

From the point of view of sustainability, contamination is primarily a problem if it makes it difficult or impossible for future generations to use a resource for its own needs. Contamination may be, but is not necessarily, a problem because of the harm it does in the present, but it is always a concern for sustainability if it complicates future use of the resource. Contamination from natural sources is common. Groundwater in many parts of the world is contaminated by arsenic that is present in rocks and soil due to the geological formation of the location and that has nothing to do with human activity. On the other hand, as with arsenic in the water supply, even natural contamination can become a pollution problem when the resource is tapped or diverted by human activity in ways that cause harm: for example, when arsenic-contaminated well water is used as a source for drinking water. When contamination does cause harm and is due to human agency, then it crosses another definitional line and also becomes "pollution," as described in the next section.

Contamination means the act of making something impure, in the sense of adding something that does not belong there. The word comes from the Latin *contaminare* meaning "to defile" (related to *con* "with" plus *tangere* "to touch"), but it is a relatively late entry into the language, coming into the English lexicon in the fifteenth century. When applied to air, water, food, or some other medium, it usually means the condition of having a detectable but not large amount of an unwanted constituent. Most definitions of contamination do not require that the impurity has actually changed the properties of the medium or caused harm (although some dictionaries do equate it with pollution in that sense). However, in technical use in environmental sciences the implication of the word is that contamination could damage something valuable and might cause harm under some circumstances. It is often used metaphorically when purity is equated with virtue or integrity. Strictly speaking, however, something that is contaminated has had its composition altered in an undesirable way, but its essential properties are not necessarily changed. The notion of contamination, therefore, is that something tangible (food) or intangible (public morality) has been defiled.

In environmental health, contamination is important because it carries a risk for the future and some degree of uncertainty. Contaminated water might carry bacteria that cause diseases such as diarrhea, but that depends entirely on the nature of the contamination (many waterborne bacteria do not cause diarrhea) and the presence of a person or animal that is exposed. Contaminated food carries a risk of disease that may or may not happen. Contaminated air degrades the atmosphere from what is presumed to have been a pristine condition originally (although that is seldom the case once human settlement has occurred, especially in urban areas).

Contamination is also used is a similar sense in other fields. In social sciences, contamination implies that values and attitudes have been influenced by others and do not necessarily reflect the culture or original views of the people being studied. Sometimes this is due to simple cultural contact and communication; at other times this is due to manipulation and propaganda or advertising. A contaminated archeological or historical site is one in which artifacts have been removed or moved around. The site has been disturbed, and layers (strata) have mixed so that it is no longer possible to reconstruct relationships accurately and with confidence that they represent a true sequence or association with other artifacts. The word does not imply that human beings were responsible, since this can occur from animals, water flow, and earth movement. In each case the use of the word "contamination" leaves open the possibility of human manipulation but reserves judgment.

In medicine, a precise distinction is made between contamination and infection by microbes (mostly bacteria and viruses). If something is contaminated, it is not sterile and has the potential to spread a pathogen that can grow into an infection, which implies that the pathogen (infectious agent) both proliferates and invades tissue. Contaminated wounds are those that are not yet infected but are likely to become so. Nonsterile, contaminated instruments carry a risk of introducing bacterial (and other) contamination and subsequently bacterial growth and inflammation, which is the essence of infection. However, use of the word "contamination" in medicine implies that infection has not occurred yet but that the risk is high.

The fundamental problem with contamination in practice in the environmental sciences is making the definition operational for chemical contamination. Measurement technology has advanced so far that it is now easy to measure contaminants at a level of parts per trillion (ppt) or even less. This has raised awareness that contamination (including pollution, defined below) is global, that hardly any site on earth is free of contamination, that unwanted contaminants are unavoidable, and that all human beings, regardless of location and isolation, now carry a burden of modern pollutants (defined as contaminants that cause harm and are from human sources).

For example, when analytical chemistry only allowed chlorinated dioxin compounds to be measured at parts per million (ppm), the world seemed pure enough from dioxin contamination. These compounds were not known to be present at background levels. They were therefore viewed as uncommon hazards found only in industrial locations as unwanted byproducts of chemical reactions. However, when technology advanced to allow detection in the parts per billion (ppb) range, it was found that there were natural sources of dioxins, for example where there had been forest fires. It also became obvious that dioxins were formed as byproducts by many other industrial processes and that levels were increasing and spreading around the world, even in places where the background levels had previously been low. Thirty-five years ago it took a whole unit of blood (450 ml) to conduct an assay for dioxins, and it cost almost as much to identify the individual congeners (forms of the molecule that vary only in number and placement of chlorine atoms) as to buy a small used car. Low levels in people who did not have high exposure from their occupations were often reported as "nondetectable" because even the presence of the compounds could not be discerned. Technology is now able to identify dioxins easily in the ppt range and congener speciation is routine. Environmental scientists routinely characterize their presence in the blood and tissues of normal, healthy people and trace their origin to far distant industrial sources. This same story has been repeated for numerous organic contaminants such as pesticides, metals such as lead and mercury, and numerous trace contaminants in the atmosphere—some of which play critical roles in the chemistry of air pollution and in the scavenging of ozone in the upper atmosphere.

Purity, it seems, was an illusion. It is not difficult to find "background" levels of any number of chemical contaminants in relatively pristine locations, such as cores from the Greenland ice cap, throughout the oceans, and in the tissues of people who have had little contact with industrial sources. From the point of view of environmental sciences, therefore, contamination is not the opposite of purity, because purity is not a distinct state: it is actually unachievable, only approximated by a highly refined distillate or a crystal. Rather, contamination is a matter of degree. Something is only pure when the contaminants no longer matter because they cannot be detected, do not change the properties of the medium, or do not confer a future risk. Purity, then, only exists in a physical or chemical sense on a relative basis and is only defined when there is a reason to characterize the degree of impurity for some specific purpose.

Culturally, this is unsettling. The concepts of health and integrity, cleanliness and purity are closely interrelated, and metaphors about purity and contamination are deeply embedded in religion, morality, ethics, and culture. The idea of contamination, of something important to people's bodies, lives, and their planet being impure, is profoundly disturbing psychologically and culturally and leads to perceptions of both risk and violation—whatever contamination may mean in a literal, technical, or physical sense.

An important implication of contamination for sustainability is that reconstituting or purifying a contaminated site or medium to recover the original medium in a purer form is difficult, expensive, and sometimes impossible, especially on a community scale. For example, if a freshwater aquifer is contaminated by nitrates from irrigation (the sources being fertilizer and waste from livestock), there is no practical way to remove the contamination. If food is contaminated with disease-causing (pathogenic) bacteria, it takes energy (in the form of cooking or pasteurization) to make it safe to eat. However, if drinking water is contaminated with a chemical such as perchlorate (ClO_4-, a highly soluble compound derived from rocket fuel but also naturally occurring in alkaline soils), there is currently no practical way to remove the chemical on a large scale from groundwater and soil where it occurs. (This is the topic of active research; several technologies look promising but will be expensive.) If there are no easily available alternative sources, decisions therefore have to be made on the basis of what is an acceptable risk. On a fundamental level, this is an application of the principle of entropy: it is easy to contaminate or cause disorder, but it takes energy to restore order. On a practical level, this raises costs for future use of the resource or may make future use impractical, directly affecting one of the key elements of sustainability.

The trade-off is that once contaminated, a resource becomes more expensive if clean-up is required and uses that depend on higher levels of purity may be deferred or unavailable. A contaminated industrial site will usually require expensive cleaning and remediation before it can be considered by developers, approved by regulatory agencies, and bought and insured for use—especially if the new purpose for the property involves human occupation and liability for health risks. If wastewater in a community upstream is not adequately treated, or if treatment is unreliable, then discharge into a flowing body of water poses a risk and therefore a higher expense for communities downstream that use it as a source for their drinking water. Drinking water is easily contaminated with coliform bacteria (bacteria from the gut that enters the water from feces), and so it is imperative, for health protection, to purify it through filtration and disinfection.

Contamination, as a concept, does not depend on human agency; but pollution, by definition, does. It can be caused by human action, but it can also occur naturally. Natural contamination is widespread and a particular problem in localities. An important example is arsenic contamination of groundwater in Taiwan or Bangladesh, which can vary between wells only tens of meters apart). No human action contaminated the groundwater. The arsenic is naturally present in soils and rock. Likewise, almost every mountain stream in the Rocky Mountains is contaminated with *Giardia lamblia* and other parasites that cause human disease that naturally inhabit the gut of forest creatures. (Giardiasis, the disease, is a particularly nasty form of chronic diarrhea). This situation did not result from human action, because the contamination arose naturally from beavers and other animals. The contamination affects human beings as consumers of the water. Human beings have certainly been part of the mountain landscape for tens of thousands of years, and indigenous peoples were not unaffected by these diseases, of course, and lost many children to them, but the adults who survived tolerated or had resistance to its effects or immunity. The problem of contamination has been there as long as the wildlife, but disease risk was never a practical problem because there were few human beings present and no large settlements. The modern approach taken to manage this risk involves filtering and disinfection of water taken for drinking water, since the watershed properly belongs to the wildlife.

The Transition from Contamination to Pollution

The difference between contamination as a whole and pollution in particular rests on whether harm has occurred and the contamination is caused by human agency. The words matter a great deal in practice.

Contamination describes a condition in which the original medium is mixed with something unwanted that should not be there, at least from the point of view of human use. In the case of contamination from natural sources, such as microbial contamination of lakes and rivers by animal feces or arsenic in groundwater from natural sources in soil, one could argue that contamination is the natural state and only becomes an issue when human communities access the resource and want to use it for their own purposes.

More often, however, contamination raises the question of whether harm has occurred or is likely to occur in the future. Whether harm has already occurred, notwithstanding the risk of harm in the future, is critically important for legal action and public policy. Harm can be difficult to prove, and the absence of harm is not always accepted by those who believe they have been injured. On the one hand, residents of the community may fear that their health has been harmed. If community residents already believe firmly that their health has been affected, studies that do not demonstrate an effect are likely to be dismissed as incompetent, biased, or lacking and will not be perceived as reassuring. On the other hand, the same residents have an interest in demonstrating that harm has occurred because this qualifies them for compensation, supports legal action, and furthers the agenda of correcting the problem. The other part of the distinction—that calling contamination by the name "pollution" implies that the contamination resulted from human agency—strongly implies that someone was responsible; if so, someone carries the legal, ethical, and moral responsibility for the act of pollution and therefore may be assigned the responsibility to clean it up or compensate those affected (if they are still around).

The distinction between contamination as impurity and pollution is therefore an important one. Whether the source is due to human agency and whether harm has occurred, the essential elements that distinguish pollution in particular from more general contamination, have profound practical consequences for social justice, legal action, public policy, and sustainability. Determining whether pollution has occurred in this more rigorous sense is a critical step in characterizing public issues related to contamination. For example, the extent of contamination and risk of future harm underlies the feasibility and cost of whether a "brownfield" (contaminated industrial site, many of which are located in otherwise desirable locations) can be remediated safely and sold for use as a site for clean industry, housing, or a school, or whether the contaminated site must remain off limits or can only be sold at a low price for other industrial uses. Whether it has already caused harm (part of the definition of pollution) is always the central issue in legal actions against the prior owner for personal injury ("toxic tort," with "tort" being the legal term for injury) and liability for claims and legal actions may make the property unsalable or uninsurable. The priority score given to a hazardous waste site that qualifies for Superfund remediation on the National Priority List depends on documenting a high level of contamination and the potential for future risk to the community but also on whether harm has already taken place and is likely to continue.

A common problem with calling an event or a situation "contamination," and not calling it "pollution," is that there is so often uncertainty on an operational level about whether harm has already been caused. The US Environmental Protection Agency is probably the leading funding source for environmental health research in the world, but it cannot study everything. The Agency for Toxic Substances and Disease Control (a part of the Centers for Disease Control and Prevention (ATSDR) has the mandate to do evaluations at the community level but has insufficient resources; the reports are often inconclusive because adverse effects can be surprisingly hard to document.

Often studies do not show an adverse health effect, and the question remains whether there was no effect or just an absence of evidence for an effect that was missed. This is a particularly common problem in assessing the potential impact that a hazardous waste disposal site has on a community; communities perceive a high risk, but studies rarely show objective evidence of harm, usually because a route of exposure to the community cannot be demonstrated. Studies to find the true level of risk are usually lacking, especially for other limited, local hazards and sometimes show uncertain results. To counter this conclusion it is often argued that "harm" has already occurred in a meaningful sense even if it cannot be documented; this is because it is detrimental to the community to live with uncertainty, or that alleged adverse effects (usually health effects but sometimes ecosystem effects) have not shown up yet.

Scientific studies are not conducted on every incident of contamination or pollution. Usually, only relatively large issues are studied thoroughly and professionally, unless they involve a novel exposure that merits scientific study or qualify and are accepted for investigation by ATSDR. Smaller-scale and strictly local incidents are more likely to be studied if there is someone with an interest in the issue (e.g., faculty at a local university who may or may not be trained in this kind of research or a community activist), if there is political support (a local advocacy group or concerned politician), and a source of funding to do the work. But often such studies are not well done and they raise far more questions than they answer.

Similarly, activists sometimes conduct their own "quick and dirty" studies to assess harm Not infrequently, members of the community conduct "studies" of their own (such as counting the number of cancer cases among residents), either to challenge studies that have been done or, more often, to force action by officials. These community studies, while often well meaning, are almost always questionable in their methods and therefore uninterpretable; yet they raise emotional tensions in the community. These faux studies convey false impressions for many technical reasons: inadequate sample size, improper research methods, failure to look for the right effects, insufficient time elapsed (especially for cancer), improper subject selection, statistical significance that is almost always improperly calculated, no valid comparison population is available, common health problems, no consideration is given to population age structure, invalid case definitions result in a cluster of unrelated cases somehow constituting a single health problem and myriad other critical issues of study validity are impatiently ignored or considered mere details. The studies demonstrate little and in the end become contentious issues in their own right. Conducting a "quick and dirty" study to determine whether harm has occurred almost never satisfies the community or solves the problem. "Quick and dirty" really means "sloppy and misleading."

Pollution

The dictionary definition of "pollution" requires that contamination be present but also requires the elements of human action and harm. Thus, the essential difference between the two words, and the concept that will occupy most of this section after a brief but important digression on etymology, is that "pollution" implies that the contamination occurred as a result of human activity (not originating from natural sources) and that harm has occurred as a result of the contamination. Use of the word "contamination" alone does not imply these two important elements.

The word "pollution" has a colorful and unusual history, as well as being rich in metaphor and negative associations. The word was first used in English about a century earlier than "contamination" and derives from the Latin *polluere* meaning "to soil or defile [something]," derived from the much older root *lutum*, which means "mud," and *lues*, which means "filth" (and is also a synonym for syphilis in medicine).

For most of its early history, the word "pollution" was used in a moralistic sense, usually by religious writers; it was especially applied to nonprocreative sexual behavior. ("Self-pollution" was once a widely used synonym for masturbation.) The core meaning of the word for most of its history was that a person defiled and soiled themselves with unclean and immoral acts. Obviously, it was a nasty word that was used in a highly judgmental context.

Around 1860, when the word started to be used to mean contamination or defilement of the environment, it already came heavy with meaning and had strong negative connotations and a sense of moral outrage. The currently predominant meaning of pollution, as contamination of the environment bringing harm, is a relatively new development that emerged as recently as 1954. Obviously this was a much-needed new application of the word; usage in its current meaning has now crowded out the much older moralistic meaning, which is still occasionally found in religious tracts and sermons. Rarely has the usage of such a fraught, discomfiting, and unpleasant word changed so abruptly and entered the technical vocabulary despite such heavy cultural baggage.

The significance of this etymology is that the "pollution" comes freighted not only with the same negative meanings for perception and culture as "contamination," such as the notions of defilement and lack of purity; it also carries notions of harm already occurring and with a residual historical sense of individual responsibility (personal responsibility for debasing oneself, or corrupting others). For many people (at least, those who were writing on the subject in generations past, when the old moralistic meaning still coexisted with the word's technical meaning) the damage caused by pollution has equated to moral laxity and personal failure.

Pollution as Systems Failure

The remainder of this section will deal with a more technical definition of pollution and its implications. Pollution can be better understood in context and defined more rigorously in terms of a systems failure with adverse consequences. For the purposes of the rest of this book, "pollution" will be further defined as the accumulation by human action of a chemical, biological agent, product, or form of energy over time beyond the capacity of the system to remove, transform, or recycle it to a level that avoids adverse consequences.

Examples of contaminants that qualify as pollutants because they do harm and are the result of human action include air pollution (chemical), water pollution by sewage (biological), plastic material in the oceans where birds and fish are entrapped or injured when they try to eat it (product), and mishandling of radioactive materials (energy). Some biological hazards also have a unique characteristic that complicates pollution: they are capable of growth and proliferation. Microbial contamination, mainly bacteria and fungi in the free state and viruses in animal reservoirs (because viruses do not grow outside the body), depending on the organism, may develop into (harmful) pollution, because the microbe, being a life form, can proliferate to become a greater hazard where there is a medium that supports its growth. The many bacteria and viruses that can affect human beings from animal reservoirs are an example, or mold on damp surfaces that causes wood to disintegrate in building and when spores become airborne can cause allergies and irritation.

Pollution is best understood in a context that includes sources, behavior of the pollutant in the environment, production, and recycling or return. This requires a systems approach to the problem.

Figure 4.2 illustrates a natural cycle. It is characteristic of all natural cycles (and inherent in the definition of a "cycle") that they are closed systems: inputs equal outputs, and mass is moved around from one place (and often form) to another but not added or removed. Why this is true is easy to understand: there is no way for carbon, nitrogen, or water to leave the planet. Thus, with the trivial exception of space exploration and in science fiction, all matter moves in a loop

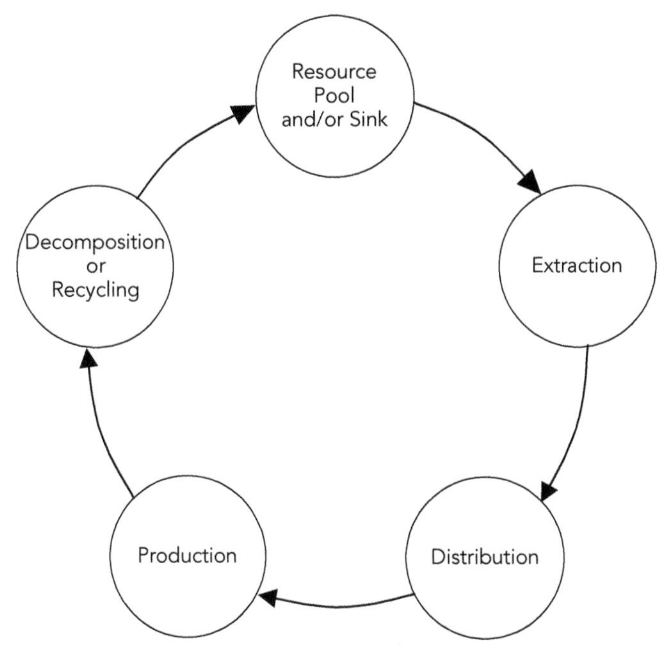

FIGURE 4.2 A natural cycle is a closed system, in relative terms, in which inputs equal outputs, and mass is conserved.

around the planet or around a self-contained ecosystem. Such cycles have similar characteristics for carbon, nitrogen, phosphorus, or water.

A natural cycle can be reduced to five basic steps or processes. The first is the "resource pool," where most of the element or material exists and that supplies new inputs of the resource to the cycle. In some way, the material is extracted from this resource pool so that it becomes available for moving through the various processes in the cycle. The "extraction" step may involve processes that change the form of the material chemically so that it is more easily available for plants and animals to take up and incorporate in their metabolism. The material is then distributed around the ecosystem, in specific ways following ecological pathways such as trophic levels (food webs) or by physical processes (for example by evaporation, diffusion, being carried in water). The "distribution" step makes the material more widely available to the species in the biological community that use it as food or, if a contaminant, that ingest it in their bodies. "Production" is the accumulated result of natural growth throughout the biological community and its ecosystem.

In ecology, "primary production" is the first step in production (first trophic level), when organic matter is synthesized from carbon derived from carbon dioxide through photosynthesis by plants. The organic matter is then cycled through trophic levels as plant matter is consumed by herbivores, (second trophic level, secondary production), which are then consumed by predators (third trophic level) that may themselves be subject to predation, all the way up to top carnivores. This organic matter, and everything else that flows through the cycle, accumulates as biomass, including nonliving structures left by the living organism such as dead trees (which serve as local pools of resources). Organisms die or leave waste, and this organic matter then decomposes through the action of fungi and bacteria, returning nutrient material to the cycle. The "decomposition" step may lead to short-term recycling through the process again, as the resource pool comes to include many pockets of accumulated resources that can be drawn upon, or material may return to the main resource pool. Sometimes, as in the case of carbon in the deep ocean and in the boreal forest, the material may become unavailable for further recycling in the short term because it is locked in chemically or physically. In this situation, the resource pool then becomes a "sink," with more material for a time going in than out until equilibrium is reached or, more often, the cycle is disrupted. In real natural cycles, the material recycles many times before it is removed into a sink and the sink eventually reaches a balance between how much of the material becomes locked up and how much is released. This equilibrium may achieve a steady state, but for the most part it is in constant flux, with some material coming in and some going out. Under different conditions, such as changes in temperature, acidity, fires, or physical disturbance, the resource pool releases more or less of the resource and makes it available again to the cycle for production, acting as a pool, or it may lock up more resource and act as a sink. For example, the deep ocean serves as a sink, rendering unavailable but storing many nutrients required by pelagic (open ocean) species. However, the "sink" becomes a "pool" when periodic ocean upwellings associated with El Niño bring nutrient-rich waters close to the surface off the coast of Peru. Mobilization of this store of nutrients supports large but temporary growth in fish populations. Likewise, forest fires in the sub-Arctic boreal forest release vast amounts of carbon dioxide from carbon that is locked in trees through photosynthesis.

The cycling that takes place in natural cycles is a consequence of the biology of natural systems and is governed by ecological principles, evolution, and population growth and decline. Natural populations are kept in check by the availability of critical resources. Whatever wastes accumulate will provide a nutrient and energy source for whatever fed on it in the first place, so the population of that consumer or decomposer increases until the resource is being consumed as rapidly as it is produced. In the long term, the ecological niches that recycling and replacement depend on will favor the in-migration of better-adapted species and ultimately the selection of organisms that can exploit the resource more efficiently.

A good example of a natural cycle is the nitrogen cycle, as in figure 4.3. (The reader should ignore the "Haber-Bosch Process" for the moment.) The atmosphere consists mostly of nitrogen (78 percent) in a molecular form (N_2) that is inert, meaning that, unlike oxygen, it resists combining with other chemicals and therefore does not normally participate in chemical reactions, including those that support life. The reason for this is that the chemical bond between the two nitrogen atoms is so strong that it takes a large amount of energy to break. Nitrogen can be "fixed," however, in a form that can be taken up and used by living organisms in two ways: by conversion to nitrogenous compounds by lightning or by fixation by certain types of bacteria that are either free-living or symbiotic (symbionts), living within the root nodules of plants. These bacteria have special enzymes (mostly molybdenum-based nitrogenases) that convert N_2 into ammonium (NH_4^+) by reducing the energy required to break the bond and thereby facilitating the split in the molecule. The ammonia may be taken up directly by plants or converted into another biologically usable form, nitrite (NO_2^-), by other bacteria which then convert the nitrogen into nitrate (NO_3^-). The nitrogen is now available as a nutrient that can be assimilated into the plant and it becomes incorporated into proteins, chlorophyll, and other macromolecules and metabolic constituents that require nitrogen in the structure of the plant cell. The plant may then be eaten by an herbivore, which converts the plant protein nitrogen into animal protein nitrogen, and the herbivore may be eaten by a predatory carnivore that converts the herbivore's animal protein nitrogen into its own. The original plant, the herbivore, or the carnivore may die and their bodies will decompose through the action of bacteria and fungi. The decomposition process will create humus and other organic debris that is rich in nitrogen and supports anther plant. It will return some ammonium back to the conversion process as well. At some point, the bioavailable nitrogen may encounter bacteria playing a very different role and undergo de-nitrification, converting back to molecular nitrogen, which as a gas rises and returns the nitrogen to the atmosphere. Something quite similar happens in the oceans, with nitrogen fixed by bacteria and passing through plankton and the marine trophic levels, sometimes sinking to the deep ocean, where de-nitrification can take place or the nitrogen can accumulate or return to the pelagic ocean carried by vertical mixing and upwellings. Thus, the atmosphere is the ultimate "pool" for nitrogen resources, but on both land and in the oceans there are also circulating nutrients in various soil, plant, and animal compartments. Each nitrogen atom cycles countless times, and the availability of sufficient nitrogen is a major limiting factor for production (biomass) in natural terrestrial and ocean ecosystems.

Natural cycles apply to mass but there is no "energy cycle." In the natural world, solar energy is a constant input and energy dissipates, so energy does not move in a closed loop. Rather, it moves through a series of transformations (solar energy, light, chemical energy fixed through photosynthesis, chemical energy released in the bodies of animals and plants, and up through trophic levels of consumption and predation).

At each stage, energy is consumed, and uncaptured solar energy is radiated back out into space. This is called an "open cycle," because the loop is not self-contained and outputs on one end do not become inputs at the other. The continuation of the process depends on a constant input of energy from outside the "cycle."

The idea of an "open cycle" may appear to be an oxymoron (a contradiction in terms) and so it is, because the term does violate the basic meaning of "cycle," which simply means "circle." In this usage the word "cycle" is best understood as a repetitive process rather than a circular system. This use of the word "cycle" to mean a repetitive process, rather than an intact circle, is well established in physics and technology (as in a "thermodynamic cycle," which requires a constant input of energy from outside the system) and in the concept of a product's "life cycle."

Sometimes, human activity enhances or interferes with a natural cycle. In the nitrogen cycle presented in Figure 4.3, the "Haber-Bosch" process, to the upper right of the schema, represents just such a human intervention. It is a chemical process that is used to fix nitrogen chemically in order to produce ammonia for fertilizer and chemical manufacturing. This was one of the most important developments in the history of chemistry and has profoundly affected world history. At the end of the nineteenth century, the world was facing a crisis in nitrogen availability for both fertilizer and chemical production. Nitrogen gas (N_2) is inert (nonreactive) under normal circumstances, so although it comprises 78 percent of the atmosphere, it had been impossible to extract from air by chemical reactions. Recycling of nitrogen-rich organic wastes and manure on a local level was already routine but was insufficient to increase the food supply at a pace that would keep up with world population. Agriculture had become increasingly and dangerously dependent on stores of nitrogen-rich saltpeter (sodium nitrate) found in Chile (and only recently acquired by that nation in a war against Peru and Bolivia). These deposits were being depleted at an alarming rate. The Haber-Bosch process, developed in Germany in 1909, made it possible, for the first time, to "fix" (derive in a form that could be used for chemical synthesis) nitrogen from air, and to do it anywhere in the world where there was an energy source and investment to build a plant. A looming catastrophic "Malthusian" crisis of population and food supply (see chapter 3) was thereby averted.

Since 1913, the process has also been used to produce feedstock for the modern chemical industry, with the effect (well recognized at the time) that explosives and munitions could also be produced much more cheaply. This technology, which without question saved thousands from starvation in the early twentieth century, also contributed to the escalation of arms in the First World War and afterward. It also, quite incidentally, collapsed the economy of Chile and kept that country in economic depression for many years by displacing its now value-less primary export commodity. Over time, the cheap fertilizer the process made available was intensively used and overused, often in situations where run-off drained into bodies of water, creating a serious problem in many waterways of

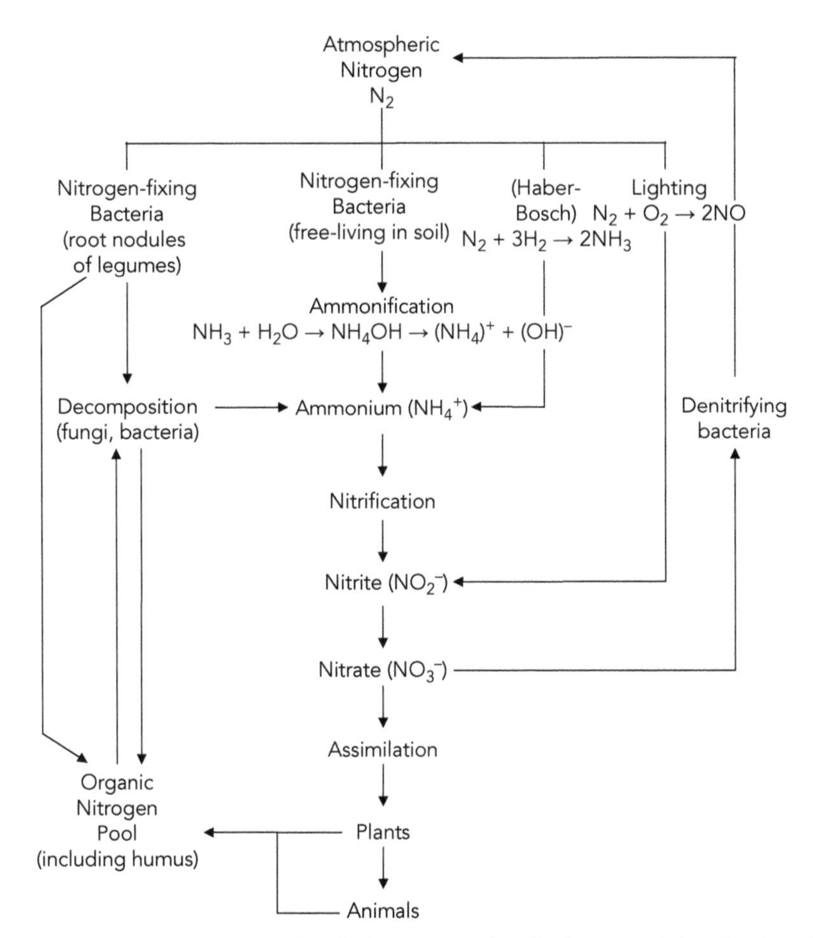

FIGURE 4.3 The nitrogen cycle, which is an example of both a natural closed cycle and, since 1910, of an anthropogenic, open cycle. The Haber-Bosch process turned it into an open cycle through the industrial production of fixed nitrogen, as ammonia.

"eutrophication" (excessive nutrients leading to overgrowth of species, particularly algae, and damage to the aquatic ecosystem). The excess nitrogen, in other words, did not feed back into a closed cycle, as it did in the natural nitrogen cycle. Instead, the cycle was "open" and driven by production in excess of the natural system's capacity to return it to the resource base.

Open cycles are characteristic of human economic activity. Figure 4.4 schematically demonstrates a human or economic cycle, identifying the same basic steps as for a natural cycle. For a renewable resource, the resource pool might be a farm, a forest, or the ocean. For a nonrenewable resource, the resource pool may be a mine, an oil well, or an inventory of materials that have been extracted historically (such as copper) and are now available for use and recycling. Similar to the natural cycle, the material is extracted (conceptually, by mining), distributed (usually

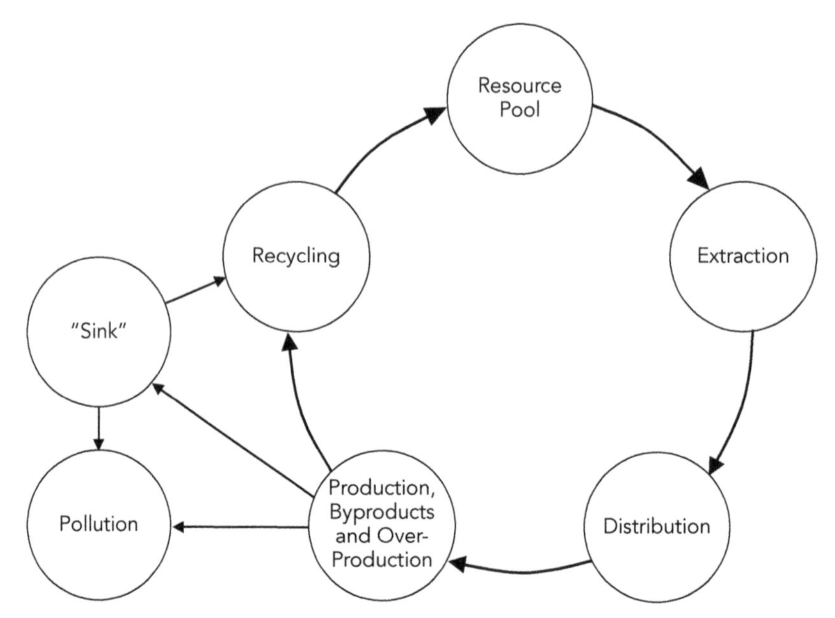

FIGURE 4.4 An anthropogenic, economic cycle, which subsumes the same basic steps as for a natural cycle and uses the resource for production and distribution for human benefit but that also produces pollution.

by energy-consuming transportation), and then produced, but here the "production" step may be disconnected in whole or in part from the next step of "recycling." "Production" also involves production of unwanted byproducts (which are much more often the source of pollution than the primary product) and overproduction of materials that accumulate (such as plastic bags). If the end product of the production step has value, the material may be recycled and returned to the resource pool. But it is just as often disposed of in a waste disposal facility such as a landfill or hazardous waste facility, which acts as an artificial "sink" that does not return resources to the resource pool. Instead, the sink makes the resource unavailable to the cycle for recycling and over time may contribute to the pollution problem (for example, by groundwater contamination associated with poorly managed landfills).

Human or economic open cycles have no intrinsic feedback or balancing mechanisms to ensure balance, and so corrective measures (dotted lines) have to be created through social and political mechanisms. For natural cycles, the products of the production step would be in some balance because of its natural tendency to balance inputs and outputs. In a human or economic open cycle, there is no self-effecting mechanism to balance inputs and outputs, and so unwanted outputs accumulate because they are overproduced and are not removed and recycled. To restore some closure to the cycle requires public policies that discourage overproduction and that link the "sink" back to the resource pool, which is the definition of recycling.

The critical elements of this model of pollution are therefore that overproduction and byproducts of production by human agency lead to the accumulation of pollutants to levels that can result in harm. The harm may affect human health or to the ecosystem. It can be reversible or irreversible. In general terms, however, harm is proportionate to the degree that the cycle is kept open and how long this unnatural situation continues.

To create a closed cycle for human activity requires an engineered or managed effort, which may take the form of reducing or mitigating pollution, changing the process of production, or returning resources from the "sink" to the resource pool through some form of recycling. In many ways this is the essence of sustainability. It may also involve pollution abatement, remediation, economic incentives, and all the policy options currently used for environmental management.

Figure 4.5 focuses on the harm caused to human health from both human and economic open cycles, which are the primary concern, and from natural cycles, when harm comes from natural contamination. As pollution is released

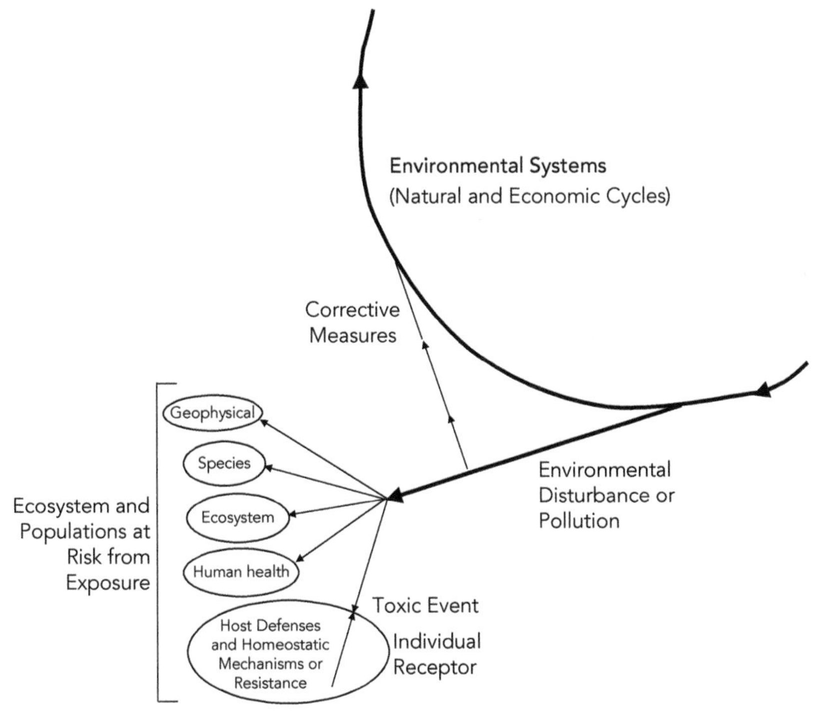

FIGURE 4.5 Harm may be caused to human health from either pollution from anthropogenic and economic open cycles or nonanthropogenic contamination associated with natural cycles. Individuals vary in their level of exposure, even when they are members of a population at known risk of exposure. People also vary, within limits, in their susceptibility to the effects of the natural contaminant or pollutant.

and exerts an effect, it may affect various individuals differently for two reasons. The first, and usually the most important, is that different people are exposed to different levels of pollution, even in the same environment. The second is that the body's natural defenses against pollution (tolerance, host defenses, homeostatic mechanisms, "resistance," and the ability to adapt) vary from individual to individual, and so the total capacity to oppose the adverse effects of the pollutant also varies. There are both genetically determined and random biological variation in strength among people in their capacity to tolerate, resist, or adapt to the toxic effects of pollutants. Past medical history and state of health also play a role in whether a person is more or less susceptible to the effects of pollution.

Ordinarily, the most susceptible people to the adverse effects of pollution are, predictably, the very young, the very old, the chronically ill (for example, with asthma), and the expectant mother and fetus. However, there are many exceptions. One is that some people are genetically predisposed to react in certain ways. For example, people with atopy (a genetic predisposition to allergy and asthma affecting about 10 percent of the population) are more likely to have respiratory problems from any airborne irritating exposure than people without atopy (roughly 90 percent of the population). There are, however, even exceptions to the generalization about susceptible subgroups. For example, young people tend to be particularly susceptible to the effects of ozone in air pollution, a consequence of the normal function of certain receptor cells in the human lung. Older people tend to be more resistant to ozone. Given this variation, it is important to think in terms of the distribution of adverse effects outcomes in a population, rather than just the effect on a single individual. For example, some people will have more eye irritation than others during a severe smog incident, while a person with asthma may have a severe attack and a high school athlete may not be able to perform at his or her usual best. A model for understanding the effects of pollution has to take all this into account, as well as variation among individuals in a population.

Figure 4.6 presents an overview of the scientific disciplines that are involved in studying the adverse effects of pollution. The main cycle that generates pollution is the object of study of ecology (which is mostly concerned with natural communities but also documents disruption and perturbation of ecosystems by pollution), economics (there is a small subspecialty of environmental economics within the discipline), demographics (because of the amplifying effects of increasing population), and environmental chemistry. The type and level of pollution is determined by many scientists but those who specialize in the doing so usually identify themselves with exposure science (a relatively recently developed discipline), exposure assessment (the general term for determining the circumstances, profile over time, concentrations, and distribution of exposure to a pollutant), and specifically in the workplace, industrial and occupational hygiene ("industrial hygiene" is more common in the United States, "occupational hygiene"

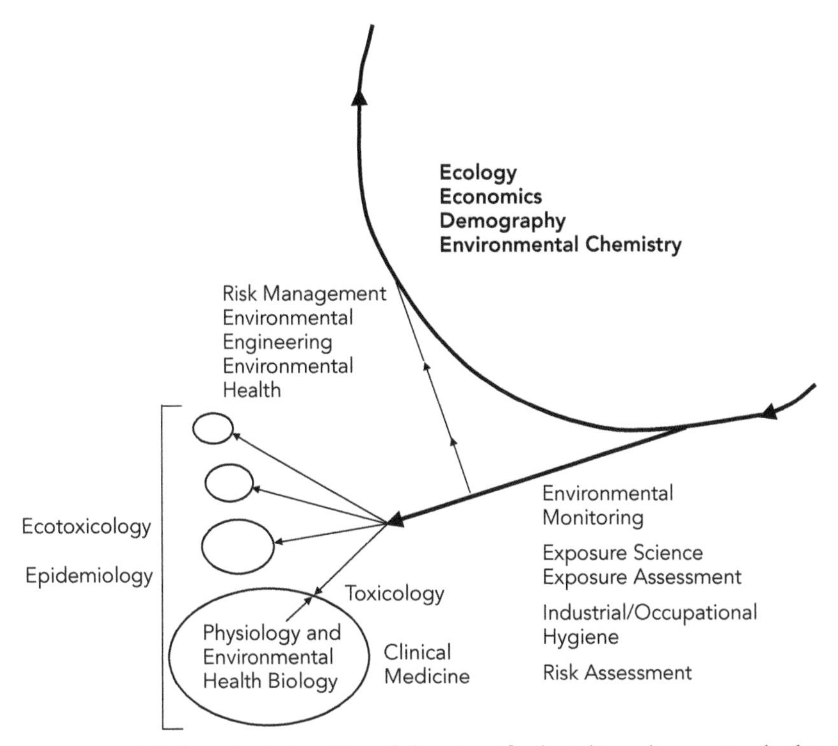

FIGURE 4.6 Role and relationships of the scientific disciplines that are involved in studying the adverse effects of pollution and natural contamination.

in the rest of the world). Many of the effects result in specific diseases that are studied by physicians practicing clinical medicine (including but not limited to clinical toxicologists and specialists in occupational and environmental medicine). Scientists who study the effects of toxic substances and the body's response are toxicologists. The capacity of the human body to resist and to adapt to toxic exposures is also a concern of toxicologists but has been the subject of investigation by many physiologists and biochemists who may not identify themselves with toxicology. Epidemiologists use statistical methods to identify patterns of disease and risk factors associated with them in populations, both human and animal; environmental epidemiology is a primary source for identifying and estimating the degree of risk associated with exposures to pollutants.

"Environmental health" is a broad professional field of public health that emphasizes the identification, assessment, and control of environmental hazards, of which pollution is a major part but not the only concern; it is focused on solutions to environmental problems that present human and animal health risks. Environmental engineering is a branch of civil engineering that addresses systems for managing environmental hazards, such as water treatment, sewage, solid waste disposal, and air pollution control. Environmental engineers are thus critical in designing and operating the systems that restore open human and economic cycles to some degree of similarity to closed natural cycles. Clearly this is a long list of professions and academic disciplines. Fragmentation and interdisciplinary communication remains a challenge in environmental health and sustainability; but cooperation among scientists from different disciplines is probably further along than in most other complicated fields.

Types of Pollution

There are three basic types of pollutants, broadly speaking:

- Microbial, in which the pollutant introduces a microorganism into a medium, usually water or food, that can infect a person or animal and cause disease
- Chemical, which is discussed in detail in the next chapter
- Energy, of which the most common forms of public health concern are noise, heat, and radiation

Table 4.2 demonstrates how these three basic types of pollution apply to the important ambient media: air (ambient and indoor/built environment), water (surface, groundwater, ocean), soil, food, and consumer products. Table 4.2 also indicates which problems are principal concerns of environmental health protection. Not every combination occurs in the real world, and not every combination that occurs is important as the others as a practical matter in shaping sustainability and public health.

Microbial Pollution

The earliest, most dangerous, and still by far the greatest global threat to human health from pollution comes from microbial pollution, the contamination leading to health risk of any medium from microorganisms. The microorganisms primarily involved in human health risk are bacteria, viruses, parasites, and fungi. There are other types of microorganisms important in clinical medicine but they are uncommon as microbial pollutants and do not have the same level of importance to public health. The media involved are primarily water, food, and the built (indoor) environment.

The most important and common microbial pollution problem is fecal contamination, which will be the chief topic of this subsection. Fecal contamination means contamination with the waste products of human and animal elimination. (Feces is another word for excrement, also known as "ordure," "dung," or "manure" when it comes from animals, and vulgarly known as "poop" or "shit" when it comes

Table 4.2 Types of pollution, by medium, with examples. (Problems that are common and of practical importance in environmental health are in bold type.)

Medium	Microbial	Chemical	Energy
Global atmosphere	Distant transport of pathogens (e.g., fungi)*	**Global atmospheric change, resulting in climate change and other atmospheric effects**	**Climate change has secondary effects on heat in the atmosphere**
Ambient air	Airborne pathogens	**Air pollution (ambient)** **Air toxics**	**Noise**
Built environment	**Transmission of air- and waterborne infectious diseases indoors, e.g., *Legionella*** **Occupational infectious disease**	**Indoor air quality** **Occupational chemical health hazards**	**Occupational hazards (e.g., noise)** Electromagnetic fields* Radiation hazards
Water	**Water pollution (microbial), particularly fecal** **Cross-contamination of food**	**Water pollution (chemical): surface waters, groundwater**	Thermal pollution (excess heat) from discharge from cooling units of power plants, industrial sites Radionuclides in water
Ocean	**Ocean pollution (microbial) near sewer outfalls**	**Release and transport into ocean** **Ocean acidification due to ↑carbon dioxide in atmosphere**	Noise/sound (affects ocean species) Very low frequency electromagnetic radiation (used for submarine communication)*

(continued)

Table 4.2 Continued

Medium	Microbial	Chemical	Energy
Soil	Contamination of soil with anthrax (local and uncommon problem)	**Contaminated sites (e.g. brownfields); migration to groundwater** **Nitrogen runoff and eutrophication**	(Not applicable)
Groundwater	Contamination of aquifer can occur	**Groundwater contamination (chemical)**	**Radiation (from radon)**
Food	**Microbial contamination; "food poisoning" risk**	**Chemical contamination, often by pesticides**	Radiation (not a major problem)
Consumer products	Uncommon (except in foods)	**Contamination of products, usually with lead; potentially toxic constituents of products**	Radiation, noise
Space, ether	(Not applicable)	(Not applicable)	Light pollution (reduces visibility of sky at night)

*Not established to be an important health hazard.

from humans. Sometimes in the environmental health sciences it is necessary to get that basic when communicating with people about risks to health.) "Fecal-oral contamination" occurs when microbes in feces are conveyed to the mouth and are ingested, causing disease in the host. In microbial pollution of drinking water, this occurs when there is contamination with fecally contaminated water. In food, it occurs most often when someone preparing food has not washed their hands after using the restroom. It can also occur when the feces of animals enter water and when intestinal contents of animals contaminate food while it is being prepared.

Many infectious diseases are specific for certain species. Human beings are of course most susceptible to human diseases, so that the pathogens most threatening to human health come from other humans, as a rule. However, human beings can and do contract some infectious diseases from animals, including some of the more serious and common microbial diseases associated with pollution. (See chapter 6.)

Many pathogens are present in the gut (small and large intestine) of people in any population at any given time, in people who are currently ill, in those who have an infection that is so low grade that they do not notice it, and in those who are entirely asymptomatic but carry the disease. Historically, one of the most important asymptomatic carrier states was the persistence of the organism that causes typhoid fever (*Salmonella enterica* variant enterica, subvariant typhi) in the carrier's bile duct, where it would periodically be released into the gut. Such people sometimes had to be quarantined and even detained in jail to prevent the spread of disease in the days before antibiotic treatment. Disease outbreaks from fecal-oral contamination therefore depend importantly on what pathogens are present in the human population in a given area at a given time and what opportunities there are for spread from the animal population. Very few people today are carriers of typhoid. The diseases spread by feces in developed countries today are usually forms of diarrhea caused by viruses or viral hepatitis.

Water

Contamination of water by human feces is a dangerous but common problem worldwide. Because of the risk of contamination wherever people are found, the frequency of the problem around the world, the potential for fecal contamination in virtually all water systems if conditions are not maintained, the debilitating effects when an outbreak occurs, and the contribution of the problem to infant mortality, microbial pollution of water ranks with malaria and only a few other individual public health issues as threats to sustainability. In the developing world, microbial water pollution is a substantial risk to life and a drain on productivity, economic development, and sustainable development. Prevention of waterborne disease was for most of history (and in many parts of the world still is) a constant fact of daily life.

The cause of most microbial water pollution today, as historically, is fecal contamination, as described above. Transmission by the fecal-oral pathway through water remains one of the principal public health threats in the developed as well as the developing world.

Fecal microbial water pollution primarily carries a risk of viral diarrhea, which is a particular threat to children, and other diseases, including cholera (itself causing an extreme form of diarrhea), typhoid, hepatitis, polio, and other serious and often lethal diseases. The fear of these diseases, disgust at the thought of human excrement, and the social approbation and dismay associated with visible and odorous human waste motivated profound changes in urban design, advanced civil engineering, and raised expectations for cleanliness to their present level in polite society.

Animal fecal contamination also plays an important role in fecal microbial pollution of water but for parasitic diseases or particular strains of bacterial infection. Viruses that are present in feces are more species specific for human infection, so the risk for viral diarrhea from microbial contamination comes almost exclusively from the vast reservoir of human infection.

Were it not for the aggregation of human dwellings into communities above the small village level and the growing size of human populations producing waste that discharges larger volumes into the finite resource of flowing bodies of water, the risk of microbial pollution of water would be much less. Prior to about 1600, the practical size of cities was limited to under a million largely because of waste disposal issues. Cities were able to grow larger and more efficient after the invention of the flush toilet and capital investment in sewerage systems. ("Sewage" refers to the waste; "sewerage" refers to the infrastructure that removes the waste.) Before the turn of the twentieth century, no disinfection of drinking water took place (although some degree of filtration was not uncommon), waterborne disease outbreaks such as cholera and typhoid (for which records are best) were sporadic and spread with the movement of individuals (called "carriers," as noted above) and ships, making the worst outbreaks of these diseases in port cities and towns.

Human activity is critical in fecal contamination from domestic animals and wildlife. All animals have their own gut flora, most of which consists of bacteria that do not affect human beings. However, some animals carry certain pathogens that can cross the species barrier with devastating results. These include the bacterial pathogens referred to below in the discussion of foodborne microbial pollution and parasites that are common in the gut of wildlife species, of which the most important is *Giardia lamblia*, a protozoan that forms a cyst that is difficult to disinfect and causes "beaver fever" (giardiasis, a difficult diarrheal disease) and *Cryptosporidium parvii*, another protozoan infection that resists disinfection and which causes a refractory infection resulting in watery diarrhea).

Even in the modern era of reduced pathogen load in wastewater, fecal contamination of drinking water continues to carry an unacceptably high risk of disease. Minimal levels of fecal contamination may be associated with a sufficient load of virus to cause disease even in recreational water (water used for swimming, boating, fishing, and bathing, but not drinking). That is why swimming pools are chlorinated and why the same disinfected water supply provides both drinking water and bathwater.

Drinking water treatment and monitoring are based on technologies that provide protection for the entire community on a continuous basis only when they function continuously, without interruption, and consistently meet standards. For over a century, the first line of defense in drinking water has been chlorination, using as chlorine gas, chlorine dioxide (which is easier and safer to handle at the plant), or a combination of chlorine and ammonia called chloramine.

Chlorination is effective in killing bacteria, viruses, and most (unfortunately not all) parasites at the point where it is introduced into the system—but so are other, alternative disinfection systems based on ozone, ultraviolet light, and membrane ultrafiltration. The advantage of chlorine is that it leaves a trace in the water known as "free chlorine residue," which is enough to kill most bacteria and enteric viruses. ("Enteric" refers to intestinal.) This free chlorine provides a margin of safety downstream because as it is carried in the distribution system, the chlorine residue in water remains bactericidal in the water lines, providing critical, reliable protection in an imperfect distribution system. Because of this "passive protection" downstream, chlorination has not been replaced by other methods that singly or in combination may be effective at the point of disinfection but that provide no protective effect downstream.

The horrible example of Walkerton, Ontario, is an example of how badly things can go wrong when there are lapses in disinfection. Walkerton is a relatively affluent, stable middle-class farming and recreation-centered community of almost 5,000 in southern Ontario (Canada) within Bruce County, not far from where many city residents from the Toronto area have summer vacation homes ("cottages"). An outbreak of bloody diarrhea occurred in Walkerton in May 2000, caused by the failure of two municipal employees (who happened to be brothers) to do their job in loading chlorine into the disinfection system and monitoring levels of disinfection. They also falsified reports to cover their tracks. A heavy rainstorm and the presence of cattle and of their feces in the area where the water was taken in led to pollution by a particularly dangerous strain of gut bacteria. Unusually, for such events, the bacterial load was so large and the rainfall so intense that groundwater was contaminated. The bacterium, *E. coli* O157:H7, is not a common fecal pathogen in cattle and does not make them sick. In humans, it has been more often associated with foodborne outbreaks, especially from contaminated ground beef. *E. coli* O157:H7, can cause liver and kidney failure as well as severe diarrhea. The pathogen exerts its effects by releasing a toxin (a chemical agent produced by the bacterium), rather than by direct infection. The toxin causes serious and even lethal complications that may occur despite antibiotic treatment (which paradoxically in this condition may even make conditions worse due to side effects). Seven people died in Walkerton of an entirely preventable disease, and roughly half of the population got sick with bloody diarrhea. This was a "perfect storm" of heavy contamination, serious lapses in disinfection, and criminal negligence in mismanagement. Such extreme malfeasance and contamination are not necessary for serious public health threats to emerge, however; much lesser lapses are quite enough.

The public health triad, which will be described later and mentioned many more times in this book, refers to the three essential elements for a disease outbreak: the pathogenic agent, a medium that conveys the agent to the host, and a susceptible host. Applying the public health triad to the Walkerton episode: a novel pathogen (agent) was allowed by negligence to be conveyed (environment) to a susceptible population (host) with fatal results. Walkerton therefore illustrates all the basic principles of microbial pollution. However, in another sense the Walkerton example is misleading because it was so severe and had its roots in the irresponsibility of negligent operators, who could be blamed as individuals. At first, the government authorities did just that, but an investigation determined that Walkerton was also a systems failure that could have happened elsewhere. Decision makers then had to face up to the need for regulatory reform, strengthening the public health (as opposed to the medical care) system, inspection and monitoring of performance, better training of the people entrusted with water treatment, and investment in better technology and water quality monitoring.

Chlorination interruptions actually happen fairly frequently, but they are usually very short, immediately corrected, and some protection is provided by chlorine already in the system. More typical breakdowns in drinking water quality treatment and monitoring result in more subtle and delayed effects in the modern world, where there are relatively few carriers of infectious disease in the general population. However, experiences of disease outbreaks during war, civil unrest, and natural disasters provide illustrations of the short-term effects of removing environmental health services.

There is now a small but troubling movement against the basic principles of drinking water safety. In the name of avoiding chemical hazards, a small but (in the United States) extremely vocal group of activists in Vermont, Pennsylvania, and California have attacked the idea of disinfection, especially disinfection with chloramines, which is the alternative disinfection preferred to chlorine for urban drinking water distribution systems. Their objections focus on alleged irritant effects of chloramines, speculative toxicology, and the unanticipated consequences of disinfection by chloramines in promoting lead release from lead service lines in older North American cities, particularly Washington, DC (where a major and highly visible problem with lead in drinking water began in 2001). One social consequence of this problem is that there is emerging in some quarters an activist movement to ban chloramines and chlorination or any disinfection treatment that leaves a residual and a nihilistic attitude against disinfection in favor of only filtering water. If this movement were to actually gain influence, it could result in catastrophic health consequences.

Food

By comparison to microbial water pollution, microbial contamination of food is a more restricted problem with fewer implications for global sustainability or sustainable development. However, food contamination (the public health terminoloty for the issue is "food safety") has major implications for the sustainability of some enterprises and of the food distribution system. Because food is usually spoken of as a commodity, rather than as a medium, the usual phraseology is "microbial contamination" of food rather than "microbial pollution of food," even though the effects on health can be severe.

Foodborne diseases are very common, leading to approximately 76 million cases per year in the United States alone, most of which require little or no treatment. About 350,000 cases a year require hospitalization, with 5,000 deaths. Obviously, most of these diseases are minor or short-term and indispose rather than incapacitate or disable the host. However, a few of the pathogens that cause microbial foodborne diseases are lethal and can cause serious illness especially in persons with compromised immune systems. Foodborne illness may easily be mistaken for other illnesses, and so unless special studies are done to identify true cases (as in the sources for the figures cited) public health statistics are always a serious undercount.

Table 4.3 presents a list of products and pathogens that have been responsible for large outbreaks of foodborne disease in recent years. Food can be contaminated by virtually any class of microorganisms, but some pathogens are more common than others:

- Bacteria (principally but by no means exclusively *Campylobacter spp.*, staphylococci, Shigella, *Listeria monocytogenes,* and *E. coli* which comes in "toxigenic" strains that produce toxins that cause acute food poisoning and "nontoxigenic" strains that cause infectious diarrhea)
- Viruses (hepatitis A being the most common; also, norovirus and many others)
- Protozoan parasites (*Giardia lamblia, Cryptosporidium, Cyclospora Microsporidia*)
- Metazoan (multicellular) parasites (*Trichinella* spp., liver flukes, especially in rural Asia)

Table 4.3 Foods and pathogens responsible for large recent outbreaks of foodborne disease

Food	Pathogen(s)	Disease
Beef, poultry, raw or undercooked eggs	*Salmonella* spp.	Diarrhea and severe gastroenteritis
Seafood	Hepatitis A and B, *Vibrio cholera*	Hepatitis, cholera
Unpasteurized milk and cheese, processed meats	Many, including *Listeria monocytogenes*	Several; Listeriosis (caused by *L. monocytogenes*) in particular may be fatal to people with chronic disease affecting the immune system
Undercooked game, pork	*Trichinella*	Trichinosis, an inflammatory condition of muscle due to infection
Milk	*Campylobacter*, norovirus	Severe diarrhea
Unpasteurized apple cider	*Escherichia coli* O157:H7 (this is a particularly dangerous strain of *E. coli*)	Severe systemic disease with risk of liver and kidney failure
Improperly canned preserves and meats and infused oils (usually garlic in oil)	*Botulinum*, producing "botulinum toxin"	Paralysis; fatal without specific treatment
Strawberries, and other fruits that are difficult to clean	Hepatitis A	Hepatitis
Any food (a frequent problem in institutions and on cruise ships)	Norovirus	Diarrhea

- Toxins derived from microorganisms than can remain on food despite the absence of viable organisms that can grow (particularly toxigenic *E. coli, Shigella*)
- Prions (abnormal proteins that act like infectious particles, principally the one causing "variant" Creuzfeld-Jacob disease, popularly known as "mad cow disease")

Any microorganism that can contaminate water can also contaminate food. Microbial contamination may also be transmitted on any surface, at least in the short term.

Contamination may take place at any time in the cultivation and preparation of food. The food itself may be contaminated to begin with (for example, fecal contamination with *Salmonella* in chicken meat and eggs); there may be superficial contamination in handling the food (for example, cross-contamination when, say, meat or vegetables are cut on a cutting board previously used for chicken that has not been thoroughly cleaned); and there may be spoilage. The pathogen may multiply or at least persist in the food, but if it produces a toxin it can cause acute (and often severe) illness even with no viable microorganisms surviving.

Most foodborne outbreaks associated with fecal-oral transmission are due to failure of food handlers to wash their hands between going to the restroom and handling food. They are easily prevented by hand washing but not by wearing gloves, because the gloves may become contaminated by contact.

Fecal pathogens may grow on food, particularly if it is kept warm (but not hot), moist, and is rich in nutrients for bacteria. Every student of public health learns early that egg and chicken salad are particularly good media for bacterial growth and that seafood is especially problematical.

Foodborne outbreaks, like many forms of infectious disease risk, disproportionately affect people who have conditions or medical requirements that compromise the immune systems, such as kidney dialysis, chemotherapy for cancer treatment, HIV/AIDS, malnutrition, and advanced age. The elderly are particularly vulnerable, both because their immune system is not as strong and because once an outbreak of many of these pathogens begins from foodborne outbreak, person-to-person transmission of the same pathogens occur easily in close quarters. This is particularly true for norovirus and *Camplylobacter*.

Travelers are often affected by foodborne and waterborne pathogens and develop "traveler's diarrhea" while visiting unfamiliar places. This is partly because sanitary standards may be different from what they are used to but also because they may be exposed to unfamiliar strains of bacteria, to which they are not immune or tolerant. Most traveler's diarrhea is caused by toxigenic *E. coli*.

Prevention of foodborne outbreaks depends on close monitoring of the food supply from "farm to fork." At the agricultural production level, prevention is primarily concerned with preventing fecal and chemical contamination of products. Inspection, as a means of preventing foodborne illness, is not very efficient and is not completely effective in the processing stages. Microbial contamination cannot be seen, and testing takes time. This approach has therefore been largely replaced in the food processing stage by a system of hazard analysis and critical control points (called by that name by the US Department of Agriculture).

Microbial contamination of processed food generally occurs during either the production stage or the preparation stage. Contamination during the production phase most frequently occurs at specific "critical points" in the process, which are or should be tightly controlled (for example, temperature and cleaning protocols), with performance closely monitored for noncompliance. Oversight to ensure that restaurants and other food services are preventing contamination during food preparation is the function of the local public health agency, which does rely on inspection. Key issues are avoidance of cross-contamination, the facilities (such as double sinks, to ensure that cross-contamination from cleaning contaminated equipment and hand washing in the same sink does not occur), maintenance of temperatures that inhibit bacterial growth during refrigeration, adequate temperature for an adequate duration during cooking and reheating, and hand washing. Local health departments normally have the authority to immediately shut down any enterprise that processes food or serves food if it that is found to be out of compliance with regulations. This is necessary and appropriate to protect the health of the public, but it often leads to tense relationships with local restaurants. Many local health departments have grading systems to alert the public to an establishment's compliance with the rules: restaurants that do not earn an "A" grade usually do not stay in business for long.

Therefore, the implications of food safety for business sustainability, particularly for small farms, food establishments, institutions such as schools and nursing homes are immense. An entire enterprise, even a highly profitable business, can be forced to a halt overnight by a failure involving microbial food contamination or neglect of oversight to ensure basic sanitation. An outbreak is sometimes accompanied by serious illnesses or deaths among susceptible patrons of the establishment, which results in exposure to legal liability. When modern molecular and genomic techniques can identify the strain of the pathogen with relative ease and demonstrate where it came from, it is difficult to defend these cases against lawsuits for personal injury or wrongful death.

One example was a popular three-restaurant chain in San Diego, California, called Gulliver's. Very popular with tourists in 1981 because they were situated in strategic locations near tourist attraction, the restaurants were highly profitable for their owners and the chain had plans to expand across the country. Unfortunately, the managers neglected to train their workers, many of whom did not speak English, in the importance of hand washing. An outbreak of hepatitis B (a serious kind of hepatitis transmitted by food) that eventually affected hundreds of diners was traced to a salad bar at one location on a particular day and to a specific employee who did not wash his hands. Media coverage and the resulting public outcry was intense, and the company was blamed for serious illness in several patrons. The owners made the decision not to reopen any of their locations, and the chain went out of business overnight.

Energy Pollution

The inclusion of energy as a pollutant presents a conceptual problem for several reasons. One normally thinks of pollution as involving chemicals, including radionuclides; but there has also been long and accepted usage of the term for "noise pollution," "light pollution" (a particular concern of astronomers), "heat pollution" (particularly regarding the effluent of power generating stations), and occasionally "electromagnetic field pollution" (a more recent concern). The problem with considering energy pollution in the same way as chemical and biological pollution is that energy dissipates quickly in almost every environmental situation when the source is shut off, and so is unlike other forms of pollution that accumulate and stay elevated for a period of time even when the pollutant (a contaminant causing harm) is no longer released from the source.

Health Effects

The "public health triangle" (see figure 4.7) is a fundamental and versatile concept in public health, derived from infectious disease and vector-borne diseases. This conceptual framework is a useful way of thinking about public health problems because it has implications for designing interventions and control, as well as for understanding the problem. The concept works equally well for microbial pollution and for chemical pollution. Although the discussion is grounded in human health, it should be understood that the same approach applies to animal health and ecosystem effects. However, in ecotoxicology the hosts or receptors may be any important species in the biological community or the ecosystem as a whole. Ecosystem effects tend to reflect the consequences of effects on the most susceptible species, the weakest link in the chain.

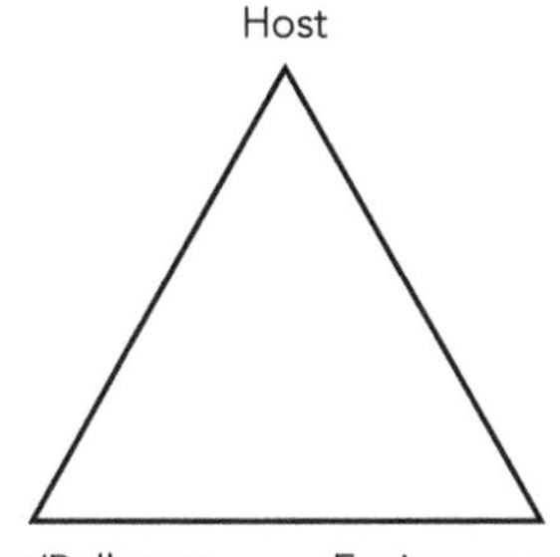

FIGURE 4.7 The "epidemiological triangle" was developed with infectious disease in mind, but it is also a helpful framework for understanding toxic exposures.

The concept of the public health triad is that for a disease outbreak to occur there must always be three elements in place:

- An "agent" capable of causing the disease, which may be a pathogen in infectious disease or a pollutant
- An "environment" that brings the two together in a manner that allows transmission of the agent or exposure to the chemical, which for some communicable diseases might be a vector and its habitat but may be and in the case of pollution would involve a "medium" (air, water, soil, food)
- A "host" that is susceptible to the disease (that is, not immune to the pathogen or resistant to the chemical hazard)

The triad is often expressed graphically in textbooks as a triangle (Figure 4.7); however, this representation is misleading because there is a definite sequence of events (pathogen or pollutant → environment/medium → susceptible host → disease that is not represented in the triangle configuration).

The term "public health triangle" is occasionally used to refer another, equally fundamental three-element "triad" in analytical epidemiology: "person, place, time." However, references to "the triangle" or "the triad" will normally be understood in public health circles to mean "agent, environment/habitat, host."

If any of the three factors is absent, the disease or toxic effect cannot occur. Therefore there are three basic strategies of public health to prevent disease and toxic effects, and they are as follows:

- Prevent exposure to the pathogen or toxic chemical, which can be achieved by eradicating it (removing it altogether), eliminating it (removing it from proximity to the host and the possibility of exposure, always with the potential risk of re-introduction in the future) or by hazard control (capturing pollutants before they are released).
- Change the host to render him or her resistant to the pathogen, for example by immunization or the use of personal protection such as respirators to prevent inhalation of hazards, which is central to protection in the workplace but impractical for most ambient environmental chemical exposures.
- Change the environment so that transmission or absorption does not occur, for example by controlling vectors of disease (addressed in chapter 6), cleaning up media such as polluted air or water, or diverting polluted media (such as sewage) so that people or animals are not exposed.

The most obvious applications of the public health triangle have been made to microbial pollution, which is a close fit to the model of tropical infectious disease (an early inspiration for the triad concept). Microbial pollution was discussed in greater detail earlier in this chapter. It is has been less often recognized that the public health triad works equally well for chemical pollution. The human health effects of chemical pollution necessarily require a hazardous pollutant, an environment in which the capacity to transform or eliminate the pollutant is slow or saturated, and a person (the "host" or "receptor") who is susceptible to the effects of pollution. For chemicals, the effects are also proportionate to exposure (dose or concentration), which gives rise to endless arguments over whether a reliably "safe" exposure level can ever be found. The model fits energy pollution less well.

All three elements must be in place for disease to occur. Ideally, all three should be identified if an investigation of the disease is to result in a complete understanding of why it occurred and, if it is a new outbreak, what changed to cause it. However, the implication of the triad is that it is not always necessary to address all of the elements in order to stop a disease outbreak. An intervention that blocks any element of the triangle (triad), or "breaks off any of the corners of the triangle," will stop the outbreak. This feature of the triad allows public health professionals to choose the best intervention strategy in a given situation (which may involve immunization against a pathogen or interrupting transmission), to use the same strategy to target more than one disease (for example, mosquito control for dengue and malaria), and to get better results by a strategy of working on more than one element at a time when a single intervention is only partially successful.

BOX 4.1

Contamination or Pollution?

"Emerging contaminants" are chemicals that have not been regulated nationally and have not traditionally been recognized as water pollutants. Traces of chemicals present in beverages, tobacco products, pharmaceuticals, cosmetics and other personal care products, and industrial chemicals can be detected in low concentrations in surface waters that provide drinking water of many American cities. They constitute a new class of "emerging contaminants" for which water quality standards usually do not exist, are incomplete, or cannot yet be based on reliable science. They are called contaminants rather than pollutants because to date there is no evidence of harm.

Environmental chemists have become aware of the presence of these emerging contaminants after monitoring programs were designed to look for them (particularly those conducted by the US Geological Service) using better measurement technology. Environmental chemistry is now sophisticated enough to find them at very low concentrations, in the parts per billion (ppb) range. Applying these more accurate and sensitive methods, environmental chemists were able to find trace contaminants in 80 percent of samples from 139 streams in thirty states, on average seven per sample.

Some of the chemicals that have been found so far in surface waters in the United States and Europe include:

- Caffeine, obviously derived from coffee but many other beverages contain this and closely related chemicals
- Nicotine, obviously from cigarettes and other tobacco products
- Prescription drugs reflecting medical prescribing patterns across the country, particularly analgesics, antidepressants, statins (drugs used to lower cholesterol), antihypertensives, anticonvulsants (anti-seizure medications), and anticoagulants (blood thinning agents)
- Antibiotics from both prescription drugs and agricultural used in livestock (which may have an effect on bacterial selection and resistance)
- Steroids and synthetic hormones used in birth control pills (which exert hormonelike endocrine activities and so may have effects on wildlife disproportionate to their concentration)
- Chemicals used in consumer and industrial products, including plastics, pesticides, fragrances, and solvents
- Silver nanoparticles (which are bactericidal—they are an even greater problem in wastewater and may soon be as ubiquitous as trace organics)
- N-nitrosodimethylamine (which has characteristics similar to the fuel additive MTBE (methyl *tert*-butyl ether), which caused huge problems in groundwater pollution); this chemical may also become widespread as a disinfection byproduct)
- Perfluorinated compounds (including C8, used in making polymers)
- Perchlorate (most often associated with rocket fuel, but which can also occur naturally)
- PBDE and PBBs (out-of-production fire retardants, not to be confused with PCBs)

The first six contaminant classes listed above are present in consumer products, and so their discharge in wastewater reflects population density. The other five appear to come mostly or entirely from "point sources," where they originate or are discharged locally. The localized emerging contaminants of greatest interest at the moment are often related in the United States to Superfund National Priority List (NPL) sites or local industrial sources, including military installations.

When people drink coffee or tea or smoke cigarettes, the caffeine or nicotine is absorbed by their bodies and ultimately excreted in urine and goes into the wastewater stream for the community. A person may not drink all of the coffee and may pour the leftover coffee down the sink. A cigarette but may be left on the street, contributing nicotine to surface water runoff, or may go into an ashtray and then be discarded in a landfill where, if it is not secure, it may reach groundwater. In each situation, it is easy to see how the total loading of these chemicals discharged in the water reflects consumption, such as coffee preferences or the number of people who smoke.

Pharmaceutical agents are a particular problem because as chemicals they are designed to have an effect on the human or other mammalian body at a particular dose, and so are likely to have toxic effects for humans and animals at lower levels than other chemicals. When people take drugs, some of the drugs and their metabolic products are excreted in urine and feces. As many as 100 pharmaceutical agents have been found in surface water. A person may not take all of the prescription, often flushing drugs down the toilet but sometimes putting them in the trash, where they may find their way into surface run-off or groundwater.

Not surprisingly, these chemicals derived from pharmaceutical and personal products are present in greater amounts in water the larger the population upstream. They are then present in the source water that becomes the drinking water supply for the next towns and cities downstream.

Municipal water utilities in North America do not routinely monitor for these chemicals because it is too expensive and impractical and because there is currently no regulatory requirement to do so. Removing these chemicals from the water supply would be very expensive and, given current technology, impractical. The European Union, which is more densely populated along its rivers than much of the United States, has been particularly active on this issue in monitoring, research, and pushing advances in innovative technologies for treating problem contaminants.

On the discharge side, where communities upstream treat and release their wastewater, conventional sewage treatment plants do not remove these emerging contaminants efficiently, so they pass through the treatment plant and into surface water. Even the best contemporary wastewater treatment does not break down these emerging contaminants to any great extent.

On the intake side, when a community downstream takes in source water to treat for drinking water, the standard water treatment technologies used for purification and disinfection, including coagulation, filtration and chlorination, only seem to reduce many organics a little, but not completely, and leave most of them in the drinking water supply. Some are only removed by highly specific and expensive technologies, such as ultraviolet irradiation, that cannot provide reliable microbial disinfection and that are not particularly useful in removing any other chemicals: these would have to be added to conventional treatment at considerable expense.

No human health effects have been identified from these emerging contaminants so far, but that does not mean that there is no risk, especially as concentrations rise in the future. But studies are ongoing, so this may change. The concern for people is that if the concentrations continue to rise they might eventually reach a level where sensitive people in the population could be affected by the agents.

A more immediate concern is that many of these chemicals, specifically those with endocrine disruption activity, may have significant effects on fish and other aquatic organisms and possibly on wildlife up the food chain (trophic levels) out of proportion to their low concentrations: this is because they mimic or block the effect of hormones. This represents a situation where ecotoxicology is of greater immediate concern than human risk in a contamination problem.

Currently, these contaminants are present in very low concentrations (ppb). These levels are probably not enough to affect human health; but if levels rise further this possibility cannot be ruled out. For species that live in water and that may bioaccumulate some of these agents, however, there is a real possibility of environmental impacts and ecosystem effects. This is especially true for agents that exert hormonelike effects (so-called "endocrine mimics").

Pharmaceutical agents are a big part of the problem. The fact is that there are only three ways to reduce concentrations of contaminant chemicals such as pharmaceutical agents in drinking water. First, reducing the load by reducing the release into wastewater is the best way; but people will certainly continue to take drugs and excrete them, and many upriver communities are growing, therefore the problem is increasing. "Take-back" programs to divert drugs from being flushed down the toilet or drain would have a marginal effect compared to the effect of a much greater quantity of drugs passing through the body. The second, reducing the intake by human beings through upgraded drinking water treatment may be good for human health, but the greater problem is probably ecosystem toxicity. Treating drinking water does nothing for effects on aquatic species, such as the endocrine effect on fish documented in many bodies of water. The third way, reducing the effluent into bodies of water by upgraded wastewater treatment is really the only solution; and it depends on infrastructure support, capital to upgrade, and technology.

The problem is national, expensive to manage, and crosses state and local boundaries. It therefore requires a national (in the United States, federal) and even global effort to achieve a solution before a serious health hazard emerges. Such an effort might consist of a combination of the following, not relying on any one strategy:

1. A national commitment to and comprehensive programs of watershed protection and upstream source protection, including land-use planning to ensure compatible uses in watershed areas
2. "Take-back programs" that allow pharmaceuticals to be returned to the point of purchase of convenient, safe disposal sites and that discourage disposal down the toilet or into trash destined for a landfill (note that this is only a partial solution for pharmaceuticals)
3. Well-designed monitoring programs to determine national trends for the increase or decrease of levels in source water. Operational monitoring for each and every utility will probably not be cost effective for a long time, and more accurate data for particular locations is unlikely to force a speed-up in technology for upgrading treatment
4. Research programs to develop robust but cost-effective water treatment technologies that are "multivalent": that is, that technology that will break down or remove a broad spectrum of contaminants, not just one or a few
5. Targeted development of control and remediation technology for specific contaminant sources that are localized, such as perchlorates
6. Further toxicological investigation to support risk assessment, in order to determine the level of risk they present

Even when the technology, to remove contaminants on the intake side becomes available, deployment will probably be best done as part of scheduled infrastructure upgrading and maintenance rather than as a crash retrofit program. There are many possible unintended consequences of choosing the wrong or an expensive technology that cities cannot afford in the long run. The risk or threat from the contamination does not seem to justify a disruptive effort that might divert resources away from upgrading basic water treatment and source protection. Most observers agree that for this problem, is better to get it right than to get it done quickly.

5

Chemical Pollution and Health

CHEMICAL POLLUTION, TOGETHER with ecosystem degradation and climate change, are arguably the most urgent and universal problems of sustainability that directly affect health. The simplest and most straightforward relationship between health and sustainability is the risk to health from chemical pollution. In the previous chapter, pollution was defined as contamination that resulted from human (anthropogenic) activity resulting in the release of the contaminant (in this case a chemical) in amounts and ways that exceed the system's ability to handle it (for example, by degradation), such that it accumulates to levels that have a human or animal health effect (toxicity) or an ecological effect (ecotoxicity).

Contamination and pollution, as they affect sustainability, have been covered in the previous chapter, which discussed some types of pollution in detail (mostly microbial) but only briefly talked about chemical pollution. Chemical pollution require much more extensive and focused discussion.

The previous chapter outlined the many ways that pollution in general compromises sustainability, following the order in the list of elements proposed in chapter 1 for the definition of sustainability. Table 4.1 described how characteristics of contamination and pollution in general fit into the cluster of issues defining sustainability. The list in table 5.1 is specific for chemical pollution:

Ambient Pollution and Toxics

Chemical pollution issues each have their own characteristics that reflect the medium, the exposure situation, and the toxicity of the chemical, but they tend to fall into two main patterns: "ambient pollution" and "toxics," which are not mutually exclusive. Ambient pollution issues deal with release of chemicals into the general environment, often involve mixtures, and tend to be managed by interventions that control multiple sources of pollution in order to reduce exposure to several chemical at the same time. Examples include management of air quality by the US Environmental Protection Agency (EPA) through the approach of vehicular

Table 5.1 Sustainability value table: Pollution

	Sustainability Value	Sustainability Element
I.	Long-term continuity	Chemical pollution, unabated, may cause serious damage to health, degrade materials, compromise ecological services, and degrade the environment such that an otherwise sustainable ecosystem cannot continue indefinitely.
II.	Do no/minimal harm	Chemical pollution deeply harms the environment, by causing toxic effects on natural species that become destabilizing to the biological communities that are cornerstones of the ecosystem.
III.	Conservation of resources	Chemical pollution may compromise conservation of resources when the pollutants involved are persistent and their effect is sufficiently serious to raise the cost or deny availability of an important resource. (For example, fisheries in which key species are contaminated with mercury, much of which is from human sources.)
IV.	Preserve social structures	Chemical pollution, as a sustainability issue, is particularly prone to create intense social conflict between (perceived or real) polluters and community residents, and results in profound inequities in "environmental justice."
V.	Maintain health, quality of life	Chemical pollution compromises health and degrades the quality of life.
VI.	Performance optimization	Chemical pollution reduces economic, social, and environmental performance by imposing "diseconomies" on the society (costs that are not paid by the enterprise doing the polluting) for health, loss of use or enjoyment of resources, costs of remediation (often in the form of higher taxes), costs of mitigation at the source because the original design did not effectively control emissions (often in the form of higher prices), costs of measures to mitigate effects (for example user fees for purifying air or water or to mitigate polluted sites for other uses), and the overhead cost of regulation and documentation of effects required to keep emissions under control.

(continued)

Table 5.1 Continued

	Sustainability Value	Sustainability Element
VII.	Avoid catastrophic disruption	At its extreme, chemical pollution can cause catastrophic events, such as the Great London Fog of 1952.
VIII.	Compliant with regulation	Chemical, and other types of pollution are subject to regulation and noncompliance is illegal; there is in the United States an elaborate system for reporting on emissions from various sources and this information quickly becomes public when there is an issue.
IX.	Stewardship	Chemical pollution is, as is all pollution, obviously contrary to stewardship and opens the polluter to legal liability. (Injury from pollution can be construed as a tort, or personal injury, when the polluter can be identified.)
X.	Momentum	Chemical pollution does not go away by itself unless the source is controlled, and so it is not self-correcting. Depending on the pollutant and the medium, it may or may not dissipate in the environment (and if it does it may take a very long time). Effective mitigation requires design and intervention to control emissions at the source. It may be impossible to clean up chemical pollution downstream.

emissions standards, and management of water under the Clean Water Act (1972). "Toxics" issues tend to involve one or a few chemicals, usually not found (at least in the same concentrations) in general ambient pollution, which tend to be managed one chemical at a time through specific interventions. The distinction is specific to the situation. A chemical or a class of chemicals, such as volatile organic compounds, can be part of ambient pollution in one context (for example, urban air pollution) and a toxic in another (uncontrolled emissions from an industrial source or hazardous waste site). The distinction is still useful because the two general types of pollution carry different implications for sustainability and management and are managed by different risk management options. In the discussion to follow, generalizations about exposure refer to concentrations encountered in the environment under "normal" circumstances, not extreme levels as might occur in an uncontrolled release, occupational exposure, or a heavily polluted site.

Ambient pollution is illustrated by fine particulate air pollution and microbial water pollution. In both cases, the effect is large in terms of number of people affected and is directly related to certain key social drivers: development, population growth, urbanization, and technological change. Fine particulate air pollution, for example, affects thousands of individuals in the most extreme and easily measured of health outcomes (i.e., excess deaths). Even so, individual risk is small, and the effect barely registers as a contribution to population health status. However, the small effect is multiplied over a very large population, often in the millions, and so the total effect can be enormous in absolute terms. Ambient pollution tends to work as a contributing risk factor for health outcomes rather than a driver. In other words, there is no characteristic disease in individuals with identifiable symptoms that one may call "air pollution disease." The effects of air pollution are expressed in populations as statistical increases in more common health outcomes such as heart attacks. From the standpoint of population health, the effects of ambient pollution are easily hidden by the beneficial effects of rising income, improved medical care, and also hidden (and confounded) by other adverse health risks in the community such as smoking.

Fine particulate air pollution as a cause of significant cardiovascular mortality, for example, was not only undiscovered but unsuspected for many years because society had, in effect, grown up with it; as a health risk, it had blended into the background. Once it was recognized, however, the death toll from pollution could be studied and quantified. It became clear that fine particulate air pollution (mostly from old, "dirty" diesel technology) and photochemical air pollution (from vehicular exhaust) were recognized as serious and widespread health risks and that then-current levels had imposed a continuing cost on society that was intolerable once it was recognized, avoidable, and costly.

BOX 5.1

Fine Particulate Air Pollution

There is a broad consensus in the scientific community that fine particulate matter air pollution is a major hazard to human health. The World Health Organization attributes 28,000 premature deaths in North America and 800,000 worldwide to ambient particulate matter each year, although more recent studies suggest that the true public health burden might be even greater. Exposure to ambient fine ("$PM_{2.5}$") particulate matter air pollution is now recognized, in the words of the American Heart Association, as a "modifiable factor that contributes to cardiovascular morbidity and mortality." ($PM_{2.5}$ means, literally, particulate matter in which the individual particles behave as if they were little spheres no larger than 2.5 μm in diameter, regardless of the true size or shape.)

The demonstration of the health effects of $PM_{2.5}$ in the 1980s and 1990s was a scientific triumph nearly on a level with the analysis of the human genome. However, when such studies were first undertaken, there was deep skepticism in the scientific community. The very small size of the particles results in a very small total mass that reaches the lung. It did not seem feasible that such a tiny mass of matter could produce serious health effects when larger particles, carrying much more total mass had much more modest effects. It seemed that fine particulate matter was violating a basic principle of toxicology, namely that the health effect is proportional to the dose of the toxic material absorbed into the body.

These studies were difficult to conduct because health outcomes are tangled up and related to one another (for example, heart disease and lung disease often go together) and because several air pollutants (such as particulate matter, ozone, nitrogen oxides, and sulfur oxides) and weather conditions move up or down closely together. It took years of observation, careful analysis, and replication at many different sites, first in the United States and later in the rest of the world, to isolate and characterize the individual effects of $PM_{2.5}$ and to separate it from, say, ozone or "synoptic" weather patterns characterized by heat and humidity. This science was conducted by many investigators in the United States who figured out the problem and by thousands of investigators around the world who have studied the problem in diverse settings to establish its generalizability. In the end the evidence is overwhelming. The body of evidence that supports this conclusion is the product of decades of intensive research conducted with stringent oversight, double- and triple-checking results, reanalysis to confirm every important finding, and laboratory validation of observations in human populations.

Many of these large studies have looked at the long-term health effects of ambient particles, and over the years they yielded highly consistent results. The first of these studies was the Harvard Six Cities study, which followed 8111 men and women living in six US cities for fourteen to sixteen years. The researchers found that over a sixteen-year period, adults who lived in the most polluted of the six cities had a 26 percent higher rate of death as compared to those in the least polluted city. Several other studies have found similar results including the American Cancer Society Cancer Prevention Study II, the California Seventh-day Adventists cohort study, and a recent national study of 66,000 participants from the Women's Health Initiative (WHI) Observational Study. These studies provide evidence linking long-term exposure to ambient particulate matter and all-cause mortality, cardiovascular mortality, and nonfatal cardiovascular events.

The impact of particulate air pollution on life expectancy is substantial. Scientists recently looked at changes in life expectancy in 200 counties in the United States and calculated that reductions in fine particle air pollution between 1980 and 2000 increased the average lifespan in these counties by approximately five months. Importantly, the greatest increase in life expectancy was seen in those counties showing the greatest reduction in fine particle air pollution during this time.

Hundreds of studies in the United States and around the world have now confirmed and extended the finding that elevations in particulate matter are associated with an increased risk of premature death, cardiovascular death, hospitalization for respiratory and cardiovascular diseases, and respiratory symptoms within days.

These scientific studies have linked particulate matter exposure to a variety of problems, including

- aggravated asthma in children
- increased emergency department visits and hospital admissions
- higher risk of hospitalization for congestive heart failure
- stroke and myocardial infarction (heart attacks)
- increased risk of premature death
- more frequent dangerous irregularities of the heartbeat
- more frequent deaths, second heart attacks, and hospital admissions for people who have already experienced one heart attack

Particulate pollution can cause health problems for anyone, but certain people are especially susceptible. Children and teenagers, the elderly, and people who already have cardiovascular disease, chronic lung disease or diabetes are among the groups most at risk. Even healthy adults who work or exercise outdoors may face higher risk. The majority of people, but by no means all, who are affected by fine particulate air pollution are older and may already be ill. However, fine particular air pollution also takes demonstrable but smaller numbers of young people. It seems likely, based on current understanding of the role of inflammation in the process, that people pass into and out of conditions where they are more or less susceptible to the effects of fine particulates, possibly related to the threshold for blood to coagulate. Even younger and healthier people may be transiently susceptible. The effects are seen disproportionally in individuals with low socioeconomic status or lower educational levels because of where they live and their health status. (Environmental justice is described at length in chapter 11.)

The mechanism for the health effects of exposure to fine particles was a mystery at first. As careful experiments began to demonstrate disturbance in heart rhythm and adverse effects of fine particles at the cellular level, it became apparent why $PM_{2.5}$ is so potent in its adverse effect on the human body. The size range of $PM_{2.5}$, being so tiny, insinuates itself in places where larger particles cannot go and presents the body, in the aggregate, with a geometrically much larger surface and larger number of particles that would be the case with large, coarse particles. This has effects on oxidation within the cell (upsetting the balance between oxidants and free radicals and the natural processes that oppose their effects), promotes coagulation (and therefore risk of stroke and heart attack), interferes with electrically active tissue such as conduction pathways in the heart, and interferes with the host defense cells that protect the lung.

Many studies showing the same effect on mortality and disease have been done in cities that were in compliance of the US National Ambient Air Quality Standards in 2012. Thus, the harmful effects of particulate matter can clearly be seen even at pollution levels well below the regulatory standards that are in effect now (at the time of this writing).

In any study involving populations and using statistical methodology, a decision point has to be reached as to whether the relationship between exposure to the pollutant and the magnitude of the effects that are seen is only an "association," meaning that they are statistically correlated but one does not necessarily cause the other, or a "causal relationship," in which the pollutant is a cause, which precedes and actually sets in motion the events the leads to the illness or health-related effect. For fine particulate air pollution, not only were essential features of a causal relationship met (such as an exposure-response relationship), but there was even evidence from quasi-experimental situations in human population studies in which air pollution dropped and then rose again and this was correlated with a parallel dip and then a return in mortality. Such situations are uncommon and constitute an unusual and very compelling validation.

Furthermore, epidemiology is not the only way of knowing that $PM_{2.5}$ has an effect on the human body. Studies of the effect of fine particulate matter in tissues, in animal experiments, and in human volunteer research has clearly shown that even low levels of $PM_{2.5}$ are associated with abnormalities of the heart conduction system, coagulation of blood, and airways.

In the scientific review of the 2009 Integrated Science Assessment for Particulate Matter, the external panel of independent scientists that make up the Clean Air Scientific Advisory Committee and the EPA scientists reviewed the evidence and concluded, without reservation, that a "causal relationship" exists between ambient fine particulate matter and both mortality and cardiovascular effects and that "a likely causal" relationship exists between ambient fine particulate matter and respiratory effects. This language is very conservative. There is no reasonable doubt that the relationship is causal.

At the time of this writing (2014), a bill has passed the US House of Representatives (H.R. 4012) in response to this work that would force the agency or authors of any study relied upon by the Environmental Protection Agency to make studies broadly available ("transparent and reproducible"), including the original data with the possibility of identifying individual research subjects, regardless of whether the studies have been peer-reviewed and validated or reproduced by other studies. In fact, the findings on fine particulate air pollution from the team that developed the Six Cities Study and its successors have not only been replicated, but the data from the original studies have already been reanalyzed at least twice (once by a statistical analysis group in Canada), by an agency partially funded by Congress (the Health Effects Institute, a model of arms-length objective research) and the and the findings were confirmed. That has not stopped politicians opposed to regulation from questioning its findings and implying that the work is junk science (see chapter 2).

The term "toxics" is environmental (not toxicological) shorthand for issues involving the release of individual chemicals that are generally more potently toxic than ambient pollutants at levels encountered in the environment but that are limited in the scope of their emissions and distribution. (This generalization is not absolute. Ozone, for example, still has toxic effects at low concentrations and is an important ambient air pollutant.) Toxic chemicals are defined by their health risks. Toxic effects in the individual are specific to the chemical, or at least characteristic to the agent.

Despite the huge number of chemicals released into the environment, there are only a relatively small number of chemicals that are potently toxic at levels commonly encountered in the environment. Relatively few of these, including a few pesticides, are carcinogenic, although cancer risk drives public concern in most cases. The evaluation and control of toxic chemicals is in crisis at the moment because many more chemicals are introduced into commerce than can be evaluated systematically for risk and regulation (especially in the United States) exempts many chemicals from thorough analysis.

Although toxics generally tend to affect small numbers of people compared to ambient air pollution, they do so in ways that are often recognizable (e.g., acute pesticide toxicity) and that target specific vulnerable groups depending on the setting: workers, children, the fetus, local community residents, fish eaters, etc. Thus, there is inevitably an issue of equity or environmental justice involved where toxics are concerned.

Some toxics become widespread in distribution and persistent in the ecosystem and present risks both locally and distant from their source. Mercury, for example, is

Table 5.2 Persistent organic pollutants (POPs). Items one through twelve constitute the original "dirty dozen"

Pollutant	Description	Status	Current situation
1. Aldrin	Organochlorine pesticide	Annex A (elimination)	Out of production and no longer registered in the United States
2. Chlordane	Organochlorine pesticide	Annex A (elimination)	Out of production and no longer registered in the United States
3. DDT	Organochlorine pesticide	Annex A (elimination); Annex B (restriction)	Out of production and no longer registered in the United States; use for public health only is allowed under the Stockholm Convention; new releases continue as a contaminant of the pesticide dicofol
4. Dieldrin	Organochlorine pesticide	Annex A (elimination)	Out of production and no longer registered in the United States
5. Endrin	Organochlorine pesticide	Annex A (elimination)	Out of production and no longer registered in the United States
6. Heptachlor	Organochlorine pesticide	Annex A (elimination)	Out of production and no longer registered in the United States
7. Hexachlorobenzene	Fungicide and chemical feedstock, unintentional byproduct and impurity of some chemical processes	Annex A (elimination); Annex A (reduce unintentional production)	Out of production and no longer registered in the United States
8. Mirex	Organochlorine pesticide and fire retardant with a uniquely toxic structure, closely related to Kepone	Annex A (elimination); Annex A (reduce unintentional production)	Out of production and no longer registered in the US
9. Polychlorinated biphenyls (PCBs)	Class of 209 industrial chemicals formerly with many uses	Annex A (elimination)	Out of production and no longer registered in the United States

10. Polychlorinated dibenzo-p-dioxins ("dioxins")	Class of 135 chemicals, unintentional byproducts of industrial processes and of combustion	Annex A (elimination); Annex A (reduce unintentional production)	Regulated as a pollutant; no intentional industrial production
11. Polychlorinated dibenza-p-furans ("furans")	Class of 135 chemicals, unintentional byproducts of industrial processes and of combustion, closely related to dioxins	Annex A (elimination); Annex A (reduce unintentional production)	Regulated as a pollutant; no intentional industrial production
12. Toxaphene	Organochlorine pesticide mixture	Annex A (elimination)	Out of production and no longer registered in the United States
13. Endosulfan (technical grade is a mixture of two isomers)	Organochlorine pesticide	Annex A (elimination)	Allowed in certain countries where alternatives for certain pests are not considered feasible at present
14. Alpha, beta hexachlorocyclohexane	Unintentional byproducts in chemical reactions, including manufacture of lindane	Annex A (elimination)	Out of production and no longer registered in the United States
15. Chlordecone	Organochlorine pesticide, Kepone®	Annex A (elimination)	Out of production and no longer registered in the United States
16. Hexabromobiphenyl	Polybrominated biphenyls (PBBs) are a class of 209 potential compounds, of which some are used as flame retardants; hexa-BBP most important	Annex A (elimination)	Out of production, banned in EU
17. Polybrominated diphenyl ethers, PBDEs	Fire retardants; suspected of endocrine effects, neurotoxicity; controversy because of fire hazard reduction	Annex A (intent to eliminate), (hexa-, hepta-); Appendix B (restriction), (tetra- and penta-)	Allowed for use as flame retardant additives for organic materials
18. Lindane	Organochlorine pharmaceutical used on head lice and scabies	Annex A (elimination)	No new production

(continued)

Table 5.2 Continued

Pollutant	Description	Status	Current situation
19. Pentachlorobenzene	Chemical intermediate in production of PCBs, fungicide, flame retardant, and unintentional byproduct of some reactions	Annex A (elimination); Annex A (reduce unintentional production)	Out of production and no longer registered in the United States
20. Perfluorooctane sulfonic acid (PFOS), its salts, and PFOS fluoride	Intermediary in production of several products, including electronic components, firefighting foam, photo imaging	Annex B (restriction)	Limited allowed uses, including production (in closed loops) of aviation hydraulic fluid, electronic semiconductor photoresist components, medical devices; certain pesticide applications
21. Hexabromododecane	Halogenated chemical capable of forming unusual geometric structures; fire retardant, polymer plasticizer	Proposed for a ban under POPs treaty	Concentrates in breast milk: proposal is also based on risk to children when breast-fed due to evidence of effects on fertility
22. Chlorinated paraffins (short-chained)	Mixtures, used principally in flame retardants and plasticizers for polymers, additives to metal working fluids and as solvents	Proposed for a ban under POPs treaty	Importation banned by EPA, restricted in EU; highly ecotoxic, especially to fish
23. Polychlorinated naphthalenes	Class of seventy-five possible chemicals, used as mixtures for electrical wire insulation, plasticizers, lubricants	Proposed for a ban under POPs treaty	Proposal is also based on health risk because also found to have substantial occupational health hazard
24. Hexachlorobutadiene	Intermediate and byproduct in making synthetic rubber and as solvent for chlorinated compounds; historically used as herbicide	Proposed for a ban under POPs treaty	Suspected carcinogen
25. Pentachlorophenol	Wood preservative and pesticide	Proposed for a ban under POPs treaty	Restricted by EPA to use in treating utility poles and railroad ties; suspected carcinogen

EPA = US Environmental Protection Agency, EU = European Union, POPs = Persistent organic pollutants.

released from many sources, especially stationary sources burning fossil fuels (particularly coal-fired power plants), travels long distances in air, deposits on land or in water, undergoes biotransformation in water to more toxic and bioavailable forms (organomercury compounds), accumulates in aquatic species, and biomagnifies with trophic level so that some species of fish reach levels that pose a potential hazard to human health at plausible levels of consumption. For this reason, control of mercury emissions has been a priority for the US EPA, with new source-control standards introduced in 2010.

There are certain particularly sensitive and harmful organic pollutants that persist in the environment because they resist physical destruction and move freely and widely through physical pathways and in the bodies of organisms such as fish. They present both a special risk to human beings and to other species in many ecosystems, some of which are far away from their point of release. These chemicals, which are variously called "persistent organic pollutants" (POPs is the most common acronym) or "toxic organic micro-pollutants" (TOMPs), and are included (together with alkly lead and mercury) and two nonchlorinated organic compounds in EPA's category of "PBTs" (standing for "Persistent, Bioaccumulative, and Toxic"), have the following characteristics:

- They have a high solubility in lipids (fats) and a low solubility in water, which means that they readily accumulate and persist in tissue.
- They have a relatively high molecular weight (>236), which is associated with toxicity for many organochlorine compounds.
- They are more volatile (evaporate more readily) than other chemicals of their class.
- They can be transported very long distances.
- They are halogenated, meaning that their molecular composition contains chlorine or bromine atoms; the more halogen, the more resistant the chemical tends to be to chemical change in metabolism ("biomodification").
- There is strong evidence of toxic effects due to exposure and often evidence that the effects caused in tissue are irreversible.
- The chemicals travel very long distances and are now found around the world in places where they were never manufactured or used, especially in the Arctic.

In order to curb production and eventually to eliminate POPs, the United Nations Environmental Programme (UNEP) developed a list of the twelve worst chemicals (called the "dirty dozen"). Most human exposure to the "dirty dozen" is by ingesting animal fat in food. UNEP ultimately convened a conference that produced the Stockholm Convention on Persistent Organic Pollutants (2001), widely considered one of the most effective environmental treaties in history. (The United States signed but has not ratified the treaty.) The treaty binds countries that are parties to it to the elimination of chemicals listed in annexes to the treaty. (See table 5.2.)

Ecosystem toxic Effects and Ecotoxicology

Chemical pollution does not just affect human and animal species directly. It can affect entire ecosystems and can migrate and transform through ecosystem pathways, the study of which is called "ecotoxicology." Ecotoxicology draws from ecology, as a critical science, and toxicology, and it applies the knowledge of both fields to the study of health effects on the ecosystem. Ecotoxicology is one of the foundational sciences underlying sustainability studies. Ecotoxicology provides a framework and the methods to study contaminants and pollutants as they circulate within ecosystems and among species and to document the degree of harm and the permanency of the damage.

The three dimensions of ecotoxicology and ecosystem dynamics that are critical to sustainability are as follows:

- The behavior, in terms of migration and modification, of the chemical that occurs when contaminants enter and pollutants are released into the environment
- The effects of the chemical on the ecosystem in the short term, including its effects on biological productivity and whether pollution precludes use or raises the cost (in money or energy) of accessing renewable resources in the short term
- The effects of the chemical on the ecosystem and beyond, including the risk to human health and to the cost and availability of the resource for future use

Biologists speak of ecology as the science of relationships among species and for those species to the physical environment and often speak of "biological communities," which are highly structured networks of relationships in which nutrients, energy, and mutually dependent biological functions create a "web of life" that is much more than the sum of the individual species and their numbers. In nature, stable biological communities are the product of many years of population growth, decline, or stabilization of the individual species, occupation of specific ecological niches by species best able to exploit them, and the availability of nutrients and energy sources. Biological communities are not static, however, and perpetuate themselves by adapting to stresses—such as climate change, drought or fire—through a process called "succession" in which new species may enter and establish themselves in new ecological niches that are created. Over a long period of time (although sometimes with accelerated bursts) natural selection acts on species by favoring individuals with traits that more efficiently exploit the ecological niche and that are better suited to conditions. Over time, this increases the representation of forms (including physical traits, metabolic pathways, energy efficiency, and so forth) that are significantly better adapted to the new environment, ultimately resulting in the accumulation of sufficient differences to define a new species.

The term "ecotoxicology" was first used in a publication in 1976 by French toxicologist René Truhaut (b. 1909–d. 1994) to name a new discipline that applied toxicological principles to all species, initially including human beings, and to the ecosystem as a whole. This covered a lot of intellectual territory and if the field had stayed so broad it would have subsumed well-established disciplines such as conventional toxicology (human and experimental) and veterinary toxicology, wildlife toxicology, and much of plant science. However, over time the term came to be used more narrowly for the behavior of contaminants of both natural and human origin in the environment after they are released, their effects on nonhuman species (animal, plant, and microbial), their effects on population density and dynamics of individual species, their effects on the relationship between critical species (such as herbivore-producer, predator-prey, host-parasite, host-commensual, interspecies competitors), and the overall effect on the ecosystem. "Ecotoxicology" is now well established as the name for the field of study that examines the fate and disposition of contaminants, including pollutants and their effects, in the environment.

The field of ecotoxicology is devoted especially to characterizing six stages of the behavior of contaminants and pollutants in biological communities:

- Release into the medium, of pollutants from human sources and of other contaminants from natural sources
- Bioavailability, the proportion of a contaminant that is available to an organism to be absorbed (the contaminant may be largely bound to a surface, such as a clay particle, or may be poorly soluble)
- Pathways of migration and accumulation
- Biomodification, which refers to chemical transformation, either by physical means (such as exposure to light) or by metabolism in the bodies of organisms in the biological community
- Removal, degradation or chemical precipitation, which eliminate the pollutant from activity in the ecosystem
- Ecosystem health and integrity, as reflected in population structure, viability of key species, the health of individual animals or plants that are sampled, and damage to the nutrient and energy flows that characterize a viable ecosystem

Ecotoxicology follows the same physical and chemical principles that are the foundation of conventional toxicology, not surprisingly. Table 5.3 compares ecotoxicology and human and animal toxicology, highlighting the similarities and differences.

Table 5.3 Ecotoxicology and toxicology (human and animal) compared

Toxicology	Ecotoxicology
Xenobiotic	Contaminant or pollutant
Route of entry	Physical pathways and movement between media (e.g., air and water)
Absorption	Uptake by species, bioconcentration (in water)
Bioavailability	Bioavailability
Distribution, including: delivery to organs, tissue affinity and concentration, partitioning into tissue	Fate and disposition, including: trophic levels, biomagnification, partitioning into tissue of free organisms (e.g. fats) or physical environment (e.g., adsorption onto clay particles)
Binding sites on proteins	Binding sites in clay soil, sediment
Sequestration (storage sites in body, e.g. bone)	Sequestration (storage sites in physical environment)
Metabolism in tissue	Biomodification (by bacteria, metabolically active species)
Organs with metabolic activity (predominantly liver)	Species with metabolic activity
Metabolizing enzyme systems	Microbial metabolism
Excretion	No exact counterpart (because chemicals and their metabolites have nowhere else to go)
Elimination (removal by metabolism, excretion, or sequestration)	Elimination (removal by physical processes, metabolism by microorganisms, or physical sequestration)
Target organ effect	Species-specific effects, Population dynamics (survival, reproductive rate, growth or decline)
Host defense mechanisms	Interactions among species that mitigate effects, pH buffering and other physical effects
Susceptibility states of the individual	Relative susceptibility of species to the pollutant

(continued)

Table 5.3 Continued

Toxicology	Ecotoxicology
Cumulative exposure (dose)	Biomagnification
Single or multiple health outcomes	Outcomes always multiple
Organ-system health effects	Effects on species
Clinical outcomes	Species viability
Effects on individuals sum to effects on populations	Effects on populations primary (also the driver for natural selection)
Populations	Species in a biological community
Epidemiology	Ecology
Biomarkers	Indicators
Health of one organism in clinical toxicology and medical intervention	Ecosystem as level of analysis and conservation biology
Individual tolerance	Ecological succession (short term) Evolution (long term)
Toxidrome	Ecosystem "health" and integrity

The fundamental pathways and mechanisms of ecotoxicology closely resemble those for human and animal toxicology, which will be described later in this chapter, because they are based on the same physical principles. In the ecosystem as a whole, different species and elements of the ecosystem play the same role in xenobiotic distribution and metabolism that organs would play in the human being or animal body. The principles of ecotoxicology, therefore, closely resemble the principles of organismic toxicology.

Movement of Contaminants in the Ecosystem

Once released into air, water, or soil, contaminants in general and pollutants in particular follow well-defined pathways of movement through the ecosystem, as seen below:

- Airborne pollutants follow wind movements.
- Waterborne pollutants follow water movement, which is governed by the adage that water always flows downhill (or in the direction of a pressure gradient)
- Pollutants in biota in bodies of water follow the movement or migration of species that take up and retain the pollutant from the surrounding water, which is a phenomenon called "bioconcentration."

- Pollutants deposited on soil follow surface water movement, the run-off following surface flows, or percolates down to the water table (aquifer), where it contaminates groundwater.
- Environmental pollutants that end up in food (as opposed to contaminants introduced during food processing) may involve deposition or uptake by plants but more often follows pathways of distribution and storage that are determined by trophic levels.
- Pollutants in oceans follow similar water movement and pathways that reflect migration of species, but movement is generally much more complicated than in fresh water and more often in three dimensions.
- Pollutants can and do move between media but in specific ways, always following physicochemical principles.

BOX 5.2

Physical and Biological Principles Fundamental to Ecotoxicology and Toxicology

- *Water solubility* refers to the ability of a chemical to dissolve in water, and is measured as the maximum amount of the substance that can be dissolved in a given amount of pure water. Environmental conditions, such as temperature and pH, can influence a chemical's solubility, which in turn also affects a contaminant's volatilization from water. Solubility provides an important indication of a contaminant's ability to migrate in the environment in surface and groundwater. A synonym for water solubility is "hydrophilicity"; a contaminant that is hydrophilic usually has low solubility in oils and fats (i.e., it is not lipophilic).
- *Lipid solubility* refers to the ability of a chemical to dissolve in fats and oils. The relative solubility of a chemical in fats and water will determine whether the contaminant concentrates in lipid-rich tissues of living organisms. When a contaminant moves from water to lipid, it is said to "partition" to the lipid, and this property is described by a simple measurement comparing distribution between a solvent that dissolve fats (octanol) and water, which gives the "partition ratio." The partition ratio predicts how contaminants will distribute in the environment. A contaminant that distributes in such as way as to partition mainly in water is said to be "hydrophilic" and sometimes "lipophobic" (rarely used in practice); a contaminant that distributes mainly into the organic solvent is said to be "hydrophobic" and "lipophilic." Highly lipid-soluble contaminants will tend to partition into living organisms

and move out of surface and groundwater. This property is also critically important in human and animal toxicology because it also governs absorption into the body and distribution among the tissues. A synonym for lipid solubility is "lipophilicity;" a contaminant that is lipophilic usually has low solubility in water (i.e., it is not hydrophilic).

- *Density of liquid* refers to a liquid's mass per volume. For liquids that are insoluble in water (or immiscible with water), liquid density plays a critical role. In groundwater, liquids with a higher density than water (called dense nonaqueous phase liquids or DNAPL) may penetrate and preferentially settle to the base of an aquifer, while less dense liquids (called light nonaqueous phase liquids or LNAPL) rise to a higher layer and may float.
- *The adsorption coefficient for organic carbon* (K_{oc}) describes the adsorption (adsorption means that the chemical is stuck to the surface, not to be confused with absorption, which means that it is taken into the absorbing substance). This number describes the affinity a chemical has for organic carbon and consequently its tendency to be adsorbed onto particles of organic-rich soil and organic sediment. The higher the number, the more tightly the chemical sticks to particles of organic matter, for example particulate matter in air pollution, and the less of it that is available to move as a free compound into groundwater or surface water.
- *Vapor pressure* is a measure of how quickly contaminants will evaporate and the pressure that results from the chemical's tendency to volatilize. Vapor pressure predicts how rapidly the chemical will diffuse into the air and is part of the equation for how rapidly it will leave water if it is dissolved. Vapor pressure increases with temperature.
- *Henry's Law Constant* is a measure of the tendency for a chemical to pass from an aqueous solution to the vapor phase. It is a function of molecular weight, solubility, and vapor pressure. A high Henry's Law Constant corresponds to a greater tendency for a chemical to leave water and volatilize to air. This tendency is often called "fugacity."
- *Bioconcentration* occurs when there is chemical partitioning between the tissues of an organism, such as fish or plant tissue, and water, so that the organism takes up contaminants from the surrounding water. (This term is properly applied only to concentration in tissues of aquatic species.)
- *Biomagnification* occurs when the bioconcentrated and accumulated contaminant in one organism is consumed by another organism at a higher trophic level, such as an herbivore eating a plant, a predator eating an herbivore, or a predator eating another animal higher on the food chain. A fraction of the contaminant from each organism consumed accumulates in the tissues of the organism at the higher trophic

level, which has the effect of magnifying the exposure level up the food chain. Not all contaminants do this.

- *Biomodification* is the modification of a contaminant by a chemical process such as hydrolysis, oxidation, photolysis, and metabolism, which in turn can result in biotransformation (being transformed in tissues of the organism into a metabolite, usually with different properties and greater solubility in water) and biodegradation (being broken down by metabolism).
- *Physical modification* is the modification of a contaminant by a chemical process, such as photodegradation by sunlight or breakdown due to temperature.
- *Half-life* is a measure of how long it takes to reduce by half the concentration of a contaminant by whatever process it is eliminated.

Source: Adapted and expanded from a list of commonly cited chemical and physical properties prepared by the Agency for Toxic Substances and Disease Registry.

However, as in all things related to the ecosystem, there is more to it than following simple physical pathways. Because of dilution pollutants to become less concentrated as they flow through media (such air or water), but they can also become more concentrated as they pass through living tissue in the form of species in the ecosystem that take them up and may accumulate them. Box 5.2 provides definitions and summarizes familiar principles of ecotoxicology, which should already be well known to any student of environmental sciences and biology.

As a pollutant (or contaminant) moves through a medium undisturbed, it tends to spread out, and the concentration is reduced through dilution and diffusion at the margins. A "bolus" (single quantity or "slug") of pollution released into a free-flowing medium such as air or water does not stay together for very long. As a generalization, as the pollutant flows with the medium, it also diffuses in all directions. Diffusion is a basic physical process driven by molecular movement and is not affected by the movement of the bulk of the pollutant. Most of the pollutant moves "forward" through transport or flow (in the direction of lower pressure or pulled by gravity) and creates a "front" or leading edge of the wave of pollutant concentration. Behind it is the remaining pollutant, which is called the "plume." However, on the molecular level the pollutant diffuses in all three spatial dimensions at the same time and in all directions, including "backward" toward the origin, so that the initial bolus soon spreads out in a distorted cone shape (bounded by surface features, thermal layers, or buoyancy if there is a temperature difference) and becomes less concentrated. When concentrations are monitored it appears more like an indistinct smear than a sharp front with a distinct boundary.

If the pollutant continues to be discharged at the same rate and conditions remain the same, a more or less stable and continuous plume forms so that pollutant levels maintain their elevated concentration. However, discharge rates and conditions rarely remain constant. Turbulence and bulk movement of air and water through currents also change the geometry of the plume and promote mixing.

Airborne pollutants do travel with wind and air currents, but their pathway may also be affected by temperature, as when warmer emissions to colder air at the source carry pollutants far upward. Their pathway is strongly affected by the height at which the release takes place and by mixing, which in turn is affected by wind speed, updrafts, temperature, and the roughness of the terrain over which the wind passes. If the release is from a tall stack, wind speed is slow and steady, and the terrain is smooth, the plume will start to spread as a result of diffusion and local mixing; it will begin to appear cone shaped—with the apex at the source (top of the stack) and widening, with diminishing concentration of the air pollutant—until the circumference of the cone reaches the height of the transition layer in the atmosphere (where temperature changes) and the ground, which will limit further spread up and down, although concentration of the pollutant will continue to diminish with distance. If wind speed is fast or terrain is rough, mixing occurs rapidly and so the concentration of the pollutant in air rapidly diminishes with distance from the source. This tendency for the concentration of the pollutant to fall with distance because of diffusion and mixing is called "dispersion." If the source is at ground level, the plume will tend to move along the ground and expose receptors nearby to higher concentrations, which is called "fumigation." The downwind concentration of an air pollutant is routinely estimated by computer models, some of them very elaborate. Point source release models, for example for a gas well, are relatively simple but distributed-source models for highways, entire airsheds, and movement of pollutants in the upper atmosphere can be enormously complicated. The most difficult challenge is estimating exposures in interrupted terrain such as cities, where wind movement forms eddies, conditions of laminar flow (when air is channeled to move briskly in a straight line and so does not appreciably mix or become diluted), and air movement is interrupted by structures. Modeling estimated concentrations at the receptor becomes extremely complicated in these cases.

In surface water, contaminants are carried by water currents and flow downstream by the action of gravity. However, contaminants may be taken up and retained by species in the water, such as fish or microorganisms, a process called "bioconcentration" (which refers only to the uptake and retention of contaminants from water and should not be confused with biomagnification). When this happens, they are carried by the organism that retains them and the pathways they follow are movement and migration patterns of the species carrying them. If that species is at the bottom of the trophic pyramid (e.g., an invertebrate that is food for a fish), then the contaminant enters the body of the next organism up the food chain, where it may be retained as well. Some pollutants, such as mercury and organochlorine pesticides, are retained in the tissues of aquatic species with such avidity that they accumulate to high concentrations, which is a process called "biomagnification" (not to be confused with "bioconcentration"). If the pollutant is not metabolized and eliminated efficiently, it will be progressively accumulated all the way up the trophic levels, achieving the highest concentrations in top predators such as birds that feed on freshwater fish.

Pollutants that are buried in the ground, applied to the soil, or that are carried by run-off water into soil follow the pathways of water, at least after whatever volatile fraction there may be has evaporated. There may be direct contact with pollutants from ingesting soil (for example, when dirt is retained on the roots of vegetables) or the pollutant may be taken up by plants. For some pollutants, such as heavy metals, this may cause a hazard in food crops. However, the principal pathway for soluble pollutants is to follow the water as it penetrates the top layer of soil, percolates down through the soil by gravity, and passes through various strata of soil, sand, gravel or porous rock—whatever lies underground. Eventually, pollutants reach the aquifer. An aquifer is an underground body of water held in porous rock or sand and kept there because the rock underneath it is impermeable. Along the way, many pollutants will adsorb (adhere to the surface) onto soil and particularly clay particles, depending on the chemical characteristics of the pollutant and the surface characteristics of the clay. This means that not all pollutants reach the aquifer at the same time and in their original concentration in surface water. The groundwater in an aquifer may flow in one direction (typically toward a lower elevation and nearby body of surface water), but some groundwater is essentially stagnant. Once the pollutant reaches the aquifer it contaminates the groundwater, and unless the water is flowing unusually quickly (for example, through sand) it takes a long time to clear, sometimes decades.

Pollutants that leak or are deposited onto impermeable surfaces such as pavement become surface water or soil pollution problems quickly because of run-off and drainage. In most cities, municipal "storm drains" channel run-off water directly for disposal in bodies of water such as rivers without being treated. The pollutants in the water are carried into the water with the run-off.

Pollutants can end up in food in many ways. Chemical pollution can come from the residue of what is applied to a crop (such as pesticide), can be incorporated into the food through bioconcentration (from water or uptake by plants), can be present in the tissues of fish or animals as a result of biomagnification (up the food chain), or can arise from the way the food is treated (burned meat and overcooked grilled foods contain significant quantities of potentially carcinogenic combustion products). In addition, many natural contaminants that would not be considered pollutants can be present in the foods themselves. These include plant toxins, breakdown products (a cause of certain types of fish-related food poisoning), mold products, and bacterial products, some of which are present even when the food products are not spoiled or unfit to eat. However, the more common problem of chemical pollution in foods is contamination after or during processing and even this pales in significance compared to the much more common problem of microbial contamination of food.

Pathways of pollution in the oceans is particularly complicated and occur in three dimensions. Pollutants in a free state or taken up by phytoplankton (the numerous invertebrate species that make up the first trophic level of the ocean ecosystem) may follow ocean currents, sink with cold water or rise with upwellings, and follow pathways of distribution and storage that are determined by trophic levels and migration patterns of the fish and other animals that retain them, the tendency of the chemical to accumulate in tissue from the surrounding water (bioconcentration), and increasing accumulation up the trophic pyramid (biomagnification).

Even these patterns of movement do not exhaust the possibilities for pollutant migration. Pollutants may move between different media. For example, acid-forming air pollutants such as sulfur and nitrogen compounds, mostly from industrial sources in the Midwest, have deposited "acid precipitation" (often called "acid rain") over vast areas of the northeast in North America, causing changes in the chemistry of lakes and streams that resulted in serious ecological damage. This problem was caused in part because, instead of reducing emissions at the source, stationary sources of emissions were reducing local and nearby concentrations at ground level by raising the height of their stacks, which propelled the pollution into higher altitudes where air currents took it longer distances. This problem was brought under relatively good control in 1991 with an international treaty (the US-Canada Air Quality Agreement), which resulted in measures to curb the emissions and subsequent long-range transport of acid-forming pollutants. (The agreement was expanded in 2000 to cover transboundary air pollution of other types, such as ozone.)

Perhaps the most significant example of how pollutants move among different media is the progressive concentration of pollutants (particularly organochlorine compounds such as some pesticides, PCBs, and dioxins) in the Arctic ecosystem from sources in temperate latitudes. These pollutants, once released into the ecosystem, move seasonally, as the temperature changes. Warmer weather favors their migration, as they volatilize (although their volatility is usually relatively low) and are carried in air. Colder weather favors their deposition on surfaces (such as trees and other plant matter) and storage in water (although their solubility is low). The end result is an episodic, seasonal "jumping" movement of these pollutants away from tropical and temperate zones and toward colder regions where they tend to accumulate in the environment. Certain species, particularly mosses and lichens, also accumulate them with great efficiency. They are also lipophilic, tending to concentrate in the fatty tissues of arthropods, fish, and animals. Because the Arctic ecosystem has few plant species that function as primary producers, food chains tend to be short and heavily reliant on energy from fats, which results in biomagnifications of these pollutants. The result has been extremely high levels of persistent organic pollutants, mostly organochlorines, in Arctic species including birds, fish, mammals (including polar bears) and indigenous peoples who live off the land. In the human population, this phenomenon has been associated with several possible health effects, particularly otitis media (ear infections) in young children, which is thought to result from reduced effectiveness of the immune system and inflammatory response; this has been confirmed in laboratory experiments as an effect of these chemicals at high exposure levels. In animal populations the same exposures have been associated with a wide range of problems associated with endocrine effects due to the action of some of these compounds in mimicking or blocking hormone activity.

The effect of chemical pollution on ecosystems, considered as biological communities, varies with the nature of the pollution, of course. The effect can be assessed expeditiously by examining the population density of various species, which reflects survival and replacement by reproduction. Critical species that can be used as indicators for ecosystem health, viability, and degradation include fish and predators, but more information is usually obtained by examining the key invertebrate species that form the first trophic level of the food chain, and upon which the sustainability of the biological community depends. One very popular indicator species for the assessment of ecotoxicity is *Daphnia magna*, a lake-dwelling arthropod (related to the flea) that serves as food for many species of fish and so is a cornerstone species in many freshwater ecosystems. It normally reproduces quickly, hatching a new brood every other day or so under ideal conditions and is easy to culture in the laboratory. *Daphnia magna* serves as an indicator of toxicity because reproduction is impaired quickly by toxic pollutants. It accumulates pollutants from eating bacteria in water. The survival and reproduction of this organism has long been used as an indicator of toxicity in rapid assays for screening chemicals for ecotoxicity or assessing the hazard to the ecosystem of run-off from contaminated sites.

Ecotoxicology (in the narrower sense it is commonly used today) is complementary to human and veterinary toxicology. From the human point of view, ecotoxicology represents the effect of pollution all around human communities, on the environmental services on which human communities depend, on the availability of food resources, and it is an indication of the sustainability in biological terms of natural and modified ecosystems. However, it is also a way of characterizing the fate and disposition of pollutants in the environment before they reach human receptors.

Pathways and Routes of Exposure

As will be discussed later in this chapter, the primary routes of exposure for human beings to pollutants and other contaminants of human origin in the environment are inhalation, ingestion, and absorption across the skin or mucous membranes. These correspond to the natural pathways of human and animal exposure to contaminants in the environment: air, water, and contact with contaminated soil or water.

Ecotoxicology describes the pathways of contaminant movement in the environment until it reaches the human or animal. Any organism or population or organisms (human, animal, plant) that is exposed to the contaminant, once it has been delivered, is called a "receptor." (This meaning of the word "receptor" is specific to ecology and ecotoxicology. In conventional toxicology, as in pharmacology and medicine, the word means the specific site on or structure within the cell that interacts with a pollutant or drug or other active molecule to produce a response and is often a specific and known protein bound to the surface of the cell.)

The standard model used in environmental health for thinking about the effects of pollution is sometimes called the "Source-Effect Model," which involves tracing contaminants from their source through transport, to the exposure opportunity, to actual exposure of the receptor, and finally to the effect, a simple theme with many variations. Figure 5.1 (top) illustrates the basic Source-Effect Model. Although the Source-Effect Model may seem obvious and is often taken

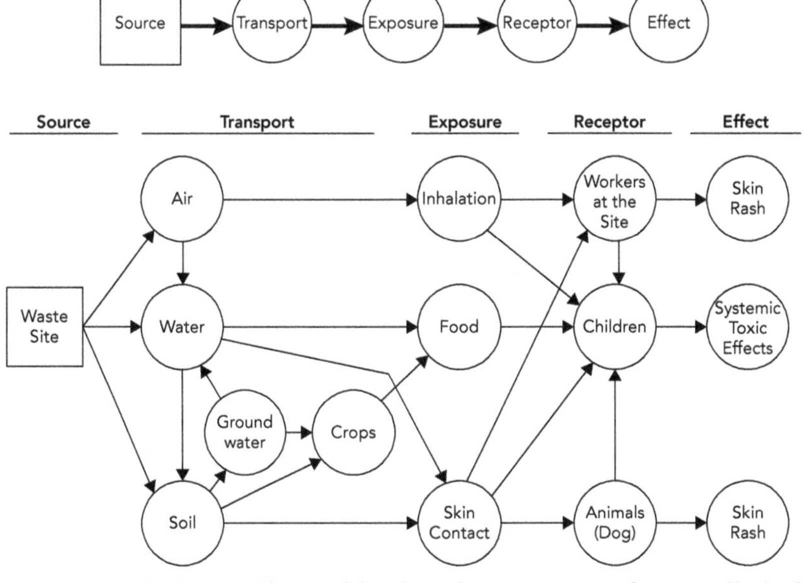

FIGURE 5.1 The Source-Effect Model is a logical, systematic way of sequentially thinking through exposure-related health risks. The top is the basic sequence of elements in the model, from the source of the hazard, transport, or distribution through a medium, the exposure opportunity and concentration in the environment or affecting a given receptor (human or animal), absorption by receptor, and the health effects that result. At the bottom is an example of the Source-Effect Model applied to the exposures and routes in the example in Figure 5.2.

for granted, it is helpful to lay out explicitly when planning studies, to simplify the organization of complicated studies (the model can accommodate many branches, interactions, and pathways, as long as the steps are in the same order for each), to teach, to communicate concepts of exposure to the public, to explain complicated situations and reasons for proposed interventions to decision makers, and to have a template for risk assessment and communication.

For human beings or animals or any other receptor to be affected by exposure to a pollutant, there must be an intact pathway of exposure by which the pollutant is transported to the receptor, and the receptor actually comes into contact with it. The pathway must be unbroken and must have sufficient capacity to transport enough of the contaminant to produce a toxic effect. If there is no "intact pathway of exposure," there is no possibility of a toxic effect. That is why the first step in almost every site assessment conducted for human health risk is to determine whether there is an intact pathway of exposure and to document that exposure is actually taking place.

Figure 5.1 (bottom) is an example of how the Source-Effect Model applies to a situation of human exposure, in the example of the scenario presented in figure 5.2. Imagine an unmanaged hazardous waste disposal site on one side of a fence. There are open or corroded drums of chemicals that have spilled out onto the ground. The

FIGURE 5.2 Hypothetical example of an exposure-related health-risk scenario involving children living at a home using well water in a rural area near a hazardous waste dump. Chemicals may reach and expose the children through air (if volatile), surface run-off, groundwater contamination, contact of dirt on skin, or contamination of food. This is an extreme situation, obviously, drawn only to illustrate many possible routes of exposure.

contents of these drums were originally the usual contents found in hazardous waste sites: dirty used solvents, metal-containing plating solutions, acid solutions, waste oils, and possibly some discarded pesticides. Over time, the composition of the mix found at the site may change as some of the chemicals evaporate. The chemicals that are left will be less volatile, and some residues in the drums will be chemically or physically modified from the original composition. On the other side of the fence, the people living in the house get their water from the aquifer (note the pump, for bringing up underground water) and have a vegetable garden. The receptors of concern are two children who live in the house. A situation like this may be unrealistic today under modern regulatory policies, but such situations did occur in North America, well into the 1980s, and still occur often in poorer countries with weak regulations and enforcement. This scenario is focused on the children residing adjacent to the site as the receptors. However, it should be understood that exposure may also occur to more distant community residents, to workers on the site, to other users of the aquifer, to other consumers of food from the garden, to pets and livestock, and to wildlife in the area. Toxic effects on animals can be an early warning of risk to human beings. Once the pollutants enter the environment, they contribute to the ecosystem and even planetary burden if they are persistent and accumulate.

Although this scenario would clearly be an unacceptable situation in real life, there would only be a risk to the children living in the house on the near side of the fence if the pollutants from the waste site were to reach them. If the pollutants did not leave the site, or if transport were in the opposite direction, there would be no exposure and therefore no effect. There must be an intact pathway of exposure for effects to occur. The caption summarizes the major ones.

Human and Animal Toxicity

Human beings are exposed to many other chemicals than those that occur as contaminants in the environment. They are intentionally exposed in the form of medications, cosmetics, personal care products (such as shampoos), and unintentionally to many natural chemical products in and on food, food contaminants from natural sources (of which there are many), and other chemicals of natural origin in, for example, drinking water. These chemicals, as well as nutrients, all follow the same principles of chemistry and biology already discussed in the context of ecotoxicology.

Toxicology is often described as "the science of poisons," but it is much more than that. It is really the science of how chemicals behave in the body, how and in what form they affect the body, and how the body copes with the effects they cause. For centuries, poisons were considered to be a special class of chemicals, and the toxicity of poisons were understood to be intrinsic properties of the chemical, or even magic. In the late Middle Ages, it came to be understood that any chemical could be poisonous (including water and salt) if the dose were high enough, and that any chemical could be safe (including cyanide and the most potent plant toxins) if the dose were low enough.

Toxicology engages many other sciences, including pharmacology, human and veterinary medicine, chemistry, epidemiology and exposure science. Many applications of toxicology are forensic, meaning that toxicologists use medical clues and chemical characteristics ("signatures") to identify which hazardous chemicals are responsible for an adverse event (a crime or an incident of pollution) and where the chemical came from. Because the delineation and acceptance of "safe" levels of exposure, for purposes of regulation and control, assume a socially determined level of acceptable risk (which is implicit in the definition of "safety"), toxicology has a deep connection with the social sciences, even though it is normally thought of as a "hard science" based on chemistry and experimental biology.

Toxicology has been adapted and refined, together with environmental epidemiology and exposure assessment, to support policy decision-making, in the form of a daughter discipline called quantitative "risk assessment" which has become the indispensable framework for regulation of chemical and other hazards. Much of the information required for this analysis comes from epidemiology. Some has to be extrapolated from animal studies in experimental toxicology. Exposure science supplies estimates of the level of exposure that occurs and how exposure varies in the population. Using a framework largely derived from basic toxicology, estimates of risk can be determined that guide the proposal of protective standards and provide the support that decision makers need in adopting standards.

Toxicology, as a discipline, starts from the point at which a person (or animal) is exposed to the chemical. For convenience in terminology, toxicologists often refer to all substances not normally present in the body and that are introduced from an outside source as "xenobiotics" (from the Greek "*xeno-*," meaning foreign). Xenobiotics may be environmental chemical exposures (including contaminants that may not cause harm as well as pollutants), drugs, cosmetics or personal care products, food constituents, natural chemical exposures, or even venoms and plant poisons. All chemicals that are extrinsic to and enter the body are considered xenobiotics, and they are all dealt with by the same chemical, physical, and physiological mechanisms.

The science of toxicology can be divided into "toxicokinetics," which is the study and description of how xenobiotics enter and are handled by the body, and "toxicodynamics," the study and description of what the xenobiotic does to the body. In public health and environmental studies, and by extension sustainability studies, there is a tendency to jump directly to the suspected health effects of a chemical without considering what is happening in the body that makes the effect take place. This is usually a mistake, because the behavior of chemicals in the environment and the risks they present to human beings and animals cannot be understood without an appreciation for their behavior in the body, which (as demonstrated in the last section) closely mirrors the behavior of chemicals in the ecosystem, because this behavior is based on the same principles.

"Toxicokinetics" refers to the behavior of xenobiotics in the body and is the analogue of pharmacokinetics in drug development and medicine. Four terms describe the disposition of axenobiotic, whatever it is: absorption, distribution, metabolism, and excretion. The latter two define "elimination," which is removal of the xenobiotic from the body and is reflected in a decline in body burden and concentration in blood. Figure 5.3 illustrates these four steps and how they define the behavior of xenobiotics in the body:

- Absorption describes the entry of the xenobiotic (in this context, a pollutant) into the body.
- Distribution describes the movement of the xenobiotic around and among various tissues and places where it accumulates (called "compartments") inside the body and is governed by the same principles that govern absorption.
- Metabolism describes the chemical transformation of the xenobiotic, which may activate chemicals (especially carcinogens) causing greater toxicity but more often reduces their toxicity and, critically, makes the metabolite easy to eliminate from the body.
- Excretion describes the removal from the body of the xenobiotic.

Elimination is the rate of disappearance of a xenobiotic from the body. It is a function of both metabolism and excretion and describes the disappearance of the xenobiotic from the body because of transformation and excretion. "Toxicokinetic models" are representations of the movement and concentrations in blood of a xenobiotic based on blood levels and how xenobiotics move into and out of tissues of the body. Because tissue levels depend on transport to the target organ and the degree to which the xenobiotic partitions (distributes itself between blood and the tissue) or is sequestered into the tissue, the kinetics of the xenobiotics in blood, especially, determines the presentation of the xenobiotic to the target organ at the tissue and ultimately cellular levels, where the toxic effect occurs.

Toxicodynamics

The mechanisms of toxic injury are too numerous to generalize over all the possible ways that tissue and normal bodily functions can be affected by chemicals. There are, however, a few general principles that are important.

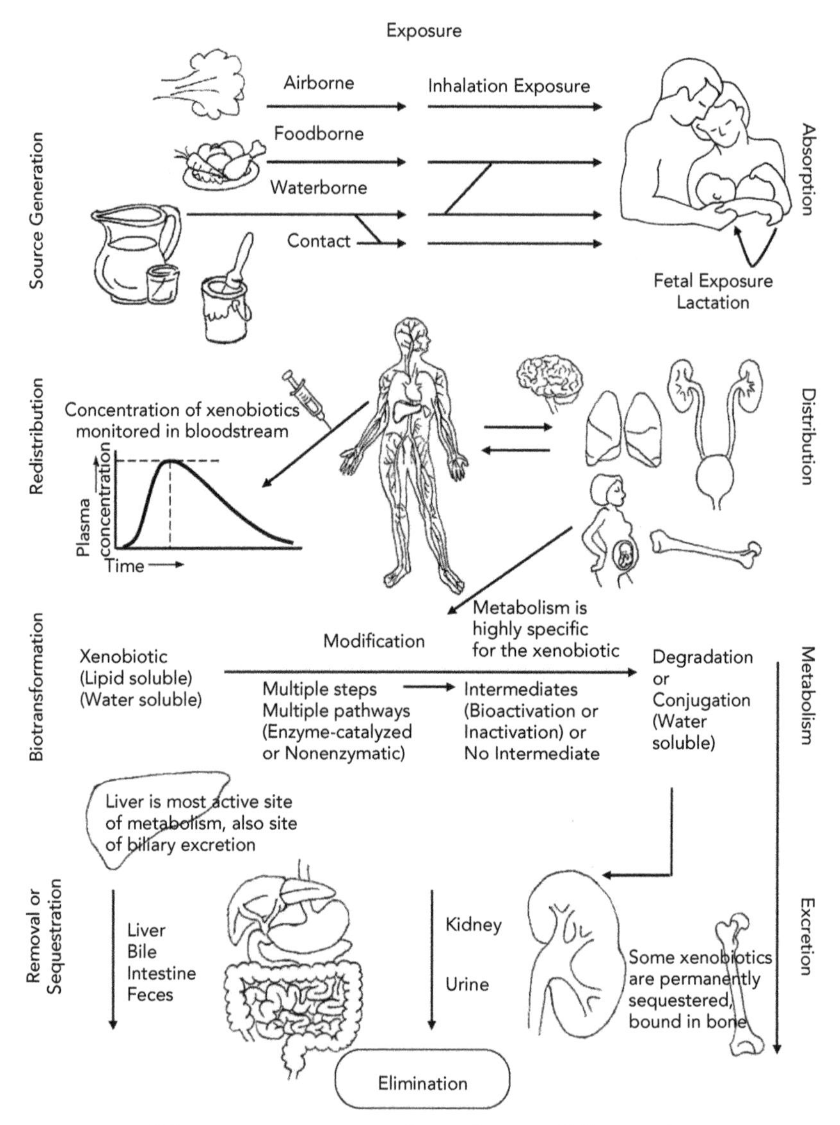

FIGURE 5.3 Schematic representation of toxicokinetics, following the exposure event and proceeding to elimination. The four stages of toxiccokinetics are absorption, distribution, metabolism, and excretion. Major routes of elimination of xenobiotics and their products are through the kidney (urine), through excretion into bile and subsequent elimination (feces), through the lung (exhaled or "expired" air); sweat glands are not a significant route of elimination. Organs may "sequester" or store xenobiotics for which they have a particular affinity or special binding sites, particularly bone, liver, and kidney.

Xenobiotics exert toxic effects by interfering with the normal functions of the body. These effects occur at the molecular and cellular level. Thus, an understanding of normal function and biochemistry is essential for understanding toxicodynamics. The toxic effect is an interaction between the xenobiotic and the cellular and biochemical mechanism. Certain organs of the body are more vulnerable than others to toxic effects: (1) the first to encounter a toxic exposure, (2) receive a large blood flow, (3) are highly metabolically active and so vulnerable to disruption, (4) actively metabolize xenobiotics themselves, (5) concentrate toxic substances or their metabolites, (6) have biochemical characteristics that render them vulnerable. The liver, kidney, lungs, skin, and bladder are particularly susceptible to toxic effects, and certain cancers of these organs are well recognized to be associated with environmental causes.

Although there are as many potential mechanisms of toxic effects as there are reactions in biochemistry and functions in physiology, there are a few processes that are particularly common and important. These processes include the following:

- *Inflammation*. The body has natural mechanisms to repair and limit damage. Many xenobiotics are irritating to human tissues and induce local inflammation.
- *Immune responses*. Immunity is a critical part of defense mechanisms of the body. When the immune response is hypereactive, as in allergies, or dysfunctional, as in autoimmune diseases, the out-of-control immune system results in diseases collectively called "immunopathies."
- *Carcinogenesis*. Cancer is defined by unrestricted growth and the ability of cancer cells to metastasize (migrate from their original site in the body). Cancer caused by environmental hazards is generally caused by somatic mutation (mutations acquired in the genome of body cells that are not involved in reproduction, rather than mutations inherited in the genetic code and passed through the germ line), although inheritance plays an important role in susceptibility. Cancer is a disorder of genetically determined control of cell division and growth that, once it begins, proceeds by its own biological determinanism. While cancer can arise in any living tissue (although it is rare in many) and manifest itself in many tissue types, the actual number of genetic mechanisms is limited, probably only to a few dozen distinct pathways.
- *Endocrine mimics*. Many xenobiotics, both organic and metal, interact with hormonal receptors: sometimes by simulating the effect of hormones and sometimes by inhibiting them.

Toxicity and Poisoning

To a toxicologist, "toxicity" is not the same as "poisoning." "Toxicity" refers to a broader spectrum of effects, from subtle "subclinical" effects (those that are physically inapparent and cannot be medically detected) to overt, symptomatic illness characteristic of exposure to a high level to an agent—which can properly can be called "poisoning, The overt clinical symptoms and signs that occur in a defined pattern in affected persons or animals is called the "toxidrome. " Toxidromes may or may not be unique or diagnostic (most are quite nonspecific) but they are always characteristic of a toxic agent. The toxidrome corresponds to toxic effects at a mid- to high level of exposure (relative to the threshold of toxicity of the agent) and defines poisoning. The term "toxicity" includes poisoning and manifestations of the toxidrome but also extends to the more subtle role of toxic substances as one risk factor among others in multifactor models of risk in a population.

In toxicology, therefore, toxicity can be widespread but poisoning is usually rare or at least infrequent, highly specific, and occurs in a context of unusually high exposure. In epidemiology, poisonings would be seen as sporadic distinct cases. that fit a particular "case definition" (based on the toxidrome) but toxicity may be manifested as changes in rates for other diseases or symptoms in a community or population.

The term "poisoning" is thus usually reserved for individual cases that demonstrate the toxidrome. In such situations, exposure is the principal factor determining the frequency of disease in a community or population and is the only important factor in the incidence of new cases, even though there may be other risk factors and causes for the condition. The exposure can be said to be "driving" the number of cases in those exposed. In the case of cancer and other outcomes that have a background rate in the population, the toxic exposure is causing at least the "excess" cases above the expected rate for the population.

In epidemiology, identifying the relative contribution of various presumed causes to the experience of a population is called "attribution." However, population data may or may not be valid for an individual within the population that has been studied or from another population; that individual may easily differ in important respects or by chance from the average, mean, or mode of the population that was studied. For most or the typical individuals in the population, the population mean may be a best estimate of the risk of that individual but for some individuals it will be way off. In medicine (particularly in occupational medicine and workers' compensation) and law, identifying the most likely causes in an individual case is called "apportionment" (because it is sometimes used to apportion the share of responsibility among parties to a legal action or insurance claim). When the potential causes are multiple, they can be estimated for the individual using the literature for populations, knowledge of the individual's personal and family history, and the context of the case can be used to apportion the relative contribution of all the causes possible in an individual case. Apportionment is of great practical importance, for example, for settling insurance claims or legal actions, although not usually for making a clinical diagnosis. Unlike attribution, which can be precise, apportionment is inexact and often rests on judgment; since it is not possible to be accurate, the system seeks at least to be fair.

For a relatively simple example (because the toxidrome is one disease), many environmental (most commonly occupational) exposures are assocated with an increased rate of lung cancer. In a typical urban population, exposure to carcinogens in air pollution contribute to the risk of lung cancer but is not the major factor determining rates in the population: that would be cigarette smoking, which is really driving lung cancer rates. For workers in some high-risk occupations, carcinogens in the work environment may play a role in elevating rates above what would normally be found (for either smoking or nonsmoking residents) and in a few occupations occupational carcinogens may drive (meaning, be the single most important determining factor) lung cancer rates within that occupation or workplace. All this occurs against a low background rate of lung cancer risk that occurs in people who do not smoke, some of whom are exposed involuntarily and passively to sidestream smoke but a tiny fraction of whom get lung cancer for no discernable reason other than biological chance. To assess the risk contributed by air pollution alone, an epidemiologist has to figure out study designs that subtract out the much larger contributions from cigarette smoking, either subtract out or overwhelm with large numbers the possibility of occupational exposures among subgroups in the population, and find ways (using inferential statistics and deriving ratios rather than calculating absolute risk) to factor out random chance and biologically-determined background rates. Then, the epidemiologist can determine relative contribution of various risk factors (assumed to be causal or indicative of a cause) to risk in the population, a process called "attribution." While this is not really difficult, once the data are available, attribution only applies to large populations. The risk experienced by individuals could be completely different. In any given individual but especially those that do not match the characteristics of the population, the risk that is conferred by that individual's exposure to particular risk factors (in this case air pollution) can be much higher or much lower that for members of their community as a whole, depending on their personal exposure and their biological make-up; there are several inherited conditions and a few that are not inherited that make a person more susceptible to lung cancer. Therefore cause in an individual case often cannot be known with certainty. Where apportionment is important, as in insurance and law, the system should at least be fair because it is impossible to be accurate or certain in many, perhaps most cases.

At intermediate levels of toxicity and when the toxidrome is more complicated than the occurrence of a well-defined disease, the toxidrome may be incomplete or not completely expressed in an individual case. It may only be possible to detect that something is wrong by epidemiological studies, comparing the health of an exposed population with an unexposed reference population and correlating symptoms, signs (medical findings on examination), and indicators (test results) with estimates of exposure. In such situations, there may be more than one cause of the illness that applies in an individual case, requiring informed judgment to decide on legal grounds (see chapter 1) if there is no definitive test that can be performed. In such situations, the toxic exposure is one among several (usually not many) contributing causes, each of which might be responsible for an individual case. At other times, there may be interactions in which the combination of exposures results in a higher rate of disease than would be predicted by the sum of the effects of each cause (positive interaction, or synergy) or a lower rate due to inhibition (negative interaction), which may be less common. Sometimes these causes may interact, as in synergistic effects of multiple exposures (often a problem with cigarette smoking and cancer). In some cases, such as heart disease, there are so many interacting factors that the disease model is called "multifactoral." In epidemiological terms, the toxic exposure is responsible for a proportion of the cases in the population, but these calculated proportions often add up to more than 100% because of interactions (synergy) among the causes.

At still lower levels of toxicity, the affected person or animal may have no specific, individual symptoms but may still be affected subclinically: that is, at a level below what is medically observable. For example, an exposure may predispose a person or animal to have a higher risk of cancer, or reduce the competence of the immune system, or confer a change in behavior that is not obvious except in large population studies. These effects are almost always missed on medical examination because routine medical tests and examinations were designed to detect illness and for the evaluation of sick people (or, in veterinary medicine, animals). They are therefore insensitive and unreliable in detecting early signs of toxicity. In such situations, the toxic exposure is only one among many risk factors; each contributes to the probability of an outcome but does not determine that outcome. Toxic exposure may not be driving the overall frequency in the population, but it is a contributing factor in a multifactoral model. One example that is extremely important in environmental health is the neurobehavioral effect of lead in children, in which lead is an important determinant, the driving factor at a population level (at levels previously considered to be acceptable) but at lower levels of exposure one contributing factor among many of comparable effects, together with arsenic, nutrition, organochlorines, endocrine mimics, early childhood education and stimulation, anemia, and many other factors.

At lower levels still, it is quite common for xenobiotics (reverting back to the terminology of toxicology) to provoke an adaptive response that does not cause obvious harm but simply represents the body's mechanisms of resilience, resistance or adaptation to the agent, such as induction of metabolizing enzyme activity or the decrease in cholinesterase activity following modest exposure to organophosphate pesticides, which may occur without apparent illness.

Exposure-Response Relationships

The dose-response relationship is the most fundamental idea in toxicology: the more of a toxic chemical one is exposed to, the greater the effect. In environmental toxicology and epidemiology, this maxim becomes: the higher the exposure to a toxic chemical in a population of people or animals, the worse and more frequent are the effects.

This idea that a poison causes worse harm the more of it that is taken into the body may seem entirely obvious today, but it was anything but obvious through most of history. Until the fifteenth century, most people (including educated scholars) in Europe and elsewhere believed that poisons had magical (or at least special immaterial) qualities that gave them toxic properties and that they were, in effect, a different form of matter. The breakthrough came in the early 1500s, when a renegade and misanthropic alchemist and physician from Switzerland discovered that magic properties had nothing to do with toxicity. Paracelsus (b. 1493–d. 1541) was the name he used professionally, but his real and entirely suitable name was Theophrastus von Hohenheim Phillipus Areolus Bombastus; no relation to the word bombastic but it would have been very appropriate to the man's personality. Paracelsus began experimenting with poisons and came to the realization that "The dose [is what] makes the poison" ("Docis facit venenum"). Paracelsus recognized that any substance can be toxic ("poisonous") at a high enough exposure level, including water; and any substance, even the most toxic, can be tolerated at low levels even if nontoxic levels have to be extremely low not to show an effect.

"Dose" technically means the total quantity of a toxic substance that is administered. However, it is uncommon in environmental toxicology that a person or an animal absorbs a pollutant in a single dose, as if they took a pill. That type of situation, where a chemical is absorbed as a bolus, is more characteristic of food toxicology and, of course, drugs. The concept of "exposure" is usually more relevant to environmental toxicology.

"Exposure" is the level of concentration available for absorption by any or all routes at or over a given period of time (duration) and is usually the preferred term in environmental sciences, although one may speak of "dose" as the cumulative exposure over a relevant time period. For a pollutant that accumulates in the body (such as lead), exposure over time may result in accumulation of a "body burden," depending on how quickly the pollutant is eliminated from the body (by excretion or biotransformation into a metabolite). For those pollutants that do not accumulate (such as most air pollutants), constant or repeated exposure may still cause cumulative effects in tissue. Therefore the dose-response relationship in environmental toxicology is more properly called the "exposure-response" relationship. However the chemical is released into the ecosystem, exposure to the chemical by the human or animal that is exposed almost always occurs over a period of time, rather than all at once, and absorption among subjects in a population, ecosystem, herd, or biological community is often highly variable.

There are three distinct varieties of the exposure-response relationship, all closely related but different in application. They need to be distinguished conceptually in order to bring together the toxicological, clinical, and epidemiological literature. These are

- the toxicological dose-response relationship, which refers to the principle that the response at a tissue or cellular level is proportionate to the amount of the agent delivered to the tissue
- the clinical dose- or exposure-response relationship, which refers to the principle that in a given individual (human or animal), different symptoms and signs may appear as different effects predominate with increasing exposure
- the epidemiological exposure-response relationship, which refers to the principle that in a population of individuals, the cases of disease become more frequent with increasing exposure

The three different but closely related relationships are illustrated in figure 5.4.

The toxicological exposure-response relationship is most often studied in the laboratory, where the degree of injury or effect is related to a controlled level of exposure to determine how the effect changes as the exposure changes. Information derived mostly from animal experimentation has been critical to determine and define these relationships, which are then used to set exposure standards to protect people. A particularly important feature of these relationships is whether there exists a "threshold" below which no effect is seen but above which there is an exposure-response relationship. The existence of a threshold establishes a "no-effect" level that, when present or as estimated, is usually critical in setting exposure standards.

The "clinical exposure-response relationship" refers to the principle that increasing exposure in an individual (human or animal) leads to the progressive appearance of new and usually more severe health problems, in a sort of stepladder effect, and that therefore people tend to show particular combinations and sequences of effects as exposure increases. Effects may occur in different tissues at the same time at the same level of exposure, however, so there will often be more than one exposure-response relationship occurring in a person or animal at the same time. Toxicity occurs in different organ systems, each of which evolves with different exposure-response relationships, and so it is common for clinicians to observe a change in pattern of disease in the individual case as exposure changes: this is the "clinical" exposure-response

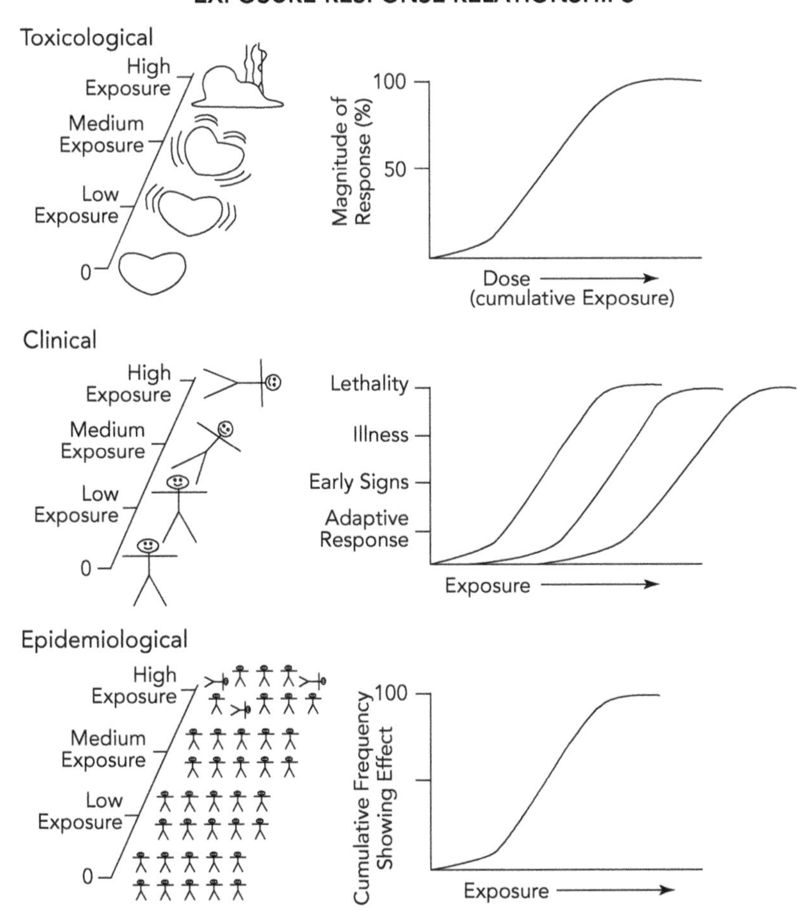

FIGURE 5.4 Three types of exposure-response relationships in toxicodynamics. At the top is the "toxicological" exposure-response relationship, in which increasing exposure leads to an increasing toxic effect in the organ or tissue, as one might measure it in a laboratory. In the middle is the "clinical" exposure-response relationship, in which increasing exposure results in more symptoms, different health outcomes, and ultimately higher risk of death, reflecting the different exposure-response relationships evolving in different tissues and tissues; this is usually presented in textbooks as a list of symptoms appearing at different levels of exposure in an individual case. At the bottom is the "epidemiological" exposure-response relationship, in which increasing exposure leads to more people (or animals) in the population showing a particular outcome, such as a gene activation, biomarker, symptom, sign, disease (by a case definition or toxidrome), or death, as in an epidemiological study. Each of the exposure-effect relationships build logically and stepwise on one another, as described in the text.

relationship. For example, an adult case of lead toxicity may present with neurotoxicity (rare today), anemia, abdominal pain, clinical gout, and muscle cramps, which will recede in that order as lead levels drop following interventions such as medically supervised chelation and "medical removal" (prescribed removal from potential exposure under occupational health regulations). This is the familiar stepladder of toxic effects seen in tables in textbooks that relate symptoms, and syndromes with levels of exposure. At a given level of exposure, one can usually expect a given constellation of symptoms and signs. This clinical exposure-response relationship depends importantly on the strength of the host defenses of the individual (which can be very variable) and whether the individual is more or less susceptible than others. In a given exposure situation, one person may show one symptom and another a different symptom, based on personal susceptibility; but there is usually a pattern in which one particular combination of symptoms predominates. This is the toxidrome.

A variation of the clinical exposure-response relationship occurs when a small exposure initiates an effect that might be considered to be beneficial but a larger exposure results in an adverse effect. Examples include trace elements such as manganese and selenium that are essential micronutrients at one level but toxic at higher levels, or the response of inducible enzyme systems to low levels of certain organic xenobiotics in which an adaptive response (increased activity resulting in rapid metabolism) may be seen but with no ill effects, compared to high levels of exposure in which the toxic effect of the organic compound is evident. When this occurs, the shape of the exposure-response curve for the key outcome (usually mortality) looks like a J or a U, with a curve downward with low exposure and a rising curve with higher exposure levels.

Some believe that this is a general principle, which has been called "hormesis," and that it should be taken into account in setting regulatory standards for exposure. The clearest demonstration of this has been with ionizing radiation, in which low levels of exposure prolongs survival in some cell systems due to induction of more active repair and compensation mechanisms, but higher doses interfere with function or kill the same types of cells. Most toxicologists do *not* believe that hormesis is a generalizable principle and consider the differences at different dose levels to be simply expressions of the different exposure-response relationships for different organ systems and functions manifesting themselves at different exposure levels. This is obscured by the measure of toxicity, since cell death or mortality is a crude indicator of toxicity and at the population level can result from different expressions of toxicity.

The third type of exposure-response relationship relates exposure levels to the frequency of the response in a population. This is the essential approach used in environmental epidemiology and yields what is usually called the "epidemiological" exposure-response relationship, with increasing count or rate associated with increasing exposure. In epidemiology, one is interested in how frequently a response is associated with a given level of exposure in a population. Recognized cases (based on the toxidrome, or some simplified "case definition") are counted as cases, and if there is an association (and if it is causal) the frequency of cases should increase with increasing exposure. An increasing number of cases cross this threshold and are observed with increasing exposure, yielding the "epidemiologic" exposure-response relationship, which relates magnitude of exposure to frequency of disease, not severity. This relationship is particularly important for disorders that are "stochastic" (arising on a probabilistic basis) rather than showing gradations of severity as a result of exposure, such as cancer, immune-mediated disorders, and infectious disease.

It is not always obvious in the literature that toxicologists, clinical investigators, and epidemiologists are each measuring the outcomes in ways that closely relate to one another. At higher levels of exposure, the exact shapes of all three of these exposure-response relationships tend to flatten out because exposure is causing higher and higher levels of response. At very low levels, the response is usually small in proportion to exposure because usually only a small proportion of the population is susceptible at lower levels of exposure, and there may be a threshold (as described above). In between, the basic shape of the curve for all three exposure-response relationships is S-shaped (sigmoid), which is not surprising because all three are related.

Stochastic Effects

The word "stochastic" has a very specific meaning today, which is "randomness." But the word has a long history that, if understood, provides an excellent object lesson and makes its meaning memorable. Stochastic derived from the word "stokhos," which is the name for a stick used by the ancient Greeks as a target for archery competitions. Archers would shoot at a pointed stick with their arrows. Some of the arrows would hit their target. The ones that did not hit the target landed around the target in a scattered distribution that today would be called "random." Hence, stochastic came to be used for a process with a random element, which could best be described in terms of probability. "Stochastic effects" are random events. They occur by random chance for any one individual. However, the probability of the event can be highly predictable in a large population or community.

Methods appropriate to random distributions are the preferred way to analyze the frequency of a stochastic health outcome: deaths, pregnancy and births, cancer incidence, incidence of infectious diseases, sensitization causing allergies.

Evaluating Sustainability Issues

This chapter opened with an exploration of how pollution in general is related to sustainability, using the elements of the definition (or meaning cluster) of

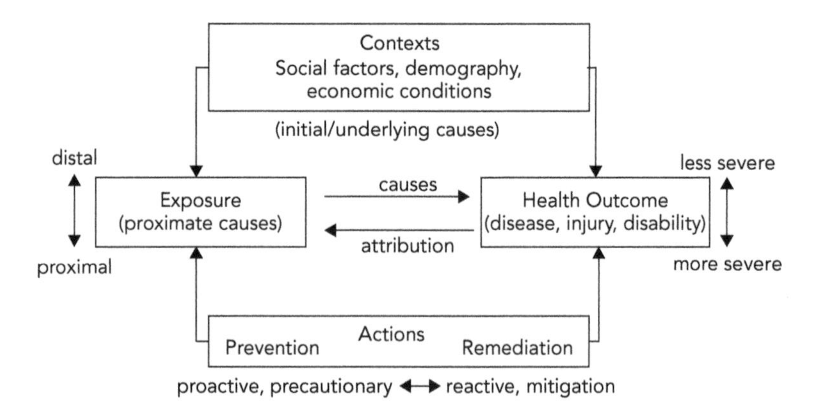

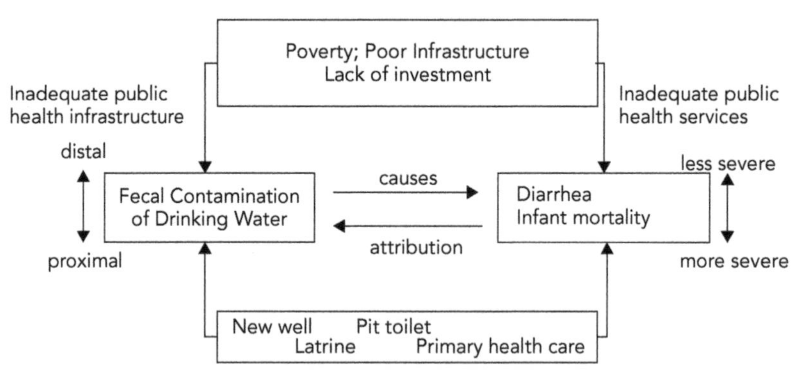

FIGURE 5.5 The Multiple Exposures Multiple Effects (MEME) model is a logical, systematic way of thinking through the context and broader implications of exposure-related health and ecosystem risk. The basic MEME model (top figure), identifies exposures and effects but also the medium or environment and social factors that allow exposure, and takes into account potential solutions (at bottom of the top figure). The example given (bottom figure) involves fecal contamination of drinking water and diarrheal disease in infants, which remains a huge problem in much of the world. "Distal" refers to higher-level causes and conditions and "proximal" refers to causes close to the community and person who is affected. MEME can be considered a combination of the "Source-Effect" Model and the epidemiological triad, adapted for deeper analysis when the social context and facilitating factors are important. It is also related to the "DPSEEA" model, which will be introduced in chapter 7.

sustainability and sustainable development. Pollution problems, as other environmental issues, are embedded in a more complicated social and economic framework. Sustainability is not just about preventing pollution but about rearranging and strengthening this framework so that intractable problems do not occur.

A useful model for understanding the context of threats to health has been developed by the World Health Organization for the analysis of problems affecting children's environmental health. It incorporates the idea of wholeness in that health problems arising from environmental proximate causes (immediate or precipitating), such as pollution or diseases associated with ecosystem modification, are recognized to have deeper causes that lie in underlying environmental conditions, ecosystem mismanagement, economic conditions, and social factors. The model is called Multiple Exposures Multiple Effects (MEMEs). Figure 5.5 illustrates the elements of this model, which relates exposures (principally exposure to pollution, but the model also fits other environmental hazards) to effects, with the assumption that both are modified by the context of the problem (including the physical environment, social conditions, economic status and level of investment, and the population) and that actions taken to mitigate the problem should act on both cause (the preventive, proactive approach; the paradigm being the public health approach, as presented in chapter 1) and effect (the responsive approach; the paradigm being medical services provided to treat preventable diseases). The lower part of figure 5.5 shows the MEMEs approach applied to infant diarrhea from microbial water pollution in a village in a less economically developed society.

Toxic effects on people, domestic animals, and species in the natural environment create unacceptable and inequitable costs and obstacles to using common resources (air, water, land). The relationship between sustainability and health is affected by pollution in other ways, such as the denial of use of resources to future generations. One example is the local contamination of property with industrial waste and, often lead, resulting in "brownfields" that often require expensive remediation before they can be used for purposes such as schools, homes, and apartment buildings. This represents a substantial encumbrance on future generations.

6

Ecosystem Change and Infectious Disease

UNSUSTAINABLE ECOSYSTEM CHANGES may produce conditions that result a realized or potential threat to human health or impede its control. This type of health issues is most obvious in the case of infectious disease, in which the change in the environment favors survival, proliferation, and transmission of a pathogen (the agent that causes the disease). To view the problem as one of sustainability requires reinterpreting the issues in a different context, looking at the problem as a broader biocentric issue in systems biology rather than narrowly as an anthropocentric issue of consequences for human beings. Table 6.1 summarizes the significance of infectious diseases to sustainability.

Infectious diseases, especially major epidemics, have in the past altered the course of societies and threatened civilizations: they have been catastrophic as defined in chapter 3 and demonstrated at the time of this writing by the Ebola outbreak in west Africa. However, as endemic or sporadic outbreaks on a smaller scale, they have more often been constituted a constant, on-going drain on resources and economic and social vitality that consumed wealth and energy, impeded progress and constrained human expression, as will be described for malaria.

As far as is known, from studies of ancient human remains, some health problems have been around for as long as the anthropological and genomic record, and provide a clear indication of the privation and harsh conditions of early and probably prehistoric life: vitamin A deficiency (night blindness), dental caries (tooth decay), mixed pneumoconiosis (an occupational disease in which silicosis dominates), coronary heart disease, cancer (which was uncommon in those days and probably lead poisoning. Many of them are infectious diseases: leprosy, tuberculosis, rabies, parasitic diseases like malaria, and common forms of pneumonia. However, these infecitious diseases would have affected individuals one at a time or appeared only in small outbreaks that probably did not spread far beyond families and clans, because there was not a large or dense enough population to sustain transmission. The concentration

Table 6.1 Sustainability values table: Infectious agents

	Sustainability Value	Sustainability Element
I.	Long-term continuity	Infectious diseases have in the past threatened economic and social continuity due to mortality and depopulation at the extreme and quarantine and travel restrictions.
II.	Do no/minimal harm	Contamination with human, animal, or plant pathogens does considerable harm to human society and can cause harm to the ecosystem, depending on the pathogen and its effects.
III.	Conservation of resources	Biological pollution may deny people, and even future generations, use of a particular resource (such as water). Some zoonotic infections are directly related to unsustainable practices, such as habitat encroachment resulting in contact with animal species carrying the disease.
IV.	Preserve social structures	The threat of infectious disease can cause and has caused social conflict and considerable social disruption, as demonstrated by the HIV/AIDS epidemic, the Ebola outbreak of 2014, and the 1918 influenza A pandemic.
V.	Maintain health, quality of life	Infectious agents cause serious disease endangering human life and health. Even minor illness, such as upper respiratory tract infections (colds, influenza), diarrhea, and urinary tract infections, interferes with quality of life and productivity.
VI.	Performance optimization	Disease, when serious, results in disruption (from death, serious illness, disability) and reduces one's ability to function at peak performance even when not incapacitating. Dealing with the consequences of biological or pathogen contamination reduces economic, social, and environmental performance directly by their impact on health, and also by loss of use or enjoyment of resources (such as contaminated beaches or bodies of water), costs of remediation (often in the form of higher taxes), costs of mitigation at the source (often in the form of higher prices or utility fees), costs of measures to mitigate effects (for example increased costs of water treatment), and the overhead cost of regulation and documentation of effects required to keep emissions under control.

(continued)

Table 6.1 Continued

	Sustainability Value	Sustainability Element
VII.	Avoid catastrophic disruption	Infectious diseases have in the past disrupted entire civilizations and threatened continuity of societies and their sustainability.
VIII.	Compliant with regulation	Water pollution and contamination of food are subject to regulation and noncompliance is illegal; most regulation and enforcement is local.
IX.	Stewardship	Biological pollution is, as is all pollution, obviously contrary to stewardship and opens the polluter to legal liability. In the case of managing infectious disease, there is an implicit responsibility to retain the effectiveness of antibiotics for future generations, which has not been respected.
X.	Momentum	Public health measures to ensure access to clean water and safe food requires continuous vigilance and must be maintained uninterrupted. Interruptions or reductions in service often lead to outbreaks of disease. Surface water pollution by bacteria and viruses, if limited, can go away by itself over time because organisms do not last forever and media such as water are, within limits, "self-cleansing."

of the human population into larger and larger communities allowed communicable diseases to be transmitted person-to-person much more efficiently, vastly increased the number of "passages" (replication in hosts allowing selection of strains better adapted to human biology), increasing opportunities for allowing mutation and new strains to form, and brought human beings together with animals who were similarly concentrated in human settlements.

The result was an explosion in human disease which seemed to grow worse as human settlements got bigger and more interconnected. Plague and smallpox stopped European civilization in its tracks more than once. Tuberculosis had a profound effect on society in the nineteenth and early twentieth centuries; called "the white plague" it affected everything from the way that buildings were designed (to promote air circulation) to the experience of immigrants in America (who often lived in tenements where they became infected). HIV/AIDS arguably set the economic and social development of southern Africa back by a generation, and in developed countries such as the United States it caused social disruption, distrust, and changed everything from medical procedures to laws on discrimination. Even lower-grade infectious diseases such as the common cold and influenza still have a considerable effect on personal performance and economic productivity.

Because pathogens, as agents of disease, are mostly specific for species, the ecosystem effect of infectious diseases that affect human beings is often limited.

The societal cost, however, can be enormous. For example, during the devastating plagues of the Middle Ages and the centuries after, much of central Europe was virtually depopulated. Ecosystems were not harmed, of course, because the diseases wiped out only humans.

This chapter will focus on infectious diseases of human beings that are of importance to sustainability and are transmissible, or communicable, from one person or animal reservoir to another person (or back to a reservoir). When the mode of transmission is from one living organism to another, the intermediary that conveys a pathogen is called a "vector." A rat, for example, can be a vector itself by conveying disease through a rat bite (rat bite fever is a rare disease, unusually variable in its presentation), by urination (which can convey the bacteria that causes leptospirosis, a particularly nasty disease), and by depositing feces (which can contaminate food that humans may ingest). Rats can also carry another vector: the flea. The flea may transmit a pathogen (the plague bacterium, *Yersinia pestis*) that causes bubonic plague or agents of other diseases. None of these are common. Rats also have and transmit their own rat diseases. Most vectors of importance to the discussion in this chapter are arthropods, such as mosquitoes, fleas, and ticks.

As described in the previous chapter, infectious diseases tend to follow a "stochastic" process, in that increasing exposure to the pathogen increases the risk of infection and development of the disease, while the disease itself is determined mostly by the biology of the pathogenic organism and the interaction between the pathogen and the host (the person infected). As will be shown below, however, this is often an oversimplification.

The emphasis in this chapter is on the "hazard," which for infectious disease is the "pathogen," and the "environment" or medium, how the hazard is affected by ecosystem change, how ecosystem change affects the opportunity for exposure, and the implications of this for sustainability. There will not be much description of outcomes or clinical diseases. Even so, it is difficult to explain the sustainability issues without some introduction to the ecological and biological complexities that help in understanding the health considerations. The details of infectious disease in medicine, public health, and biology quickly grow complicated and may be confusing for readers without some background in microbiology. They will be kept as brief as possible.

Infectious Diseases

Infectious diseases are caused by pathogens, biological agents that are contagious, which means that they can be spread from one host (infected person or animal) to another. (There are also plant pathogens. In fact, even bacteria have pathogens: small organisms called "phages.") All infectious agents are biological agents, at least so far (or at least until Internet viruses leap over the cyber-reality barrier).

It is characteristic of pathogens to infect only a limited range of hosts so that diseases they cause are usually unique or more or less restricted to one or a few related species. Human beings cannot get more than a few dog or cat diseases, for example. However, some pathogens can pass from one species to another, particularly viral diseases, some of which (mostly types of influenza) can pass through birds and pigs and ultimately can infect human beings. Sometimes, passage through another species renders the pathogen more virulent to human beings through selection, which is famously the case for certain types of bird and swine influenza.

People present a much greater risk to people, in general. Within the human species, many pathogens can pass freely. This easy passage is facilitated by many aspects of human behavior, such as the normal tendency to groom (in which a person brings their hands, which may be contaminated with a pathogen, to their mouth or nose or rubs their eyes), sexual behavior (and the risk of sexually transmitted diseases), lapses in good practice (e.g., inconsistent hand washing), and invasive procedures (such as medical interventions such as surgery or piercing). These behaviors may increase the risk of introducing pathogens into the body either directly or by reducing "host defenses" (which are the many processes and systems in the human body that protect it from infection or toxic exposure).

The pathogens that cause infectious diseases are biological agents, such as bacteria, organisms of various particular types that resemble bacteria (with names such as mycoplasma, rickettsia, chlamydia), viruses, protozoa (single-celled parasites, such as malaria), metazoa (many-celled parasites such as hookworm), fungi, and, very rarely, other forms of life, or by prions (nonliving proteins that behave like an infectious agent and are an uncommon cause of dementia).

Infectious diseases may develop in response to the proliferation or effect of the organism, or by the response of the host to the organism. Disease can be caused by the proliferation (growth in numbers) of the infectious agent itself or by an effect of the infection (such as inflammation), by a toxin produced by the infectious agent (such as cholera or staphylococcal food poisoning), by a counterproductive immune response mounted against the infection that results in damage to the host's body (such as tuberculosis), and by many other mechanisms, in great variation and with many potential complications.

Box 6.1 provides a glossary of the highly specialized vocabulary of infectious disease and introduces some of the biology behind the immune response, without which the disease effects of infection cannot be understood. However, this is not a textbook of tropical medicine, public health, or vector biology. The broad outlines of how ecosystem change affects health risk are more important in this chapter than the details of a pathogen's life cycle. This chapter, therefore, relies on the life cycle for

BOX 6.1

Useful Terminology in Infectious Diseases

Like all fields of public health and medicine, infectious disease has its own vocabulary. It is important to understand the precise meaning of the words used because they sometimes differ from common meaning in conversation.

An *infectious disease* is one that is caused by a *pathogen*, a biological agent that in some way reproduces itself inside or outside the body. (The word originally was not limited to biological agents but it is used that way now.) When it is introduced into or onto the body and grows, the presence of the pathogen is said to *colonize* the body. If it grows within the body and invades or causes disease, it is said to *infect* the body. Some pathogens also produce a poison, called a *toxin*, that can cause disease itself, with or sometimes without infection. Pathogens are mostly *parasites*, that live off of the tissue and nutrition provided by the organism it infects, which is called the *host*. (Viruses also co-opt the genetic machinery of the host in order to replicate.) Parasites that sicken or kill their host are poorly evolved; those that do not disturb the host are called *commensal* parasites. Some parasites serve a useful purpose for the host, resulting in a *mutualistic relationship*. Most bacteria in or on the body are in a mutualistic relationship or are commensals. Only a small number are or can be pathogens.

The body has a series of natural defenses, called *host defense* mechanisms. One class of host defense mechanism is called *innate immunity* and consists of mechanisms that attack germs or parasites nonspecifically, without the need to identify what it is. The white cells, or *leukocytes* in the human body are one element of innate immunity. The second class of host defense mechanism is *specific immunity*, usually just called "immunity," in which the body recognizes the germ and targets it specifically. *Lymphocytes* in blood and specialized attack proteins called *antibodies* are elements of the specific immune system. Many pathogens are incapable of infecting people with normal, or *intact* immune systems, but they can cause disease in people who for one reason or another have impairments in their immune system. Such impairments are often caused by diseases (including diabetes, autoimmune diseases, and HIV/AIDS, but to very different degrees and with different types of deficiencies), malnutrition (and iron deficiency, which is not always nutritional), or by drugs, most notably steroids, or *immunosuppressive* cancer treatment.

Most pathogens are *bacteria*, which are tiny cellular organisms without a nucleus, *viruses*, which are packets of DNA or RNA that depend on the molecular apparatus of the cell it infects to preserve and reproduce itself, *unicellular* parasites (such as malaria or amoebae) or any one of a variety of multicellular or *metazoan* parasites (such as hookworm and roundworm), which stay outside cells. There are other pathogens, including certain fungi, parasites, and dependent subtypes of bacteria known as rickettsiae (small, bacteria-like particles) and chlamydiae (tiny bacteria), both of which can only live inside cells. Even one type of algae (*Prototheca*) has been shown to cause human disease by infection, in people with immune deficiencies. One type of pathogen is not even a life form: prions are aberrant proteins, variations on a normal class of protein in the body, that cause disease and propagate by inducing a structural change in other proteins. In all, there are approximately 1,400 species of microorganisms of any type that are known to cause disease in human beings, compared to perhaps an estimated billion species of bacteria alone. The conclusion is that relatively few microorganisms, even bacteria, are pathogenic to human beings. Even fewer can only survive as human pathogens or parasites, because this is an ecological dead end limiting the species to a human distribution.

A *transmissible* pathogen is one that can be conveyed to a person or animal by any of a number of means, such as food, or water, or direct contact. Diseases that arise from a transmissible pathogen are called *communicable diseases*. Diseases that are spread by direct transmission by person-to-person contact are called *contagious* diseases.

A pathogen may secrete a biological poison, called a *toxin*, or have some other effect on the body because of how it interferes with function (falciparum malaria, for example, may affect the brain, with devastating results) or due to the inflammation it causes (tuberculosis provokes a response in the body that is greater than the damage done by the tubercle bacterium itself).

A disease that is normally present in a community or population is called an *endemic disease*. Malaria, for example, is endemic to many of the world's tropical coastal regions. Endemic diseases can be very severe and debilitating but they are sometimes overlooked or misdiagnosed. For example, in many of the regions where malaria is common, hepatitis B is also highly prevalent, but it tends to be undercounted because its symptoms overlap with malaria. An *epidemic* is simply an *outbreak* of disease at a higher rate than expected for a population. The word "epidemic" does not necessarily mean that the outbreak is widespread or lethal, but it does imply that more cases are being seen than usual. Although the two words are synonymous, "epidemic" tends to be used in reference to populations, such as "the H1N1 influenza epidemic of 2009," and "outbreak" tends to be used in reference to a community or defined group, such as "an outbreak of norovirus caused severe diarrhea among the passengers of a cruise ship."

When an outbreak occurs, one of the first things to be done by *epidemiologists*, the scientists who study the distribution of disease and health risk factors in populations, is to define the sequence of events and time course as new cases appear; this defines the *epidemic curve*. An epidemic curve that shows that everyone who got sick did so more or less all at once and suddenly usually implies that they all got it from the same place, which defines a *common-source outbreak*. Foodborne and waterborne illnesses are common source. An epidemic curve that shows that people got sick more or less one after the other and had contact with others who later became ill defines a *propagated epidemic*, demonstrating disease transmission. Sexually transmitted diseases, diseases caused by pathogens that are inhaled as fine droplets (such as tuberculosis and influenza) or on contact surfaces (such as colds and influenza) show up as propagated epidemics.

One way that diseases become transmitted is when they are carried from an infected individual to an uninfected individual by another species, called a *vector*. Although this mechanism can operate with many different species, the most common vectors of public health significance are *arthropods*, which is the biological phylum that includes insects, ticks, fleas, lice, and mites. Some organisms are just *mechanical vectors*, like the housefly, which can carry bacteria from feces. Others are passive biting vectors, such as the conenose or reduviid beetle or "chinche," which transmits a trypanosome parasite that causes Chagas' disease in tropical areas through biting a human host, usually on the face, and contamination of the bite by its feces. Others are active biological hosts, such as the mosquito in malaria.

One way to control communicable disease is by *immunization*. Immunization acts like a fake infection: it tricks the body into mounting an immune response as if the person were infected (actually, usually stronger). To do this requires a vaccine to be developed that carries molecules, *antigens*, that resemble the pathogen on a molecular level but that are not themselves infectious. Immunization is also called *vaccination* because its modern history began with intentional inoculation with cowpox (vaccinia, from the Latin vaccinus, meaning "from cows") which conferred sufficient cross-immunity to prevent infection with the much more dangerous smallpox. Immunization protects individuals well—as a strategy, it is, together with drinking water disinfection, the most effective public health measure ever invented. However, few vaccines approach 90 percent effectiveness in protection because not everyone can mount the same level of immune response or keep it at high levels for long enough. For this reason, the level (*titer*) of antibody to the antigen is tested in situations where protection is critical, such as hepatitis B and cytomegalovirus, which a health worker might transmit to patients if immunity were insufficient.

The more people in a population who are immunized against a threat, the better, excepting the few people who may be at risk for adverse effects. However, immunization rates do not have to reach 100% to be effective in protecting against infectious disease. Immunization of the large majority of the population protects even those who are not fully protected because of a phenomenon known as *herd immunity* (because it was initially described in livestock). High but subtotal immunization coverage reduces the probability of transmission until it is so low that the infection does not effectively propagate from person to person in the population. However, when immunization levels drop too low, a disease that was thought to be under control can return as a new outbreak, as has happened repeatedly in recent years in England and the United States with pertussis (whooping cough). Pertussis is a severe and occasionally fatal disease in children and adults, and when many parents who were persuaded on the basis of a medical fraud that the available vaccine might cause autism (it does not), declined to immunize their children.

Immunization requires getting the target antigen right for the vaccine. Some pathogens frequently change their antigenic sites, particularly on their surface, so that it is difficult or impossible for a vaccine to work. Malaria is an example of a pathogen that shows *antigenic variation*, with mechanisms that constantly change antigenic sites to evade detection, as well as expressing different antigens at different points in its complicated life cycle. Influenza mutates frequently so that the antigenic sites change with each flu season. The common cold mutates easily and frequently also, but there are also so many antigenically different viruses (over 200, mostly rhinoviruses) that cause the same cold syndrome that a vaccine is currently impractical. (The usual symptoms of a cold are not determined by which virus is the pathogen but by how the body responds to it and the level of *cytokines*, the body's chemical messengers, released in response.) HIV, the AIDS virus, has a mechanism called *immune escape*, in which the virus changes in predictable ways under selection pressure of the immune response. This is why there is currently no effective vaccine against these diseases. One way to get around this is to develop vaccines against parts of the pathogen that do not change, such as its DNA, RNA, or essential proteins. This is much more difficult than targeting surface antigens, but *DNA vaccines* are being developed for testing.

Sometimes it is possible to get rid of a pathogen completely, through a combination of immunization and treating people or animals that carry it. This is called *eradication*, and it is achieved by immunizing people or animals until there is a sufficient buffer of protection that halts transmission. This is only possible when the pathogen has a very limited reservoir that protects and maintains it (e.g., pathogens that only live in human beings or in cattle and not other species). There have been exactly two diseases that have been eradicated in world history: smallpox and rinderpest (a highly transmissible and fatal disease of cattle).

At the time of this writing, an otherwise almost successful global eradication campaign against polio is in crisis. The disease has been eliminated in the rest of the world through immunization campaigns conducted by the World Health Organization since 1988, but transmission of the disease persists in two remaining remote areas: northern Nigeria and border areas of northwest Pakistan and Afghanistan. There, health workers providing immunization have been repeatedly attacked and killed by extremists who accused them, falsely, of a conspiracy to sterilize people against their will with the vaccine. (An echo of eugenics.) This sad state of affairs places the entire world at risk. In mid-2014, new cases of polio were discovered in Iraq and in Syria, and poliovirus was newly detected in wastewater sampled in Israel, Egypt, and the Palestinian Authority, all three of which matched the strain in Pakistan. Poliovirus was obviously coming back, almost certainly brought by a traveler from (or returning from) Pakistan.

In northern Nigeria, rebels and religious leaders have often been persuaded that polio vaccination is part of a plot by foreigners to infect people with polio, or to make them sterile (eugenics again). This has led to harassment and even killings of members of vaccination teams in those areas (particularly around Kano). Very recently, imams (Muslim religious leaders) there have changed their minds when they saw evidence of the vaccine's effectiveness. These imams have begun preaching the benefits of vaccination and cooperating with vaccination teams and in those areas rates of immunization have increased dramatically.

Vaccination is all that is holding back another, possibly global polio epidemic. Unless and until the disease is finally eradicated from those two small pockets, it will certainly come back into the rest of the world sooner or later, appearing as paralysis among children in places where vaccination rates dip below the critical point because of neglect, lack of understanding, war, financial setbacks, intimidation, misguided parental beliefs, or public attitudes of distrust. The situation demonstrates just how tenuous protection can be from a devastating illness. It is also an example of how war and domestic unrest are devastating for health protection, even outside the arena of conflict.

When eradication is not possible, a pathogen can still be controlled insofar as the disease is totally prevented for a period of time. This is called *elimination*. The difference is that a disease that is eradicated can never come back, in theory (for smallpox there is only the possibility of reintroduction from stored laboratory samples). A disease that is eliminated may be controlled so that it does not affect an entire generation but can be reintroduced from a reservoir where it is protected. For example, cholera can be eliminated and many areas have done so, but there is always the possibility that it will be reintroduced because the bacterium lives protected in brackish water in a dormant state and sticks to various aquatic organisms. Cholera epidemics have been caused by ships emptying their ballast tanks in a harbor and dumping the bacterium out with the water. Elimination requires a huge effort but is never fully complete because the disease can come back. Eradication requires extreme, heroic effort, but once the disease is gone it is gone for good.

malaria for illustrative purposes only, to demonstrate that every infectious disease agent has its unique features and that every disease has its own characteristics.

The infectious diseases of primary concern in the context of sustainability include the following:

- The traditional great scourges of humankind, such as malaria and tuberculosis. These diseases are so dominant that they have a demographic impact on human populations and limit the social and economic potential of nations.
- The historical scourges of humankind that have shaped how we view infectious disease. These diseases, including smallpox, plague, and cholera had such a profound impact on society that they shaped our history and left perceptions and attitudes that remain today.
- "Emerging" infections such as HIV/AIDS (which of course truly emerged and is now endemic), SARS, "swine" flu (influenza A H1N1) and a host of more recent diseases that present a greater or lesser threat, such as Ebola. These diseases require constant vigilance and remind us that we are not as safe from disease as we might think.
- "Nosocomial" infections, often resistant strains of common pathogens, arising from health care and in hospitals and clinics as a result of intense selection pressure through the use of antibiotics, the concentration of people with potentially communicable diseases, potential contamination, and advances in medical care that paradoxically lead to invasive procedures or breaches in host defenses of the patient that an allow an infection to become established. At the time of this writing, the nosocomial risk of infection with Ebola virus

is receiving widespread media attention; resistant bacterial strains in hospitals have already become a huge problem, however.

- Diseases for which resistant strains have emerged by natural selection. The efficacy of antibiotics has been compromised by overuse and inappropriate prescription in human patients and by indiscriminate use in animals. The list of resistant agents is long, but one of the most troubling and difficult to treat is "methicillin-resistant *Staphylococcus aureus*" (MERSA). Many of these resistant strains emerged because of unrestrained use of antiobiotics in cases when they were unnecessary.
- The "zoonoses," diseases that are carried by animal species and can be transmitted from them to human beings. This has led to the emergence of a new concept: the OneHealth model.

These categories are not mutually exclusive. HIV/AIDS began as a zoonosis; but once it jumped the species barrier, it established itself as a communicable disease transmitted from person to person. Tuberculosis, an ancient hazard, has reemerged largely because of drug resistance and because of the susceptibility of persons with HIV/AIDS.

In discussing infectious disease, especially malaria, it is important to appreciate the critical difference between "elimination" and "eradication," as technical terms. Elimination occurs when public health interventions result in no new cases in a particular geographical area, so that the disease can be said to have been eliminated from an area; elimination has occurred many times in many countries and is the goal of effective public health measures, however for infectious diseases there is always a remaining risk of re-introduction from elsewhere. Eradication is the permanent suppression of a disease, such that there are no new cases, anywhere, ever. This can only occur when the pathogenic agent is completely removed and has only happened twice in world history, with smallpox in human beings and the serious and highly communicable disease rinderpest in cattle. At the moment, there is an epic effort underway by the World Health Organization to complete the eradication of polio form the world; if it fails, the disease could easily return but if the effort succeeds, in theory it can never return.

A Critical Example: Malaria

Malaria is critically important to sustainability for many unfortunate reasons. Considered alone, it is one of the greatest threats to global sustainability and therefore must be understood on its own terms. It is the leading cause of death in the world, especially in the case of children. It is one of the clearest and most relevant models for the relationship between ecosystem change and infectious disease risk. As a case study, it also demonstrates almost every aspect of concern in infectious disease issues that applies to sustainability. Even if progress toward a vaccine simplifies the management of malaria in the near future, the problems getting to control are an object lesson in how infectious disease relates to sustainability.

Because modern malaria is a disease of human beings and people are its only reservoir, in theory malaria could be eradicated if all human cases were to be cured or prevented. In practice, the problem is much more complicated, as will be shown, and elimination, area by area, is the practical but still ambitious goal.

Malaria is the greatest killer among diseases in the world. Every year approximately a million people die of malaria (estimates range from 700,000 to 2.7 million). Every day, up to 3,000 children die of the disease in sub-Saharan Africa alone. Malaria is one of the most significant burdens that tropical developing countries have to carry, particularly in Africa; malaria consumes a large proportion of productive capacity (over 1 percent of GDP, as noted above) that could otherwise be used in health care, education, and infrastructure improvement. The costs associated with malaria include health care but mostly involve loss of productive work capacity for people who are sick, disabled, and who die prematurely from the disease, as well as those adults who may be well but must care for sick children.

Malaria is a social problem as well as a disease, particularly in Africa. It severely weakens those who have it, decreases productivity, and therefore undermines economic security; those who have it are kept from achieving their full social capacity in their communities and families. The disease places a great burden on health care and social services, which are fragile to begin with in most malaria-endemic regions and mostly family based. At the same time, control of malaria is impeded, in part by cost, limited economic development, inadequate health services, lack of education and awareness of effective means, cultural reluctance to adopt practices that would reduce the risk, and methods for prevention that are technically complicated and often only partially effective.

Climate change has dramatically aggravated the incidence of malaria. Average warming has extended the range of potential mosquito habitat not only in area but in altitude, with increased incidence of mosquitoes climbing higher in mountainsides and highlands of Africa and South America. Warmer average temperatures increases the survival of mosquito eggs, shortening hatching time and the maturation time of mosquito larvae, as well as of the gestation of the parasite within the mosquito. Biting frequency also increases with temperature. The result is that malaria, which is bad enough, is also "emerging" and would be an even worse threat if this were not already widely recognized and the object of a concerted global effort to stop it.

Malaria is also an example of how inadequate resources in health care can lead to a distorted approach to infectious disease. Until very recently, the diagnosis of malaria used to require a time-consuming test that was not practical in developing or resource-poor countries. "Thick-film" slides were prepared so that a technician could look for the parasite in blood samples, using a microscope (which is an expensive piece of equipment in a poor country), a process which sometimes took a long time. Unfortunately, a very stressed hospital or clinic with limited resources simply could not afford to do this test on a regular basis. As a result, many people were diagnosed only from their symptoms and never had the disease confirmed.

In much of the developing world, anyone with fever was (often still is) given antimalarial medication without testing, because there is no money to pay for diagnostic tests. If the patient does not respond to treatment, it is assumed that the disease must have been something else. This strategy usually works for the individual case, because the disease usually is, in fact, malaria; however, in a lot of cases, other diseases are present but not acknowledged. Of course, many of these cases are never followed up on, because health care is expensive, and many patients cannot afford medication, especially in the poorest countries.

Simple and inexpensive but accurate "Rapid Diagnostic Tests" have recently been introduced and are becoming more widely used in developing countries, although they may still be out of reach because of the expense in very poor countries. They are several types, some of which can be used for all four types of malaria or some for falciparum, the worst type, alone. They have a common problem of poor shelf life because of sensitivity to heat, which is a serious problem in tropical countries without easy access to refrigeration. (The problem of maintaining a "cold chain" is discussed more extensively below, in relation to vaccines.)

Until fairly recently, malaria was not a focused priority target for global health and economic development. This is not because its impact on the world was not appreciated but because it is technically difficult to eliminate, no vaccine was practical, and it was not thought that a large investment in disease control would make much difference. Now, thanks largely to the Bill and Melinda Gates Foundation and programs of the World Health Organization, malaria is being directly addressed, and there is progress to show for it. (This will be discussed in greater detail in a later subsection.)

The interested reader is referred to textbooks of medicine and public health for a greater level of detail but cautioned that the medical management of malaria is a specialization in itself and will not be described in any detail.

Knowing something about the biology of malaria is important to understanding the message of this chapter and the book as a whole, but it is not critical to master the details. In particular, issues of prophylaxis, diagnosis and treatment, drug resistance and its mechanisms, medical management of complications, and drug side effects are too complicated and specialized to outline in detail here and would distract from the emphasis in this chapter. The reader only needs to know, for the purposes of this chapter, that drug treatment of malaria has been compromised by resistance, because treatment selects for those organisms that can survive exposure to the drug, especially to chloroquine and some other of the best antimalarial agents available. Even where resistance is not a factor, not everybody can take the most effective malaria drugs because of a common genetic condition that cause potentially serious side effects associated with ingestion of certain antimalarial drugs.

Controlling Malaria: Three Basic Strategies

The malaria parasites, all of the genus *Plasmodium*, are extraordinarily well adapted to their common niche in the ecosystem, circulating among the two species, mosquito and human, and avoiding detection by frequent antigen changes. As a practical matter, there are four of them, in increasing order of severity in the form of malaria they usually cause: *malariae, ovale, vivax, falciparum*. The plasmodium parasite effectively evades the immune response and so no effective vaccine has been available until now. The most recent development has been the development of a new, partially-effective vaccine that has overcome many of the intrinsic problems of immunizing against a parasite. It will be described under "changing the host."

No one strategy now available seems to be able to eliminate the disease when used alone. Integrated management and overlapping measures seem to be the only viable pathway that could lead to leading to elimination of the disease.

Control the Pathogen

Until recently, the strategy of eliminating malaria by eliminating the pathogen itself has seemed tantalizing but distant and unattainable. Malaria is vulnerable because it passes through a phase in which it is vulnerable to treatment. The life cycle of the malaria parasite depends on human beings to serve as a reservoir for the disease and incubates it in some stages; mosquitoes of the genus *Anopheles* transmit it between people and to incubate it in other stages. Therefore, one possible way to get rid of malaria would be to get rid of the malaria parasite itself in the community, through treatment of sufficient people that mosquitoes would have a low probability of picking the parasite up after biting a resident. Effective immunization would achieve the same effect, but vaccine prophylaxis is discussed later, under "changing the host."

The drugs available to treat and prevent malaria have problems, however. For many years there has been active development of new drugs for both treatment and prophylaxis (prevention of disease) because current options are not ideal. Treatment has serious limitations because of drug side effects and because the parasite becomes resistant to the drug with time. Use of some of these are limited by toxicity. Chloroquine and drugs related to it cannot be taken by people who have a particular hereditary enzyme deficiency (G6PD deficiency) that is quite common in some parts of the world. Once the best hope for a standard cure and for prophylaxis, mefloquine is effective but causes debilitating psychiatric and neurological side effects in some people. Most recently, regimens including a Chinese herbal agent, artemisinin, in the form of combination therapy, have been effective and in some settings are now the regimen of choice. These new and more effective protocols reduce the risk of resistance considerably, but they also involve multiple drugs. Because of this complicated pattern of resistance, susceptibility, and propensity to produce side effects, treatment and especially prophylaxis (treatment to prevent infection, rather than cure the disease) is complicated and specific to the local situation. A drug that can best be used in one situation may not be suitable for another, and it takes an expert to know.

Each species of malaria has its own pattern of sensitivity and resistance to drugs. The first-line drug chloroquine, which for many years was the mainstay of malaria treatment, does not work everywhere now because many strains of malaria in various parts of the world (and all strains in some regions) are resistant. This resistance arose because both prophylaxis (prevention) and treatment have been intense and widespread and so selected for survival mutant strains of malaria parasite that were resistant.

It has long been known that there is a limited form of immunity to clinical malaria that arises among residents of endemic areas where the malaria burden is high. In places where it is impossible to avoid being infected repeatedly from childhood, most people who survive develop a partial immune response that is constantly boosted by reinfection. This partial response is unreliable in protecting the individual but overall reduces the risk of clinical malaria and is partially protective against the most severe forms of falciparum malaria. This phenomenon is why not everyone who lives in highly prevalent malaria-endemic regions gets clinical malaria. This form of acquired immunity disappears when a person has been removed from that environment and is no longer reinfected, but it returns after some time when they go back and get the booster effect of repeated mosquito bites and reinfection.

Because human beings are "obligate" hosts for the malaria parasite (meaning that the plasmodium must pass through human beings to perpetuate its life cycle), this strategy would require treating (or immunizing) as many malaria-afflicted people as possible to eliminate the plasmodium and preventing new cases from being introduced into the population. If the plasmodium could be eradicated, an uninfected mosquito could not pick up the parasite from someone in the population when they bite because no one would be infected, and so there would be no pathogen available in the population to transmit to another person. Preventing new cases would have to be achieved by effective prophylaxis in the case of malaria in the absence of an available vaccine. With a reduced burden of the parasite in the community, mosquitoes are much less likely to pick up the plasmodium after biting a local resident and therefore less likely to infect someone else. Unfortunately, there would still be the potential for reintroducing it from other parts of the world through immigration and travel, however, so this approach would not be sustainable in the real world for malaria. Until the early twenty-first century, therefore, the prospect of malaria elimination by removing the pathogen through treatment or prophylaxis looked like a dead end. However, there has been a twist to this story.

Recently, malaria experts who conducted analysis of the incomplete and interrupted UN-sponsored effort to eradicate malaria in the 1960s came to a different conclusion. They compared malaria prevalence and incidence in countries that had maintained low incidence of malaria after the campaign and those in which the disease had surged back. They found something that had been largely overlooked before, which is the number of secondary cases caused on average by each case of malaria in the population of countries (described in biostatistics as the reproduction number, R_0, where $R_0 = 1$ means that for every case there is another case that results, so the disease continues at the same prevalence and if $R_0 > 1$ the outbreak propagates into an epidemic. R_0 is about two for influenza and ranges up to thirty for pertussis (whooping cough, which is one reason there have been so many epidemics in unvaccinated children recently). It turns out that for malaria R_0 is very low in some countries (on the order of 0.04); those countries were the ones that had made a big push to eliminate malaria in the population through treatment combined with controlling the mosquito vector. The clear implication is that in those countries the disease can, in theory, be eliminated by treatment alone, as long as vector control measures are adequate to stop or slow down transmission from imported cases. It helps that those who were already infected often develop partial immunity, and so tend to become resistant to malaria. This refines the strategy of elimination and suggests that even partially effective treatment and partially effective vaccines could be used strategically to achieve elimination.

Change the Environment

The second broad approach is to change the environment, so that the vector does not transmit the disease. The main strategy used in the world today is to get rid of the vector that transmits malaria through control of mosquitoes. Historically, this was done by controlling standing water by draining swamps, pouring oil on standing water (to suffocate the mosquito larvae), and keeping the ground dry so that mosquitoes would not have suitable breeding grounds. In some places, such as California, small fish (*Gambusia affinis*, the "mosquitofish") have been widely introduced to eat mosquito larvae. Recently, a promising new gene technology has been developed to sterilize male mosquitoes in order to control local populations.

One novel approach has been a new vaccine developed by Sanaria, a private company in the United States with participation by the National Institute of Allergy and Infectious Disease, is aimed at using antibodies in the immunized person's blood to attack the parasite in the mosquito foregut as soon as it first draws blood, before the parasite even enters the human body. This would provide greater individual protection and achieve the same result of eventually eliminating the pathogen in the population. Philanthropic support of these vaccines has been necessary because in poor countries the market cannot be expected to recover the cost of development.

In recent times the most common approach to controlling malaria has been eradication by using insecticides to kill the mosquitoes that carry the parasite. Unfortunately, the practice of "fogging" (spraying indiscriminately into the air), while it looks impressive, does not work effectively. A more effective way of controlling mosquitoes is by "indoor residual spraying," using an insecticide that persists on indoor surfaces like walls. The best residual insecticide for mosquito control was once DDT, which became unavailable for the purpose because of its ecological impact and bioconcentration as a persistent organic pollutant (POP). (See box 6.2.) Under the Stockholm Convention on (the "POPs Treaty") DDT was banned, except for restricted and indispensable uses. (See chapter 6.)

One of the most effective measures for protecting individuals is pesticide-impregnated bed nets, using environmentally acceptable pyrethroid insecticides, but this only protects individuals one bed and one night at a time. DEET, which is a repellant rather than a pesticide, is also effective protection for individuals; it is sprayed on clothes, not skin. The main problem with mosquito repellants is that they become ineffective about two hours after application.

One of the complications of malaria control is that timing is so important. Mosquito species do not all transmit malaria. *Culex* species, for example, bite ferociously and often. Where they are prevalent, the bite frequency and annoyance a person experiences has nothing to do with malaria risk. Only a few species of the *Anopheles* genus (and only the females of the species) transmit malaria. Those species that do transmit the disease bite with different frequencies, fly at different altitudes (higher or closer to the ground), and may bite at different times of day (although usually starting at dusk, so the highest risk is dusk to dawn). To add to the complication, the timing or peak bite frequency can change in a mosquito population as it comes under stress by elimination measures.

Change the Human Host

Changing the human host means making people more resistant to the disease. The preferred way to do this in practice is immunization. As of the early twenty-first century, an effective vaccine has been elusive, although the search may soon be over.

BOX 6.2

DDT and the Incomplete Elimination of Malaria

One of the most regrettable stories in public health is the history of how malaria was almost eliminated in much of the world but could not be completely eliminated because of the misuse of chemical pesticides for other purposes.

DDT (dichlorodiphenyltrichloroethane) is a chlorinated hydrocarbon pesticide that persists for many years without being broken down in the environment. DDT had first been synthesized as a compound in 1874 but had no obvious use until 1939, when it was found to kill insects effectively on contact, even when dry.

The main pesticide in common use in the 1940s was pyrethrum, which was manufactured from powdered chrysanthemum flowers and for which there was a shortage during wartime. That shortage was corrected by replacing pyrethrum, a natural product, with DDT. DDT was used during the Second World War to control malaria (spread by mosquitoes) and typhus (spread by lice) among troops, prisoners of war, and civilian in liberated areas, where it was applied under the clothes using "flit guns," which were a form of aerosol applicator.

DDT had excellent repellant qualities. The neurotoxic effect of the pesticide repelled the mosquito almost immediately and knocked it down before it had a chance to bite anyone else. DDT acted as an excitatory neurotoxin to the insect but not for human beings. DDT appeared to be essentially nontoxic to human beings; it was cheap to manufacture and easy to apply by hand spray. Thus, health workers did not require extensive training or safety gear. It seemed at the time to be the ideal mosquito vector control agent.

The greatest advantage of DDT, however, was that once applied to the walls in a house it persisted for years, so that it kept killing mosquitoes, even if the wall was cleaned with soap and water. Very little DDT was required but as long as the wall was not painted, it kept working. It continued to work year after year for this purpose. This procedure became known as "indoor residual spraying" (often abbreviated "IRS") and became the main strategy for malaria control because it prevented the mosquito from biting at home during the evening and night, when many species of mosquito are most active.

The World Health Organization launched its Global Malaria Control Programme in 1955 with great optimism. It relied heavily on two pesticides: one that went by the familiar initials DDT and the other dieldrin, which was later found to be too toxic for routine use. Initially, there were spectacular successes. Malaria was eliminated quickly in Taiwan, Sri Lanka (then Ceylon), India (then under British rule), Corsica, and much of the South Pacific, and elsewhere. The program collapsed less than ten years later, in part because the pesticide was no longer effective in some areas where resistance began to develop, and in part because of a wave of concern for the environmental effects of DDT.

At the same time that it was introduced for civilian public health use, DDT started to be used in massive quantities for agriculture. All the advantages of DDT for public health applications, including and especially its low price, made the pesticide very attractive to farmers. DDT became one of the most heavily used pesticides in history, applied indiscriminately and in huge quantities for crops, particularly those such as cotton that were prone to insect infestations. Usage in agriculture dwarfed its use in public health measures by a factor of thousands. Indiscriminate and massive use meant that soon strains of insect species were selected that were resistant to DDT. The once-ideal pesticide lost effectiveness.

In 1962 Rachel Carson described the effects of DDT and other pesticides on the environment in the seminal book *Silent Spring*, the most influential and one of the first books that launched the environmental movement. The effects on desirable insect species, on birds (notably peregrine falcons, osprey, and pelicans), and on fish were undeniable, and there were allegations of human health effects. Subsequently, it was found that through bioconcentration (see chapter 4) raptor birds were particularly vulnerable to an unsuspected hormonal effect of DDT interfered with eggshell formation and was devastating the symbol of America, the bald eagle, which was headed for extinction at the time, as was the California condor.

DDT was the first battle in what became the toxics front of the environmental movement of the 1960s. (Rachel Carson herself advocated the small, measured use of DDT as necessary to control malaria.) The US Environmental Protection Agency de-registered the pesticide in 1972 except for public health uses, which were rare by that time in the United States. Most other countries followed, although many countries continued to allow the manufacture of dicofol, a pesticide that is metabolized to the same active metabolite as DDT. DDT was incorporated into the "POPs" list of the Stockholm Convention in 2004 and has been in effect banned worldwide ever since. The chemical stability and characteristics of DDT that enhance its value as a pesticide also determine its propensity to remain adsorbed onto particles of soil and in sediment, its ability to migrate and concentrate in colder northern ecosystems, and its bioaccumulation in fat in the bodies of human beings and animals, The persistence of DDT has resulted in detectable but declining DDT levels throughout the world and in almost every person tested regardless of where they live. Some of the highest levels recorded have been in people and animals in the Arctic, as described in chapter 5, far distant from where the pesticide was applied.

In subsequent years the toxicity of DDT has also been reevaluated, with some concern over its role in cancer risk (particularly breast cancer) and interference with hormonal mechanisms. It is not established that DDT is a carcinogen in human beings, but it is known to affect the metabolism of many xenobiotics (see chapter 5) by inducing metabolic enzymes; it also has some effects that mimic those of hormones and is present in mixtures of other residues in tissue that are associated with diminished immune response in children. DDT travels with many other chemicals with similar characteristics, and it is difficult to disentangle effects in isolation. How important DDT exposure is for human health in the real world remains unclear, but it is unlikely to be a major determinant of health by itself.

Today, the mainstay for mosquito control are pyrethroids, which are modern synthetic derivatives of pyrethrum that are more expensive than DDT (but not by much when the lower amount applied is taken into consideration). Resistance to pyrethroids has not been a severe or widespread problem, and their use in pesticide-impregnated bed nets is proven to provide effective protection for individuals.

However, as of this writing (2014) DDT is increasingly used again for mosquito control by residual application, primarily by resource-poor countries in Africa (such as Ethiopia and Zimbabwe), in South Africa (because alternative efforts failed), in China and in North Korea. For over forty years there has been debate over whether DDT should be allowed to be used for indoor residual spraying. It is a perennial topic of debate at the World Health Congress, which is the biennial meeting of the governing body of the World Health Organization. Contrary to popular belief, however, there is no broad ban on its use for public health purposes, and it has been used on occasion, for example in India. DDT has also been used as a wedge issue against environmental activists, with allegations that strident environmental advocates deprived public health professionals of an essential tool in the fight against disease. In fact, knowledgeable sources seem to agree that the use of DDT against malaria transmission failed mostly because agricultural use rendered it nonviable and that it is no longer required or preferred except in extraordinary situations, particularly in situations where indoor residual spraying must be maintained at high levels.

The story of DDT and malaria control has many dimensions and illustrates the complexity of strategies that seek to combine health protection with sustainability. However, it is part of a much larger story of how DDT was adopted as a panacea in agriculture with an excess of enthusiasm and every incentive for overuse, and how this not only had unintended consequences for ecosystem sustainability but ultimately failed in its mission of controlling pests, as resistant strains were selected. A public health opportunity was lost forever.

Human beings are not defenseless against malaria, but the defenses are not sufficient. The human body has mechanisms (called "innate immunity," for example by release of a certain chemical factor from platelets) that are capable of fighting the parasite just enough to take the edge off the disease in adults. Resistance to malaria has also evolved, but it has done so at the expense of health in other ways.

Many people in Africa and the Middle East carry the gene (just a single mutation is responsible) that causes sickle cell anemia, a potentially serious blood disease, when two genes are inherited and others carry other genes for disorders of hemoglobin (and they can occur in the same person). The sickle trait (when only one gene is inherited) makes the person resistant to malaria because the abnormal red cells do not support the parasite. Since the parasite has to spend part of its life cycle in blood, the sickle trait prevents the infection from being fatal, but the survival adaptation of the gene results in many cases of sickle cell disease in the population, which is a high price to pay. (Unfortunately this mechanism does not work well for the worst type of malaria: falciparum.) People who carry the sickle trait do have a strong adaptive advantage, however.

Similarly, G6PD deficiency, an inherited condition that causes no problem unless a person's blood cells are stressed (as with one of the antimalarial medications mentioned), also makes a person resistant to malaria because their blood cells become damaged when they carry the parasite and so are removed by the body. Also there may be partial immunity to malaria (again, not so much for faciparum) when a person has grown up in an area where malaria is endemic and is repeatedly infected, but it is not sufficient to prevent the disease altogether and fades quickly if a person stops being bitten by malaria-carrying mosquitoes. From the biological point of view, therefore, the human body evolved with malaria as a constant threat, shows evidence of having struggled against malaria, specifically, for hundreds of thousands of years, and can be said to have beaten the disease to a stalemate. Unfortunately, this stalemate involves a lot of unavoidable disease both from malaria and the adaptations that were necessary to fight it, and only limited biological protection that proved completely inadequate to stop the surge in malaria that occurred in the last century.

Parasites such as malaria are good at becoming resistant, hiding in the host's organs, protecting their most vulnerable cellular mechanisms, and changing their surface antigenic structure to evade the immune system.

On the horizon is a vaccine that can protect people from infection with malaria, In 2014, the first malaria vaccine to complete trials for efficacy and safety reached the stage of an application for approval one vaccine. The vaccine, developed by pharma company GSK, completed trials in children and was shown to be effective in reducing malaria infections by about half, with no limiting side effects apparent. At the time of this writing, the vaccine is pending approval from the European Medicine Authority. Preliminary reports are that although it is effective early on, its protection diminishes over the following few years and may require a booster dose. However, it is a major step forward and could, conceivably, change everything in a very short time. Several other candidates are in various stages of development at the time of this writing, with various possible advantages or disadvantages yet to be documented.

A practical vaccine will face obstacles that are not insurmountable but that must be planned for. For example, if a traditional vaccine becomes available that requires refrigeration for preservation, there will be logistical issues in its delivery and especially in maintaining the "cold chain," which is simply the infrastructure of having refrigeration available and reliable electrical power to run it in poor areas, since existing vaccines need to be kept cold to prevent degradation.

Action Plans

The roots of the malaria problem in Africa, the center of the global problem, are long and convoluted and reach deep into colonial history and the economic setbacks the region experienced after decolonialization. Malaria was not eliminated by British, French, Dutch, Belgian, or Portuguese colonial governments, but it was controlled to varying degrees, particularly around centers of population, industry, and colonial administration. When the colonial powers withdrew, the resulting shortage of resources and skilled personnel resulted in rapidly worsening public health conditions. Malaria, in particular, went out of control.

A series of initiatives and action plans, mostly sponsored by the World Health Organization, assessed and tracked the problem, but the resources were not available to confront it. This was partly due to the realization that contemporary methods were inadequate to control the disease in Africa, so that a large investment was seen as unlikely to succeed, and in part because the problem of HIV/AIDS loomed as a more urgent global threat and took priority. The political situation in sub-Saharan Africa was also perceived as highly unfavorable for success in elimination because countries seemed incapable of supporting a sustained effort to achieve it.

Into this bleak vacuum of health and sustainability stepped an unexpected new player. Microsoft founder Bill Gates (b. 1955), one of the richest men in the world and a man who had made a vocation of highly selective and intelligent philanthropy, together with his wife Melinda Gates (b. 1964), became interested in the problem. In 2007 the United Nations called anew for the eradication of malaria. They were prepared to back efforts with their personal fortunes. (This is a distinct echo of the earlier success of the Rockefeller Foundation, which was able to act relatively unfettered by politics and the constraints of government, for example in eliminating hookworm-associated iron-deficient anemia from the

American South.) Being a private initiative with extraordinary resources, the new Bill and Melinda Gates Foundation has been able to sidestep political obstacles and direct sufficient resources to programs most likely to succeed. Acting in concert with WHO and a new Global Fund for fighting infectious disease set up within the UN/WHO structure and a fund established in the Office of the President in the United States, the Gates initiative took a new look at the assets available and the tools that could be applied. The first priority was a research program to develop new tools to break through the technical obstacles. That now appears to be paying off with the promise of new vaccine alternatives.

Although malaria continues to be a devastating burden for the world's most vulnerable people, and no easy answer is in sight, even if the GSK vaccine succeeds, there is a much greater and more realistic prospect for elimination than ever before. Achieving elimination will have broad social implications. Parents will increasingly expect their children to live long enough to grow up, adults will be able to live their lives with less fear of death or incapacity, workers will be more productive, families will live with less fear of disabling illness that could thrust them into poverty, and economic opportunity will improve for future generations. The future of Africa, especially, will be much brighter, and the world will then be able to tackle the next big, seemingly intractable health problem. Sustainability will become achievable and not a distant prospect.

Malaria and Ecosystem Sustainability

Climate change has aggravated malaria, as observed at the beginning of this section, and this compromises sustainability, both globally and in the regions affected directly. On the other hand, the historical methods used for control of malaria, such as draining wetlands, potentially face limits where they conflict with other priorities in sustainability. Contemporary methods for controlling malaria include trade-offs at the expense of ecosystem integrity and sustainability. Malaria illustrates that some of the control measures relied upon in the past, such as draining wetlands and polluting surface water with oil, would not be acceptable today because of their ecological consequences.

Climate change has raised concerns that malaria may expand out of its current range, which is defined by the home range of mosquito species that act as vectors and recolonize areas that are now free of the disease, such as southern Europe and the United States. (Malaria used to be found as far north as Canada and Scandinavia.)

Fortunately, there are at least three reasons why epidemic (expanded) malaria is unlikely to expand into developed countries, at least much. One is that developed countries have the resources required to continue to change the environment to make it inhospitable to the vector, in a regime where irreversible environmental change is already taking place due to climate change. The second is that malaria has no substantial human reservoir in most developed countries, so for the vector to pick up and disseminate the parasite would require reintroduction of the pathogen from an endemic area. Thousands of people do travel between malaria-free and malaria-endemic areas every day, of course, so malaria is reintroduced constantly but not in sufficient case numbers to propagate an epidemic. The third is that either drug therapy or, in the future, immunization could be deployed quickly to stamp out outbreaks in areas where malaria is reintroduced. Leaving aside issues of case recognition and diagnosis, this implies that effective drugs or a vaccination program can in fact be deployed quickly with adequate adherence to treatment for resistant malaria or cooperation with immunization. An effective malaria vaccine would provide a safer margin of protection.

Malaria is a considerable threat to human health and life in the world and a major cause of loss and instability. Removing malaria would be a major step forward in the sustainability and progress of human society.

Emerging Infections

A major and recurrent threat to sustainability is the emergence of new disease threats, the risk of significant increase in the impact of these diseases in the future, and the consequences of the disease for social cohesion, economic productivity, and diversion of resources to control it. Infectious diseases are most likely to emerge in this way, because of their ability to propagate and also because they are often exquisitely sensitive to ecological and social change and changes in limiting conditions such as evolving antibiotic resistance. Emerging infectious diseases are those that show an unfavorable change in their prevalence, incidence, or control options and that have the potential to present an increasing threat to the status quo and to sustainability going forward. This is not the formal definition, but it could be—at least from the standpoint of sustainability.

Emerging infections have come out of nowhere, gone out of control, and become serious world health hazards. A historical example is that of one of the great scourges of humankind. In 1817 a new disease that had never been described before was first noticed in India. Within a few years, it had spread around the world and afflicted millions of people, many of whom died within hours or a few days by dehydration resulting from diarrhea. Although it can now be treated symptomatically (meaning that the effects, rather than the disease itself, can be treated), outbreaks of the disease still occur and sometimes still kill, two hundred years later. That disease is cholera, which ranks in the same league as plague and tuberculosis for human misery. Cholera is still a global public health problem, but it is now much less of one, with improving sanitation and the introduction of a cheap and highly effective rehydration therapy.

More recently, the "sudden" appearance of a constellation of highly unusual symptoms in a small number of patients in the late 1970s foreshadowed the discovery of a pandemic (global epidemic) now known as HIV/AIDS. In retrospect, the first known case was in 1931 but it was not recognized. This epidemic caused human tragedy, wrenching conflict and discrimination, economic disruption (especially in Africa), and required a major medical and global public health effort to bring it to its current level of still-tenuous control.

Since then numerous other infectious threats have been discovered, a few entirely new pathogens have been identified, and a small number of these threats have shown the capacity to upset sustainable development, global health standards, and economic progress. At the time of this writing, avian influenza (influenza A H1N5), a pathogen that produces a mild disease in chickens and ducks but when conditioned to infect human beings can be lethal, is the leading candidate to propagate a globally disruptive outbreak in the near future; however, it has so far resulted in less than 400 human deaths (as of 2013). Ebola, however, at the time of this writing dominates public attention because the devastating outbreak in west Africa is taking a heart-wrenching human toll in poor countries and has resulted in cases of transmission in Europe and North America.

Emerging infectious diseases are usually understood to be diseases that have newly appeared in a population or that have been known but that are rapidly increasing in incidence or expanding their geographical range. In practice, however, emerging infectious diseases are defined differently in different contexts in the United States and internationally. The original definition, developed by the Institute of Medicine (one of the National Academies of Science in the United States) described them as "diseases of infectious origin whose incidence in humans has increased within the past two decades or threatens to increase in the near future." The Centers for Disease Control, institutionally and in its journal *Emerging Infectious Diseases*, defines emerging infectious diseases as those whose incidence in humans has increased over the previous two decades or threatens to increase in the "near future," or those that are expanding their geographic range. The World Health Organization prefers to refer to "emerging diseases" as those appearing in a population for the first time or those that may have existed previously but are rapidly increasing in incidence or geographic range. On the other hand, a "re-emerging disease" is one that was once controlled but is now on the rise again; often this is because strains resistant to previously effective antibiotics have emerged through selection.

In practice, once an infection has been put on somebody's list of emerging infectious diseases, it rarely comes off, even if its incidence stabilizes or its range becomes global. Malaria remains on the "emerging infectious disease" list at CDC and elsewhere, although it is hard to imagine a disease more firmly established in world history. HIV/AIDS is often described as an emerging infectious disease; but it was first identified in 1983, and its presence is now permanently established worldwide. After years of struggle, HIV/AIDS is declining in incidence in many parts of the world. Can it truly be said to be emerging, or has it well and truly "emerged"? Likewise, the pathogen responsible for Legionnaire's disease, *Legionella pneumophila*, was isolated in 1976; although sporadic outbreaks continue, the means to prevent them by controlling growth of the pathogen in humidifiers and hot water tanks are well known and widely applied. The reality is that there is great variation among institutions and countries regarding which diseases are recognized as emerging, with some extremely dangerous diseases still too rare or geographically restricted to make the cut on most lists. One is left with the impression that what is recognized as an "emerging infectious disease" depends on national priorities, institutional mandates, and perceptions as much as epidemiological criteria. The label of "emerging infectious disease" should therefore be understood to be a way of raising awareness so that it represents a changing threat requiring monitoring and attention (not that it is always new or even necessarily widespread or on the rise in the population).

Emerging infections, by whatever definition, tell us a lot about what we do not know. As noted, there are only about 1,400 pathogens causing disease in human beings out of millions, perhaps billions, of potential species of microorganisms that possibly could do so. Since 1980 only approximately eighty-seven new pathogens have been discovered, despite the many more "immunocompromised" people (those whose immune system does not work properly) who are alive in the world because of survival following cancer diagnosis, advances in treatment of chronic diseases that impair the immune system, transplantation (which requires immunosuppression), and the spread of HIV/AIDS.

Most so-designated emerging diseases have one or more of the following characteristics:

- The disease had never been described before and probably did not previously exist as a disease in human populations, such as the novel coronavirus associated with so-called Middle East Respiratory Syndrome (MERS) presently (2014) circulating in the Middle East.
- The disease had never been described before but probably did previously exist unrecognized in human populations, such as the fungal disease *Cryptococcus gattii*.
- The disease, usually a zoonosis, had been confined to a population or geographic area but has broken out of its historic range to threaten new populations, such as chikungunya, a widespread viral diseases that suddenly exploded in the Caribbean in 2013 and is now (2014) establishing itself in Florida.
- The disease, usually a zoonosis, had been confined to a small population or geographic area but has entered the human population because of increasing

human incursion into its habitat and the new opportunity for exposure, such as several of the hemorrhagic fevers, including Ebola.

- The disease is the direct result of a leap across a species barrier or the newly acquired ability of a disease to be transmitted to humans and human-to-human, such as avian influenza H5N1.
- The disease does not affect people with normal immunity and is newly recognized or encountered with increasing frequency in patients with compromised immune systems, such as opportunistic infections with fungi.
- The disease arises because of resistance to antibiotics that were previously effective in controlling the pathogen, such as many methicillin-resistant *Staphylococcus aureus* (MRSA) and multiple drug-resistant tuberculosis (MDR-TB); this may occur because the drug was over-used in the past or, in the case of TB, because treatment was started but interrupted or the patient was not adherent, and this allowed resistant strains to gain the advantage.
- The disease arises from a new medical procedure, product, or invasive practice, such as a recent (2012) outbreak of fungal meningitis due to contamination of vials of injectable steroids by a compounding pharmacy (one that makes drug preparations to order rather than manufacturing them) or an outbreak of the novel pathogen *Burkholderia cepacia* among patients with cystic fibrosis (patients with disease this have compromised host defenses in the lung but intact immunity) due to a contaminated nasal inhaler (2004).

Table 6.2 summarizes emerging infectious diseases and pathogens that are recognized today (in 2015) and that are under investigation by major institutions. Of the more recently discovered emerging infections since about 1995, only two coronaviruses (SARS virus in 2002 and MERS, the novel SARS-like coronavirus discovered in Saudi Arabia in 2012) appear to be completely new human pathogens, in the sense that they may not have existed prior to recent outbreaks (at least in a form capable of infecting human beings). (The SARS coronavirus began as a zoonosis of civet cats, and MERS is associated with camels.) The rest represent newly discovered infections that have probably been there all along, "old" diseases that are on the move (mostly zoonoses), and diseases arising from resistance to antibiotics.

Ecosystem changes, including climate change, have an effect on emerging infections, as in the example of malaria. Many ecosystem changes are local and reflect changes in habitat. For example, the proliferation of the deer population in North America, which has been estimated now to be higher than at the time of the first European invasion, has brought with it a myriad of problems related to deer ticks, especially a dramatic increase in Lyme disease prevalence, incidence, and geographic range. Previously unknown diseases with characteristics similar to Lyme disease but caused by different pathogens (examples include the

Table 6.2 Emerging infections, as of 2014

	New pathogen	Zoonosis	Habitat incursion	Drug resistance	Discovered recently*
Acanthamoebiasis (encephalitis and eye infection, rare), *Acanthamoeba*					
"Acute encephalitis syndrome" of India, Nepal (probably multiple pathogens, including Japanese encephalitis)					
Australian bat lyssavirus disease (very rare) Australian bat lyssavirus (similar to rabies)					
Babesiosis (a parasitic tickborne disease) *Babesia microti*, atypical					
Bacterial meningitis strain in new subgroup *Neisseria meningitides*					
Bas-Congo hemorrhagic fever Bas-Congo virus (rhabdovirus)					
British Columbia cryptococcosis *Cryptococcus gatii (fungus)*					
Bovine spongiform encephalopathy ("mad cow"), variant Creutzfeld-Jacob disease BSE prion					
Cat scratch disease *Bartonella henselae*					
Chandipura encephalitis Chadipura rhabdovirus					
Chikungunya fever Chikungunya alphavirus					

“C diff” diarrhea *Clostridium defficile*					
Coccidioidomycosis (“Valley fever”) *Coccidiodes immitis*					
Crimean-Congo hemorrhagic fever CCHF bunyavirus					
Cryptococcus gattii					
Escherichia coli O157:H7					
Ebola hemorrhagic fever Ebola viruses (several, Filovirus family)					
Ehrlichiosis (tickborne bacterial disease) *Ehrlichia* spp., esp. *chaffeensis*					
Encephalitozoon (Systemic infection in immunocompromised host) *Encephalitozoon cuniculi* (fungus)					
Eye infection in in immunocompromised host) Encephalitozoon hellem					
Enterococcus, vancomycin-R Enterococcus, usually hospital-acquired					
(Opportunistic infection in AIDS) *Enterocytozoon bieneusi*					
Hantavirus Pulmonary Syndrome Hantavirus, Sin Nombre virus					
Peptic ulcer and gastric cancer risk factor *Helicobacter pylori*					

(continued)

Table 6.2 Continued

	New pathogen	Zoonosis	Habitat incursion	Drug resistance	Discovered recently*
Hendra virus disease (respiratory, very rare) Equine morbilli virus (paramyxovirus)					
Heartland virus disease Heartland virus					
Hepatitis C					
Hepatitis E					
Herpesvirus 6 Encephalitis, Sixth disease Human herpesvirus 6 (two related viruses)					
Influenza A H1N1 "swine flu" (last epidemic in 2009)					
Influenza A H5N1 "avian influenza, bird flu" (threatened pandemic)					
Influenza A H7N9 (newly emergent in humans, bird flu type, 2013)					
Influenza A H6N1* (newly emergent in humans, bird flu type, 2013)					
Influenza A, H10N8 avian (newly emergent in humans), not typical bird flu					
Invasive non-Typhimurium Salmonella (in AIDS patients) *Salmonella typhimurium* strain					
Kaposi's sarcoma Herpesvirus 8					
Human herpesvirus B (encephalitis, very rare)					
HTLV I (human T-cell leukemia)					

HTLV II infection (disease not determined) HTLV II					
Kyasanur Forest Disease ([Indian] monkey fever} Alkhurma hemorrhagic fever virus					
Lyme disease *Borrelia burgdorferi*					
Marburg hemorrhagic fever Marburg virus					
Middle East Respiratory Syndrome MERS-CoV (coronavirus)					
Monkeypox (infects prairie dogs, rare) Monkeypox viruses (several, Filovirus family)					
Nipah encephalitis Nipah virus (paramyxovirus)					
Parvovirus infection, Fifth disease Parvovirus B19					
Polio-like (flaccid paralaysis), probably enterovirus 68, possibly other (California outbreak)					
SARS coronavirus					
"Severe fever with thrombocytopenia" SFTSV virus					
Staphylococcus, methicillin-R (MRSA)					
Staphylococcus, vancomycin-R					
Tuberculosis, multi-drug resistant					
West Nile disease West Nile virus					

* Since 1995.

uncommon diseases babesiosis and ehrlichiosis) were once rare in their restricted distributions but pose a greater and wider threat today.

Incursion of human beings into previously inaccessible or inhospitable ecosystems is one well-recognized process for introduction of disease into human populations, especially many of the less common but often exceedingly nasty "arboviruses" (arthropod-borne viruses) but also pathogens with an animal reservoir similar to human beings (such as primates). Incursion into other species' habitats is also almost certainly the root cause of how and why HIV had the opportunity to jump in the 1920's from an endemic infection among monkeys in Africa, which sporadically infected the few people who handled them, to a human outbreak in the 1930's and to a global pandemic in the 1970s. HIV is the best known example, but numerous other diseases seem have emerged in human populations in much the same way.

Population density plays a major role in some emerging infectious diseases. Antigenic changes in influenza and person-to-person transmission of novel pathogens is much more likely to occur where the human population is dense and infection rates are high. This is the reason so many strains of influenza arise from Asia and relatively few from sparsely populated regions of the world. Humankind probably experienced only sporadic outbreaks of communicable disease in prehistoric times, with transmission of communicable diseases confined to a small clan or tribe or village after a one-off introduction from another or from an animal reservoir. Historically, larger outbreaks of infectious diseases began to occur about the time that people aggregated into larger communities, such as the ancient cities. Most ancient cities (and provincial cities well into the early Middle Ages) had very poor sanitation compared to Rome and its colonies. Population growth in the richer cities of the later Middle Ages kept the risk of infectious disease at a high level regardless of prosperity, particularly for waterborne diseases such as typhoid and in port and trading cities vulnerable to the plague.

Population movement also plays a major role in emerging infectious diseases. Introduction of communicable diseases by travelers was a major (and justified) fear until modern times, although the risk of serious illness introduced by immigration was often exaggerated for political reasons. In earlier days, it took a while to get anywhere, and so travelers more often had symptomatic disease before arrival if they were infected. Once they were at their destination, the latency of the disease had passed and they were symptomatic. Travelers were also more readily distinguished from locals and recognized by all residents with whom they came into contact. In modern times, the facilitation of pathogen movement by modern transportation, the ability to treat symptoms that mask the developing infection, and the anonymity of daily life and protection of confidentiality makes it theoretically much easier for a pathogen to be introduced without warning and complicates ability to trace contact when this occurs. Of course, better monitoring and public health measures also provide greater protection.

Juxtaposition of human beings and animals is a major factor in zoonotic infections, of course, but it also facilitates a leap across the species barrier for pathogens that require additional modification to be infective or pathogenic in human beings. Many pathogens, influenza in particular, exist in animal reservoirs but are not sufficiently adapted to human biology to infect people. There may be different antigen sites or aspects of the immune response. "Passage" through an intermediate species that can serve as a host for the zoonosis, but that is more similar to human beings biologically, can facilitate the transition to a human-capable pathogen. Strains of the pathogen are selected by survival or competitive advantage in the intermediate species and some of these strains that are sufficiently adapted to human biology then go on to infect human beings. This is another reason why so many influenza strains, which tend to be abundant in birds and pigs, seem to arise so often from east Asia, where pigs, birds, and people are constantly in intimate contact both in rural areas and in urban live food markets.

A large number of emerging infections are antibiotic-resistant strains of established pathogens, such as MRSA (methicillin-resistant *Staphylococcus aureus*). It may seem to be stretching the rubric or a metaphor to call these diseases of ecosystem change, but in fact it is literally true. Microorganisms live in a highly competitive micro-environment, in which their survival and proliferation depends on adequate nutrition (in whatever medium it is growing on), its ability to resist adverse conditions, and its ability to extend and expand its own colony against competition. At the level of microorganisms, most of the action is chemical. There must be adequate water, a favorable pH, and sufficient free water; microorganisms vary in their need for free oxygen in the context of infection, but in the ambient environment oxygen is obviously abundant. The medium (food, water, soil, leaf litter, tissue, decaying matter, surfaces, whatever) must provide sufficient energy sources and essential nutrients for the microorganism to survive and grow. The chemical environment in or on the medium may or may not contain inhibitors and antibiotics produced by one set of microorganisms (antibiotics particularly by fungi) to compete against others. One consequence of the emergence of antibiotic resistance is that physicians are increasingly facing cases of serious infections for which the usual antibiotics are less effective and may not work at all. This is considered one of the most urgent problems facing clinical medicine today.

The "One Health" Concept

The idea that human health and animal health are intimately linked to one another seems obvious to anyone who has lived or worked on a farm, in wilderness or on the margins of settled communities, or in close proximity to animals for a long time. The real surprise is that many people in the world who live urban lives apart from other species except for pets do not always grasp that human health and animal health are both part of a whole and inextricably related.

The idea of "One Health" means that humans and animals live in one integrated biological community, affect one another's environment, share exposure to hazards such as microbial pathogens, share potential health risks, give each other diseases, protect one another from diseases, and have both common and comparative biological responses that make biomedical research possible for human and veterinary patients.

Animals are important to human health in many ways, including the following:

Nutrition. Animals are a source of nutrition for human beings, as well as other carnivorous animals. Being particularly rich in protein, meat (including seafood and poultry) is an almost universal part of the human diet. Substitution for meat (usually by lentils in vegan diets) may be possible with special effort or where the culture supports it (as in India), but very low levels of meat consumption are generally reliable indicators of poverty and malnutrition in a population regardless of the type of meat. However, the type of meat matters greatly to sustainability, and the reliance of most rich countries on beef as the preferred protein source has frequently been criticized because of its high cost to produce and as a source of cholesterol and other health risk factors unrelated to infectious risk.

Food safety. Infectious diseases resulting from pathogens such as *Campylobacter jejuni, Salmonella* spp., and a variety of parasites are a risk with undercooked meat, poultry, and seafood due to pathogens that colonize but usually cause no illness in the animal. The challenge of maintaining a reliably safe food supply is one of the fundamental issues in human public health. Another food safety risk involves the production of a toxin by bacteria, with the result that the individual ingesting it experiences a toxic reaction, often severe, as in the case of staphylococcal toxin.

Vectors of diseases. Many infectious diseases that can infect human beings are carried by animals, which may provide a reservoir of infection, a vehicle (vector) by which the pathogen can reach the human host, or both. Diseases that are transferred to human beings from animal reservoirs are called "zoonoses." A familiar example is Lyme disease, which has as its reservoir rodents and as its vector ticks that infest rodents and deer. (Human beings cannot transmit Lyme disease to one another, nor can dogs, which are also frequently infected in endemic areas.) As may be expected, some of these diseases are occupational risks of veterinarians and in agriculture, such as brucellosis and Q fever. At the time of this writing, influenza A H1N1, which affects several species of birds, including and especially chickens, is thought to present a serious risk for a potential pandemic (global epidemic) because it can also be transmitted both from bird to human and human to human.

Incubators of diseases. Some diseases are transformed by passage through animal species in ways that present a risk to human beings. For example, strains of influenza virus that infect birds such as ducks and chickens do not normally affect other species, but they are constantly mutating and on occasion a strain is produced that can infect a pig. When pigs and chickens are raised close together, it is possible for a subset of one of these strains to jump the species barrier and to infect a pig. Within the body of the pig, the virus multiplies but has to face a very different immune system, one that is much more like a human immune system than a bird's. Some of those substrain viruses escape immune suppression and survive in the pig and may spread to other pigs. When human beings come into contact with the infected pig, some of the virus substrains are capable of jumping the second species barrier because they are better adapted than others, and can infect the human being. Some of the virus substrains that can do that adapt so well in the body of the human host that they can be transmitted efficiently from one human being to another. In summary, this is the story of the H1N1 ("swine flu") epidemic of 2009 to 2010, which probably infected at least 22 million people and killed over 280,000 worldwide, although it was only moderately severe as influenza epidemics go. Malaria made the leap from birds to human beings by the vector of mosquitoes at least four times and possibly more; this is a subject of controversy.

Incubators of drug resistance. Antibiotic use in animal feed has clearly been shown to result in resistant strains appearing in human beings who live and work around the animals, and this is thought to contribute to drug resistance in the community. The problem is of such magnitude that in 2014 the US Food and Drug Administration was able to persuade pharmaceutical firms to place voluntary restrictions on the use of antibiotics for certain applications in animal health. For many years, animals have been given antibiotics in relatively small doses to promote their growth; why it does so is not fully understood, because the effect does not relate to the antibacterial properties of the drug. However, this practice is likely to favor selection of resistance strains of bacteria in the animal, and resistant bacteria have been shown to be carried by farm workers who handle animals receiving such treatment. The greatest problem in the emergence of resistance to antibiotics remains the inappropriate use of antibiotics in human patients but treatment of animals probably contributes greatly to the problem.

Sentinels of disease. When animals and human beings share disease risk, the most susceptible animal species is usually more sensitive than a human being. Monitoring the health of animals can identify risks for human beings and can lead to effective prevention. For example, birds of the family Corvidae are highly susceptible to West Nile virus, as are horses. When the West Nile virus was introduced into North America in1999, corvid birds such as crows and jays died abruptly in large numbers, and their mortality accurately mapped the areas of rapid spread of the disease. By comparison, there were relatively few human deaths because the disease is less severe in human beings. Animals that live in close proximity to human beings have also been studied as possible sentinels for cancer risk. Long-nosed dogs exposed to indoor air pollution and passive cigarette smoke do appear to have an increased risk of nasal and sinus cancer, which is more prevalent in dogs than lung cancer. A legendary use of animals as disease sentinels was the use of caged canaries in coal mines as a warning to miners of the accumulation of dangerous levels of carbon monoxide. Birds have a much higher ventilatory and metabolic rate than human beings, and so the gas would affect them much earlier than a person, giving the miners time to escape.

Models of disease. The achievements of biomedical science would not have been possible without experiments using animals, which are still required to explain disease at the level of mechanisms and to predict the effect of treatments and hazards. The modern field of toxicology, and the ability to identify chemicals likely to be toxic to human beings, depend on the knowledge base of studies conducted on animals. Also, veterinary medicine depends on research on animals.

Habitat incursion. A subset of zoonoses represents those species geographically or spatially isolated from human habitation, which are not normally encountered in human communities, and so rarely infect human beings under normal circumstances. The disease risk is restricted to the habitat of the animal host that serves as a reservoir, or of the vector that transmits the disease to humans. However, when people enter these places, they may encounter these diseases as individuals. If they are not readily transmitted from person to person, then the disease may be limited to a single, sporadic case, as in Lyme disease or relapsing fever (borreliosis). If the pathogen can be transmitted from person to person, a propagated outbreak of disease may occur. This is thought to have been the origin of HIV/AIDS, which is supposed to have jumped the species barrier from monkey species to human beings in the African bush and was then amplified enormously by human-to-human transmission through sexual intercourse, unsafe practices, and contact with body fluids. Periodic outbreaks of viral hemorrhagic fevers (of which there are several, Ebola being the most notorious) are similar in this respect.

Waste production. Animal waste presents a serious problem in agriculture, requiring disposal of large quantities of manure and other waste, such as dead animals and feathers. Manure, especially, may cause water pollution and eutrophication (adverse biological overgrowth because of a surplus of nutrients, specifically nitrogen and phosphorus) in waterways, and may also contain pathogens, antibiotics and other veterinary medications, and trace elements (arsenic is commonly added to chicken feed in small amounts). Concentration of animals in "confined area feeding operations" (CAFOs) is a particularly severe problem, and allowable discharge into waterways is highly regulated and very controversial.

Health protection. Animals protect human health in many ways, most of which do not involve biology but instead reflect behavioral responses. Assistance dogs for the visually impaired (previously called "seeing-eye dogs") is an obvious example. Pets are associated with lower blood pressure and show a calming influence in many studies, particularly in people who are ill, incapacitated, elderly, depressed, or lonely. Some animals, because of their biological and sensory characteristics, even have the capacity to identify disease: dogs have been trained to identify bladder cancer from urine specimens or to detect melanoma, and the African pouched rat is capable of identifying tuberculosis bacteria by smell. In return, human beings should and often do take good care of the health of animals in their care, but the reciprocity is far from complete.

In order to raise awareness that humans and animals live together in one biological universe and affect one another's health, leaders in medicine, public health, veterinary medicine, and biological sciences have begun to promote the concept of One Health within the health professions and to the public. The practical application of One Health has been to open a new dialogue between practitioners and scientists in human and veterinary medicine, with several organizations formed to advance the idea. Most notable of these is the One Health Initiative, which is a global repository of information on the relationship between animal and human health; the One Health Commission, an interorganizational forum for a dialogue between human and veterinary physicians; academic centers; and a growing number of websites and local groups, such as the One Health Academy in Washington, DC. (One Health should not be confused with an unrelated business partnership for health care that calls itself OneHealth, with no space between the words.)

The Epidemiological Triad

The relationship between the pathogen, the host (person infected) and the environmental link between the two, which seems so obvious today, is a powerful concept that was at the heart of one of the most important disputes in scientific history.

Infectious diseases involve an opportunity for the pathogen to reach the person (or animal or plant), followed by propagation in a host that cannot control it. Unlike a chemical exposure, the cause of the disease (here, the pathogen) is self-replicating and follows its own biological pathway once it is introduced and the process gets started. This leads first to colonization, and if the pathogen cannot be quickly cleared before it gets established in the body, it leads to infection. If that infection is not initially controlled, the process leads to an infectious disease with identifiable clinical characteristics. Key to the process, then, is the agent that causes disease: the means by which a person gets infected, and the capacity of the person to control the pathogen once it reaches the body.

This essential framework for studying infectious disease has traditionally been called the "epidemiological triad," reflecting the early years of epidemiology when it primarily studied infectious disease and had a close relationship with microbiology. The epidemiological triad is also often called the "public health triad," but the latter term is also used for the triad of "human, animal, disease" as used in the One Health model and in studies of animal and human health risk. To avoid confusion, the traditional term "epidemiological triad" is used throughout this book.

The triad consists of three elements that must be present, as a matter of necessity, before disease can result. Lacking any one element, the disease cannot be present. The three elements are

- an infectious agent (the pathogen)
- an environment that provides a mechanism for exposure, such as a medium (for example, water) or a vector (for example, a disease-bearing mosquito)
- a susceptible host, in the biological sense of susceptibility (such as a person who is not immune to the disease)

This concept states that for a disease to develop (in this case an infectious disease) there must be three elements present. If all three are present, the disease may occur, and it is the epidemiologist's job to identify an outbreak and to understand why it occurred. After removal of any one element, disease cannot occur. In the absence of more than one element, the disease is unlikely to recur even if reintroduced under otherwise favorable conditions. This three-strategy template for disease control was introduced in chapter 6 as it applies to pollution and will be developed in this section as a general model for infectious disease, to which it was first applied.

The mechanism for exposure in this model could be a medium such as contaminated water, for viral agents causing diarrhea, or a "vector," such as an insect that carries disease. The triad has been the framework for all success in eliminating or controlling communicable infectious disease, and it defines the explanation for failure when this could not be achieved.

The epidemiological triad not only describes causes of a disease but just as importantly defines the approaches to controlling it. The epidemiologist or public health practitioner then determines which elements can be removed to stop the outbreak.

The concept of the epidemiological triad has ancient roots, but the three elements of the triad first began appearing together as a coherent formulation in the teachings of Max Joseph von Pettenkofer (b. 1818–d. 1901, Bavaria). Von Pettenkofer was a chemist, pharmacist, and literary figure who became a hugely influential but tragic pioneer of public health (called "hygiene" in those days). Ironically, the champion of the key ideas of the epidemiological triad did not initially accept that the pathogen in infectious diseases was biological in nature. Von Pettenkofer believed that a disease would not establish itself unless the host was susceptible (for example, malnourished) and an agent was introduced by the appropriate medium. To demonstrate his point, he intentionally swallowed a culture of cholera and, for whatever reason (probably a robust level of stomach acid), did not become ill. He was lucky to survive but unlucky in drawing the wrong conclusions. Unfortunately, due to this dramatic but uncontrolled and inadequate experiment, he was confirmed in his belief, based on many false assumptions, that cholera was caused by emanations (as in the "miasma" theory of disease) from putrifying material in soil, not by proliferating germs, and that its transmission was airborne.

Von Pettenkofer was not completely wrong in his thinking, because cholera does require an environmental means of transmission and, as he inadvertently demonstrated, not everyone is susceptible all the time. He argued in opposition to Robert Koch (b. 1843–d. 1910), one of the two founders of modern medical microbiology (together with Louis Pasteur [b. 1822–d. 1895]) and in the end Koch prevailed easily. There was a political dimension to the argument, because Koch, who held important public positions in the national German government, believed that only exposure to the agent was really important and advocated state-sponsored centralized control measures directed at interrupting person-to-person and eliminating or eradicating the pathogen, such as quarantine and organized disinfection. Von Pettenkoffer, powerful in his own right in the public service of Bavaria, one of Germany's wealthiest and most important states, emphasized the environment and thought that unique local conditions were just as important in causing disease. He was committed to decentralized local control in management of public health and the overriding importance of environmental measures to control exposure. Despite his previous groundbreaking work on sanitation, prevention of lead poisoning, and control of water pollution, he was largely discredited by the end of his life and killed himself while in a deep depression. Von Pettenkoffer's broader notions of the multifactorial nature of disease later came to be recognized in early textbooks of epidemiology and public health.

The legacy of the dispute between the two men has been profound. Koch was correct that the biology of the agent and exposure of the human host is critical in causing infectious disease and his influence can be felt, as is Pasteur's in clinical medicine today. However, von Pettenkoffer's somewhat confused ideas turned out in the end to be more useful and closer to the truth in terms of how transmission occurs and how people can be protected, and so plays a larger role, to this day, in public health. His adherence to the miasma theory of disease, long after Koch and Pasteur provide the existence of germs, can be viewed as perfectly reasonable when applied to chemical pollution, just not to infectious diseases.

The epidemiological triad is of foundational importance in thinking through infectious disease, where it was first devised. Von Pettenkoffer redeemed himself in the end.

7

Socially Mediated Issues

UNSUSTAINABLE SITUATIONS MAY result in health outcomes that are mediated indirectly by social, economic, and shared behavioral responses. These issues define a class of sustainability-health interactions in which natural or artificial ecosystems are destabilized, and this has indirect consequences that are mediated by mechanisms other than direct toxicity or infectious disease risk. These mechanisms are social and usually economic.

The chapter also provides useful a useful framework for thinking about such issues, which will be called "DPSEEA+C," the acronym for its essential elements. How this broad area fits in with sustainability values is presented in table 7.1.

Social Systems Exist to Manage Risk

Social systems exist to mitigate risk: that is why human beings live in communities in the first place. People living in groups can protect themselves better from enemies or predators, share resources, and help one another beyond the immediate kinship group. Through technology, trade, and political organization, organized communities under stress can access help that will offset the local impact of ecological problems and diffuse the impact over a much larger area or population. A society in which people consider themselves to be one big family is more likely to show resilience in a disaster or prolonged stress than one in which each individual models his or her behavior on self-reliant individualism, all other things being equal. This is the fundamental idea behind insurance, extended family networks of assistance (a major safety net in traditional and immigrant societies), and disaster assistance. Therefore this class of problem often depends importantly on the level of resilience a society demonstrates under stress.

Social Behavior and Sustainability

The conventional way of looking at the "social implications" of unsustainable ecosystem change emphasizes 1) economic responses, 2) social adjustments by an

Table 7.1 Table of sustainability values: Social mediation

	Sustainability Value	Sustainability Element
I.	Long-term continuity	Social responses to sustainability issues may distort prices or access to needed resources. Reaction to these responses (such as political backlash) may impede and even stop progress toward sustainability and indirectly affect health.
II.	Do no/minimal harm	Social responses to sustainability issues that are inadequate or dysfunctional may aggravate problems of poverty and equity.
III.	Conservation of resources	Social responses to sustainability issues may be motivating for conservation of resources but may also promote hoarding or efforts to corner markets.
IV.	Preserve social structures	Social responses to sustainability issues often promote cohesion and organization in communities, but divisions over resources can also be destructive.
V.	Maintain health, quality of life	Social values generally favor maintaining health and the quality of life, but questions of equity, who benefits and who pays the initial cost of investment, may be difficult to resolve fairly.
VI.	Performance optimization	Social priorities must be balanced in the optimization or maximization of economic, social, and environmental performance, because if they are not, consensus in society will be unsustainable and both sustainability and health goals cannot be reached.
VII.	Avoid catastrophic disruption	Social stability and strong institutions confer greater resilience of society to outside stress and a lower risk of catastrophic disruption or compromise of health.
VIII.	Compliant with regulation	Social pressures that arise from concern for the future, deep understanding of sustainability problems, and unbiased decision making are more likely to value compliance with appropriate regulation, coherent policies, and best practices for sustainability. Commitment to sustainability also encourages compliance because the reasons behind the regulations are clear. Social pressures arising from misapprehension, misunderstanding, or competing priorities interfere with compliance with regulation and best practices of sustainability.

(continued)

Table 7.1 Continued

	Sustainability Value	Sustainability Element
IX.	Stewardship	Social institutions and cultural attitudes may foster stewardship and a responsibility that goes beyond ethics in dealing with other people to include protection of health, other species, life and some shared notion of nature.
X.	Momentum	Social pressure and cultural norms that favor sustainability and health protection are capable in the short term of perpetuating progress under its own momentum and stability.

individual or family, 3) collective adjustment or action by the community, and 4) individual behavioral responses, all of which collectively over time (and when experienced on a community-wide scale) may lead to 5) cultural responses. This view conforms to a framework familiar to readers and students and to the disciplines of economics, family dynamics and psychology, sociology, political science, psychology, and anthropology. Health consequences are most likely to be mediated by economic disruption, poverty, unemployment, and lifestyle models and adaptations. However, readers should not be misled into thinking that these classes of response are discrete and separable.

The social sciences as they are now defined are disciplines of convenience and historical tradition. They are compartmentalized for the convenience of teachers and to isolate problems for feasibility of analysis. Human behavior is not neatly compartmentalized. It is all one glorious mess of interconnected stimulus, response, emotion, and avoidance behavior (on the individual level, studied by psychology and family dynamics) that manifests itself in group action (in sociology and cultural anthropology) and structured communities and institutions (political science and economics) through shared understanding, values, and perceptions.

The consequences of unsustainable practices for health in these issues are often mediated by externalization of costs (e.g., through unmitigated pollution and ecosystem-related health effects) and they may be mediated by distortions in pricing, particularly in low-income societies, where pricing of food resources may affect food access and nutrition and of fuel may affect access to transportation and therefore opportunity.

Social responses, in the broadest of generalizations, reflect human psychological tendencies that favor short- over long-term decisions. People place greater priority on exigencies of the moment and expediency and therefore frequently compromise conservation of resources for the future. For example, it is quite possible for people to feel genuinely confused about recycling and packaging when they hear conflicting information about paper and plastic and do not have sufficient information about life cycles to make an informed choice; in such a situation, they are likely to do what is most convenient, an action that may lead to a solid waste problem and unsustainable landfill. It is also a natural tendency to extrapolate thinking on a small scale to a larger scale where it is inappropriate. For example, what works for local government in zoning ordinances, investment in public works, and taxation does not apply well to entire countries or to global finance and trade. Communities import goods and draw resources from far beyond their own boundaries, or as William Rees (b. 1943) has taught, communities have a "footprint" over a much larger area than their political boundaries.

Commodities, like a papaya or a drop-forged wrench may not be simply what they seem to be. For example, water is scarce in most places, abundant in some. There is often resistance to exporting water, but one can view agricultural production for export to arid countries as a way of achieving the same end, effectively exporting water. In that same sense, importing a manufactured good from a country dependent on coal-fired plants with high carbon-emissions outsources responsibility for global share of carbon emissions.

Social responses and concerns related to sustainability may place great pressure on social and political structures. Most obviously concern over local unemployment may put pressure on politicians to back off from supporting enforcement by regulatory agencies. At the moment, it is unclear how the state of West Virginia, with its long and lamentable history of abject dependence on coal, could adapt to a future without it. If that is to come to pass, it seems inevitable that some massive program of economic diversification, expansion of opportunity, and infrastructure development may be required. Similarly, but in less dire straits, the state of Iowa has pinned its hopes on corn ethanol and may find its commitment unwise in the long term but difficult to reverse. Just as in West Virginia, choices of the past lead to positions taken today that are locking the state into untenable policy choices that virtually gurantee disruptive adjustments in a few years.

Environmental justice issues (see chapter 11) and inequity, as in the "NIMBY" ("not in my backyard") phenomenon, are associated with perceptions of unequal health risks and resulting health disparities. Here there are two levels to the problem, that of actual exposure to environmental hazard and that of the social effects of unequal treatment and discrimination.

Resources

Many issues involving socially mediated health effects arise from sustainability problems that have to do with access to resources. Some basic principles of resource economics are helpful in understanding their relevance to sustainability. The effects on health of economic issues are indirect but can be quite substantial, since they affect food supply, access to resources and therefore priorities, and allocation of wealth and therefore life choices.

Economic and social responses that are dysfunctional in reaction to a threat to sustainability may result in distortions in pricing, limitations on the availability of necessities (such as food), and aggravation of poverty and unemployment, which have indirect effects on health. For example, anticipated shortages of food or commodities may result in higher prices, reallocation of resources (for example, away from education), and privation for poorer families in the community. The response to sustainability threats (or their denial) may themselves be counterproductive and even result in harm to the environment and health. The response to perceived oil shortages (oil has many problems but a supply shortage is not one of them) and air quality issues led to corn ethanol, which appears certain to be non-viable as a solution in the long term, because of its very high cost of production, consumption of conventional oil, and competition for food uses of corn.

The traditionally critical distinction in resource economics is that between "non-renewable" and "renewable" resources, which would seem to be obvious and unambiguous. However, the distinction is much less clear than may first appear.

"Nonrenewable resources" are those for which there is assumed to be a finite supply that is subject to extraction, the risk of shortage, exhaustion of sources, and subsequent scarcity. The most relevant industry is mining and the most common examples might be minerals. Nonrenewable resources may also be classified as either "augmentable," meaning that it is possible to find other sources and so to augment the supply or "non-augmentable" (not expandable), when supply is fixed.

Renewable resources are those that grow back, are regenerated, are available in continuous supply, and exist in a form from which a sustainable "harvest" might be taken without disturbing future availability. They are not exhausted or depleted with one use. Forest-derived resources, fisheries, and water are common examples of renewable resources. Renewable energy resources, such as solar energy technologies, geothermal, and biomass are discussed together, in chapter 12.

In actual practice, the distinction between nonrenewable and renewable resources is anything but distinct in operational terms. Some nonrenewable resources are effectively unlimited and act as if they were renewable, such as oil. Limestone used in making cement, for another example, is a mineral and so by definition nonrenewable but is so abundant worldwide that almost every area of every country has an indigenous industry to make concrete for construction, locally sourced because imports of cement are usually impractical due to weight. Iron, the ore of which was at one time thought to be at risk of exhaustion, is extensively recycled in the form of steel and is increasingly often replaced by lighter-weight aluminum and strong composites. On the other hand, presumed "renewable" resources often conform much more closely to the traditional nonrenewable model as their capacity to regenerate is impaired. For example, worldwide there is a diminishing yield from commercial fishing that increasingly seems to herald a global crisis, especially for high-value fish such as tuna. Extraction under these circumstances comes to resemble mining the ocean more than harvesting a renewable crop. A global shortage of fish harvesting has dire consequences for food supply for poor countries reliant on fishing, in terms of nutritional supply, protein intake, and sustainable food prices. This will be outlined in detail in the next section.

Renewable Resources

Resources that can be tapped indefinitely or that can be replenished through natural growth or replacement are called "renewable resources." As noted above, these resources are then "harvested" at levels that do not deplete the resource and allow for regeneration. Such resources include fish and forest products that are harvested for human use, as well as all agricultural commodities. Resources that are finite and could, in theory, be exhausted are called "nonrenewable resources." Some energy sources that are continually regenerated through natural processes (such as sunlight, geothermal wells, and wind) are also renewable resources, but their issues are a little different and are discussed in chapter 10.

Sustainable use of renewable resources therefore consists of at least two elements: the renewability of the resource at the level of "harvest" must be assured, and the consequences of resource cultivation or harvest must be negligible or well within the capacity of the ecosystem to accommodate. Each is almost never strictly or entirely true, except for renewable energy sources (which is another reason why they are discussed in a different chapter). Mining, agricultural, and fishing practices result in collateral damage to the environment in addition to extracting resources.

The definition of "renewable" paradoxically rests on certain assumptions and is only valid for a relevant time period. For example, agriculture may yield crops year after year, but the consequences of cultivating and harvesting the renewable resource may also be so severe that yields are diminished and extraction becomes unsustainable—and so the resource becomes "renewable" in name only. If the soil is depleted and nutrients are not replaced, the renewable resource will depend on a nonrenewable resource (the soil), and yield cannot be sustained. If nutrients are replaced, the problem of agricultural run-off and water quality may compromise sustainability. The agricultural yield may be "renewable" but it will not be sustainable. Intensive agriculture with irrigation and over-application of fertilizer, resulting in nitrate run-off and eutrophication (overgrowth of water organisms) downriver is an important and common example of how renewability and sustainability are separate, disconnected issues.

Nonrenewable Resources

The simplest system in which social responses mediate the health implications of sustainability issues is in response to shortages of a critical nonrenewable resource. In this situation, a finite resource that plays an important and even essential role in society becomes exhausted, increasingly scarce, or unavailable. The response is a rise in prices, which constrains availability and use. Much of the social response is mediated by the economic principles derived from the relationship between supply and demand. At least some of the response is magnified by perceptions of shortage and security of supply and some at least is mitigated by technological innovation and substitution of materials. Shortages that arise in issues of this type lead to secondary distribution and utilization issues that are often the clearest examples of the problem of the "commons" (see chapter 4).

One might expect the economically mediated health consequences of renewable resource shortages to be most severe when they first occur, but this is rarely the case. It is important to understand the reasons for appreciating the dynamics of sustainability and to understand why the expected effects on health are usually small or absent.

One would expect that a resource in limited supply (for example, a mineral ore) would first be exploited where it was found, and then as those deposits ran out or use increased, sought for elsewhere with exploitation of the most convenient deposits. Then one would assume that the less-rich deposits would be exploited (with increasing difficulty and therefore investment), until they became scarce and the deposits were exhausted, at which time there might be economic disruption that would compromise any enterprise or community that depended on it, leading to unemployment and poverty. In practice, there are many mining towns (such as Virginia City, Nevada, which was a silver mining boom town) that became depopulated and were marginalized after a resource ran out or drought dried up water resources. However, in general these were local issues, and effects were not seen on a more general societal level. (The Dust Bowl era of the 1920s and 1930s was a conspicuous exception. Another was the economic depression in Chile after the monopoly on nitrate harvesting was broken by industrial nitrogen fixation, as described in chapter 4.)

In the 1970s there was considerable concern expressed at the highest levels of scholarship, government, and the environmental movement over the imminent depletion of key mineral resources, such as oil, copper, uranium, cobalt, manganese, chromium, and platinum. (The last four have special strategic significance because of military applications.) In the subsequent years, the anticipated critical shortages simply did not occur. The rather obscure example of helium is a current example of an important resource that is essentially nonaugmentable and where a shortage in supply is definitely occurring without a viable substitution. Helium, an element, is in fixed, short supply but is indispensable for certain uses.

It is separated (not really produced) as a byproduct of oil and gas production, and so its supply is subject to artificial shortages due to policy incentives completely unrelated to its availability and that incentivize wasting the resource. There are surprisingly few other examples.

The rarity of situations in which critical resources are actually demonstrably exhausted without alternatives at hand has led some economists to assume that the supply of natural resources should be treated as if they were infinite or "perpetual" (defined as inexhaustible within a time scale relevant to human beings, in practice a question of centuries). This is perceived by many as a sort of dodge that ignores limits to the planetary content of any resource, apart from ease of availability. It may be preferable to differentiate resources commonly thought of as simply nonrenewable by assigning them to certain categories depending on the potential for augmentation and substitution. Clearly helium, which is in short supply and not easily replaced, is in a different category than iron, which is not only not in short supply, but owing largely to weight-related advantages has been replaced for many applications by aluminum and other even more abundant materials.

There are several reasons why the supply and availability of most renewable resources deviate from their predicted behavior. One is that, as in the case of oil, a resource may be much more augmentable than traditionally assumed, as shown in chapter 10. Quite differently, rich supplies of bauxite—the best ore from which to extract aluminum—are being depleted and are already essentially depleted in North America, but it is also true that aluminum is among the most common elements on the surface of the earth (6.1 percent by mole), and there are other, plentiful sources of aluminum (alunite), although the cost of extraction is higher.

A second reason for deviation of resource supply from predicted behavior is that many critical nonrenewable resources can be recycled with great efficiency, such as iron (as steel), copper, and aluminum. Because recycling usually begins with a richer source material than the mined resource, it is generally a much more efficient proposition than primary extraction, making it the preferred alternative in an industrial economy and taking pressure off the primary supply. (That iron and other metal prices are still high mostly reflects currently high demand for from China, which is still building its first generation of modern infrastructure, and to a lesser extent India.)

Changes in the use of resources precede actual depletion by a wide margin. A key reason is that as a resource becomes increasingly scarce as exploitation proceeds, its price increases much earlier than any equilibrium is achieved between supply and demand. (This is predicted by "Hotelling's rule," which says that a premium, or "rent," adds to the price as the resource becomes scarce, increasing at a rate much higher than the increase in cost of extracting the resource.) The increasing price favors conservation of the resource but also encourages substitution, so that a commodity price that becomes unreasonable pushes demand toward alternatives. For example, at a time when copper prices were high, electrical wiring made of aluminum was introduced and used in many homes built before 1979. (In practice, it turned out to be inferior and a fire hazard.)

Copper illustrates other aspects of resource economics. The world price of copper subsequently fell considerably. Copper was again used for wiring, but this time demand for copper was far below what it was in the 1970s and remained so for many years. Copper's use in telecommunications, has been displaced by fiber optic technology, which is technically far superior. Copper is still in high demand, however, because the expanding economies and new infrastructure of China and India created demand over the last decade and this supports the price. When the price of copper stays high, as it is at the time of this writing, one social consequence is crime and associated vandalism, because thieves steal copper wiring to sell it to recyclers, who are often in on the crime. The thieves also cause local pollution with toxics, because in order to sell the wire to recyclers criminals burn insulation off the copper wire and in so doing generate highly toxic combustion products, including dioxins (which are among the POPs, see chapter 5), usually in a way that contaminates the ground and exposes local residents downwind. It is now illegal in many jurisdictions for metal recycling dealers to buy burned copper.

Advances in technology have fostered substitution much earlier in the cycle of resource exploitation than was possible in the past. This is due to the development of silicon chip technology (with its generations of change in semiconductor innovation), materials science (particularly composites with characteristics often superior to metal structures), nanomaterials (including conductive organic materials and polymers), and information technology (efficiencies derived from which pervade modern technology). At the same time, advances in manufacturing (such as additive manufacturing) and certainly design are reducing wastage and the content of scarce materials in most products.

Embargoes and export restrictions occasionally lead to artificial resource shortages. The 1973 OPEC (Organization of Petroleum Exporting Countries) oil embargo was the most consequential and dramatic example of this rare strategy in modern times. The embargoes against South Africa during the apartheid era led to development of that country's synthetic oil industry, as inefficient and expensive as it was. However, in the long term, embargoes work poorly for the same reasons that resource shortages are usually circumvented. For example, it may be argued in response that an important exception to the previous paragraph, on technological solutions, is the content of "rare earth" elements in catalysts and in modern consumer electronics, for which created shortages have been predicted. (The "rare earths" are seventeen elements with generally similar chemical properties that tend to occur in deposits together and with uranium.) This issue recently (in 2011) received extensive media coverage as a concern for the consumer electronics industry. China has had a true monopoly (97 percent market share) on bulk supply of these elements and has imposed export restrictions that both drive up the price and conserve the resource for its own industry. However, "rare earths" are not, in fact, particularly rare (some are more common than copper) and are found in large deposits outside China. Now that there is an incentive to search for them, previously uneconomical extraction and production has been restarted in the United States, viable deposits have been found in other countries producing high-value products (even in Japan, where they constitute one of the islands' few significant mineral resources) and Israel, and historically smaller producers such as Malaysia and Australia have ramped up production. There was even a price drop in lanthanum (used in certain types of rechargeable batteries) in 2012 because of expanding supply.

Another "artificial" cause of resource shortages has been inaccessibility due to war or civil unrest (such as tin during the Second World War and cobalt during the civil war in Congo in the 1960s). Natural latex rubber plants and seeds were famously embargoed in the nineteenth century by Amazonian planters in order to maintain a Brazilian monopoly. The cartel was eventually broken by the United Kingdom in 1876 by smuggling rubber plants from Brazil to Malaya, which was under British control, and developing plantations there. When natural rubber was again restricted in supply, this time by the Japanese occupation of Malaya during the Second World War, the problem was ultimately solved by the invention of synthetic rubber and plastics, which worked well enough to reduce the world market for natural rubber. In recent years, the occupational health problem of latex allergy has further reduced the market for natural rubber in medical products in health care, one of its primary uses. Natural rubber now remains indispensable or preferred for some applications but has a much smaller market and range of uses.

Where there is interference with market supply and demand, price distortions may also accelerate depletion of resources. Even the helium problem referred to above has a social cause: the US Congress has ordered the privatization by sale of stores from the US National Helium Reserve, a federal facility that stores helium derived as a byproduct from natural gas production, without a guaranteed price. This has created a glut in the helium market and no mechanism for factoring future limits on supply or externalized costs into the price. This has made helium artificially cheap despite impending shortages and has perversely discouraged recycling and conservation.

All in all, the relationship between nonrenewable resource exploitation and health is complicated, not always as strong as one might expect and usually local in impact.

A Case Study: Resource Collapse

One unfortunate example of the interaction of unsustainable practices, ecosystem change, and health risk mediated by social effects is the collapse of the

Atlantic cod fishery in 1992 due to massive overfishing. The cod fishery of the Grand Banks lies off the coast of Newfoundland, where cod has been fished for at least 400 years. The drivers of overfishing were demand for the fish, stimulated by increasing awareness of health benefits and an increased capacity for catch due to technological change. The result was a catastrophic failure of a "renewable" resource. It is also an example of what happens politically and economically to a place with very little resilience.

For centuries, salted cod had been in high demand in Europe, particularly in England and southern Europe, where it was generally known as "bacalá." Traditionally, salt is used to preserve the cod, which allowed transport on long sea voyages back to Europe. Cod is a staple food in Spain (*bacalao*), Portugal (*bacalhau*), Italy (*bacala*), and England, where it is usually consumed as "fish and chips." Its popularity over the centuries derived from its culinary appeal and because unlike other fish it was easy to preserve, store, transport, and trade.

The Atlantic cod consists of three individual species of the genus *Gadus*, which are so similar that no distinction is usually made. Cod is a voracious predator fish that consumes many invertebrates and smaller fish species. Its prey includes capelin, a keystone species that is highly important as prey for other predator fish and is sensitive to ecological change; however, capelin of no commercial value itself. Cod is also a common name for several other more distantly related species of fish, not all of which belong to the genus *Gadus*. Common usage includes species in the order Perciformes (such as rock cod), some of which are also called grouper. Closely related species, such as pollock, whiting, and haddock, are often marketed as cod. Younger individuals of these species are often called "scrod." When cooked, they all share a tasty, flaky white flesh that goes well with other foods and is highly nutritious without excessively accumulating contaminants such as mercury. In short, cod is a desirable fish.

The Canadian province of Newfoundland has a long, complicated, and unusual history, quite different from that of the rest of the country. Newfoundland has only been a part of Canada since 1949, having previously been first an informal settlement occupied seasonally by European fishermen (principally from the French Basque country), a French colony, a British colony, a "dominion" (an autonomous state of the Commonwealth), and a failed state governed by direct rule from London. For most of its history, Newfoundland was poor, isolated, and weakly governed. The waters of the Grand Banks were for centuries the richest fishing grounds in the world but the money went elsewhere and Newfoundland itself has never been rich.

By the 1950s, there was intense and growing pressure to increase the harvest of cod in the North Atlantic. The prospects for diversifying Newfoundland's highly cod-dependent economy were essentially negligible. (Oil had been discovered, but offshore extraction was not yet viable.) The province was highly dependent on transfer payments from the federal government in Ottawa. Fishermen from other countries, especially Spain, were fishing in the same waters more frequently than in the past and were harvesting amounts that suddenly dwarfed the take by Newfoundland fishermen. (The gender-neutral form "fisher" is almost never used.)

A shift in food preference in Europe and North America, motivated in part by perceptions of heart health, also made cod and other fish more popular than in the recent past, so demand increased. Iceland, the economy of which was then dependent on fishing, reserved the waters around their island for its own fishermen and by doing so came into serious conflict with other countries, including Britain, in what came to be called the "Cod Wars." In the 1980s Spain and Portugal were admitted to the European Economic Community (EEC) with the unique proviso that for a period of years, they would not fish in the territorial waters of current member states. Barred from competing with their near neighbors for a period of time, the increasingly efficient fishing fleet of the two Iberian countries expanded their reach elsewhere. Since Spanish and Portuguese fishermen had already been fishing in the waters off Newfoundland for centuries, this was the obvious and preferred destination until they could go back to Icelandic waters.

For centuries, fishing technology had barely changed, with nets and lines preferentially harvesting larger adult fish and "inshore" fishermen searching by trial and error for places in the Grand Banks and George Bank (south of Nova Scotia) and other North Atlantic regions where cod was abundant and mature. Traditional technology could not support fishing below a certain depth, which protected cod and their prey fish in the deep ocean. These methods and traditional practices left alone the capelin and most of the small fry on which cod depended.

In the 1960s, fishing technology changed dramatically and became much more efficient. Huge increases in catch, aided by draggers pulling gill nets and bottom-trawling that scoured the ocean floor, depleted the breeding stock. These methods no longer protected smaller fish (both smaller cod fry and prey fish, including crapelin, would be caught in the gill nets). Large trawlers equipped with modern navigation devices could stay at sea much longer and more safely. Sonar could locate concentrations of fish reliably, and nets could harvest entire schools of fish with a wide sweep, picking up other species as well as cod, including prey species for the cod.

Fishery science, however, did not keep up with the technology for harvesting fish. Estimations of fish population size and health remained based on catch data, with limited direct observation in the ocean. A general trend toward smaller fish size was noted. But due to the fact that there was often variation in catch and fish maturity by ocean "season" and conditions, and because up and down changes in the fish population followed by spontaneous recovery or return to average levels are not unknown as a result of natural causes, and also because the catch remained high for many years, the reassuring conclusion was reached that the cod population could sustain intense harvesting indefinitely. The situation was further confused in the 1960s when an ocean cooling trend brought abundant fish stocks into the Grand Banks and Georges Bank areas, making it appear, falsely, that the fish populations were healthy. In reality, the scientific data available to manage the fishery were contradictory, imprecise, and controversial, and so could be conveniently ignored.

The peak harvest of cod from the North Atlantic came in 1968, after which the yield fell, and there was a dramatic crash around 1972. In 1977 exercising its right under a newly ratified UN Law of the Sea Convention, Canada declared the fishing grounds to be an "exclusive fishing zone" and actually dispatched naval vessels on several occasions to chase out trawlers from other countries. Intense fishing continued by the Canadian industry, however, on the assumption that this was a temporary population contraction, as had happened many times before. Canadian fishermen increased their yield and were able to sustain it for about ten years at record levels, putting unyielding pressure on the cod population. Finally around 1990 the yield collapsed almost completely; and compared to previous years almost no cod were being caught. Fish stocks dropped to about 1 percent of historic levels, wiping out an annual yield worth $15 billion.

In 1992 Fisheries Canada imposed a moratorium on commercial cod fishing that was supposed to last for two years. However, the recovery of the cod population has hardly been significant. The slow recovery has been thought to be due largely to the depletion of prey species and possibly also to genetic drift (because larger and more fit young adult cod were preferentially removed from the breeding pool). Two species of cod remain endangered. A slightly encouraging study in 2011 suggested that cod stocks were approaching 35 percent of pre-collapse levels, measured as biomass, but since pre-collapse levels were already greatly diminished from historic levels, let alone highs, this would amount to only about 10 percent of sustainable biomass. Thus, it appears that cod stocks are unlikely to come back to sustainable levels for at least another generation.

In the wake of the moratorium there were bitter recriminations and blame. Fisheries Canada and provincial agencies were blamed for mismanaging the fishery and for imposing the moratorium, which many fishermen still did not believe was required.

Cod generally migrate, although some small populations remain in bays and estuaries year around. Cod are more likely to recover if there are areas of refuge where they can reproduce and recover. However, the American fishery, to the immediate south, was intensively harvesting at the same time and putting pressure on some of the same migrating populations. The American harvest sustained high levels for longer and peaked in the early 1990s and then collapsed, but more slowly. In 2013 a similar but less strict moratorium was proposed for cod fishing in the Gulf of Maine and the American area of Georges Bank, off Massachusetts. Cod fisherman in Gloucester, Massachusetts, and surrounding towns are especially affected.

The collapse of this resource put an estimated 40,000 people out of work in Newfoundland, historically Canada's poorest province. Over 400 communities were directly affected, most of them geographically isolated "outports" that were already economically marginal. Unemployment soared, and special welfare programs were created to deal with the economic emergency (the Northern Cod Adjustment and Recovery Program, followed by the Atlantic Groundfish Strategy). On one level, the traditional virtues of self-reliance, self-confidence, and income opportunity were weakened, replaced by further dependence on welfare and federal transfer payments to an increasingly stricken provincial government, since provincial tax revenues were also falling.

What are the consequences of such social dislocations? The literature on unemployment and social isolation is clear. Unemployment is closely associated with rising mortality rates, neglect of health, alcohol and other substance abuse, family violence and abuse, and rising crime rates. These were all seen in Newfoundland.

The complexities of tracing these connections are formidable, however. Obviously, Newfoundland was not in great shape even before the cod fishery collapsed. Unemployment was already high, and wages were low at the time of the collapse. The prevalence of chronic disease, disability, and health risks arising from lifestyle were already elevated as in many isolated populations, and residents experienced a burden of common disorders related to genetic predisposition (relatively few families in small, isolated communities), diet (traditionally high in cholesterol), and lifestyle. As a result, Newfoundland has historically had the worst overall health indicators in Canada, including age-adjusted rates for cardiovascular disease, diabetes, hypertension, heavy alcohol consumption, and smoking, which exceed other provinces in Canada.

During this time, as always, the social safety valve was migration of young workers and their families from Newfoundland to other parts of Canada. Employment opportunities were becoming available elsewhere in Canada in places such as the western city of Fort McMurray, center of the oilsands industry. These migrations disrupted families and led to those who left were more often skilled and educated than those who stayed behind. While the rest of Canada experienced an increase in younger males of working age (twenty to forty-four years of age) from 1987 to 1997, Newfoundland experienced a decrease in all regions of the province, particularly in outports (remote fishing villages, in the past often only accessible by sea) and the Northern Peninsula, the highest at −11.4 percent, and the South Coast (−9.0 percent). Even the relatively urbanized area of the Avalon Peninsula (which includes the capital and largest city, St. John's) experienced a net low although it received extensive internal migration from the outports. Most of this massive demographic shift took place within only four years. Employment rates, among men and women, dropped as much as 31.5 percent to 33.8 percent unemployed (in Burin, a peninsula on the south coast, which will be used as a benchmark here); regions where the drop was appreciably less had high unemployment to begin with. The share of personal and family income that came from government relief (i.e., welfare) correspondingly increased, by as much as 31.6 percent of income to 34.6 percent of income (in Burin). Rates of violent crime, which are low in Newfoundland, increased immediately after the cod fishery collapse but came back to "normal" levels by 1996; however, this may be in part or in whole due to the emigration of young men, the demographic group most likely to be involved in such crimes.

Noteworthy, however, is that Newfoundland in 1990 showed much more resilience during this economic crisis than it had in 1940. It had a huge advantage compared to its earlier history in that integration with Canada gave it a fiscal lifeline in the form of intergovernmental transfer payments. The infrastructure in the province had improved with consolidation of many of the outports under Premier Joey Smallwood, a move that at the time was controversial and often resisted. Many communities along the coast had already been consolidated or effectively closed by a policy of the government to encourage resettlement in more accessible and economically viable communities where higher-quality health and social services could be delivered more efficiently. The building of facilities, such as a provincial medical school with a community health emphasis, improved access to high-quality health and human services. The historically high birth rate in Newfoundland had also been declining, as it had across North America, so the labor surplus was not as large as it once would have been, and there were fewer "mouths to feed" proportionate to gross domestic product. The fishing economy of Newfoundland is seasonal, and unemployment insurance is a mainstay for families in the off-season: so the increased reliance on government-supplied welfare was often a matter of degree rather than a new development in the economic structure. In some parts of Newfoundland, local hunting and gathering of "country foods" and wood for fuel also supplement purchased foods and supplies.

Other indicators showed less change. In a positive trend following the collapse of the cod fishery, given the province's historically low education completion and university attendance rates, students tended to stay in school longer (since there was less incentive to find employment in a depressed job market, and university education was essentially free) and earned more degrees. Property crime rates continued their previous decline, suggesting that social cohesion was not disintegrating.

From the economic development perspective, Newfoundland had a disproportionate supply of *social capital* compared to its relatively deficient *human capital*, two concepts often confused. "Social capital" is a concept in sociology that refers to the number and quality of networks that support cooperation and involves relationships, access to cooperation and mutual support (often available through extended families), organizations in society both informal (community elders) and formal (for example, through churches and schools), and shared attitudes. "Human capital," as it is used in development economics, means the total stock of education, skills, knowledge, creativity, competence, and behaviors in a population that affect economic productivity, including innovation. Social capital was not lacking in Newfoundland, where local society is tightly knit and hyperlocal, modeled and derived as much of it was from the costal and rural villages of Ireland in centuries past.

At the same time, Newfoundland was on the verge of a massive boom in offshore oil and gas, and the economy eventually rebounded a decade or so after the collapse because of the spin-off economic development fueled by oil and, more recently, by mining and tourism. Although Newfoundland is still not rich today, it is no longer far behind other provinces of Canada in income and amenities and maintains a material culture comparable to other regions of North America.

In assessing the health impact of ecological changes that are socially mediated, one could use words such as "uncertain," "multifactoral," and sometimes "ambiguous." These situations are always complicated, arguable, and therefore politicized. Underneath the complexity, however, the human health effects are there, mediated by stress, income insecurity, poor lifestyle "choices" (better understood as a choice between few good options), and social isolation.

At present the cod fishery remains moribund but there are possible opportunities for bottom-dwelling seafood harvests. Snow crab and northern shrimp have proliferated on the seabed after the decimation of their natural fish predators, and this may provide for a new but smaller and ecologically less rich fishery.

In summary, the collapse of the Northern Atlantic cod fishery is a classic example of the "tragedy of the commons" (see chapter 3). The collapse was entirely avoidable, but the signal that collapse was imminent was not heeded because it was arguable—the trend was not absolutely clear and so decisions could be made that were expedient but unwise. The pressure to overexploit the resource was irresistible, politically, against uncertainty of arguments for conservation. The impact of the collapse of the cod fishery was largely mitigated not only by the timely (one might say lucky) appearance of an economic alternative but also by enhanced resilience in the province due to government policies in the postconfederation era.

There is evidence that tuna will be the next ocean fish to experience such a collapse.

Boom Towns

In American slang, a "boom" is a period of relatively unrestricted expansion and prosperity, and a "bust" is a rapid decline in business to the point of collapse.

A boom strains existing institutions to keep up, inflates prices by creating demand, and causes severe shortages, not least in labor. A bust leaves behind major losses and excess capacity and results in unemployment. Both are particularly characteristic of economies based on commodities such as metals that are volatile in price due to fluctuations in demand and supply. Oil used to be characterized by boom and bust cycles; however, demand has been high for so long that the cycles have essentially disappeared. But gas is in such plentiful supply that the market is essentially in a bust cycle at the time of this writing (2014). The crisis in the Atlantic cod industry mentioned in the previous section was a bust caused by resource depletion. This section examines the consequences of a boom on communities with limited capacity to expand together with demand.

An important social phenomenon on an area or local level associated with unsustainable development is the "boom town." For example, many small communities throughout the upper Great Plains, particularly in North Dakota, are currently experiencing explosive growth with the development of shale gas. Some of these same communities experienced a previous boom thirty years earlier that then collapsed. Boom towns have important implications for sustainability, both as almost inevitable consequences of unrestricted growth and as case studies of tensions in the community that arise between opportunity and sustainability, and the relative weakness of both planning and market forces in moderating these strains.

Boom towns are complicated social phenomena that are characteristic of mining, oil and gas, and other nonrenewable resource exploitation, although they can occur with other industrial activities. A boom town is usually a relatively isolated community that experiences a sudden influx of people, direct investment, and expansion of the local economy in an uncontrolled manner. This places abrupt and severe demands on the local infrastructure, particularly housing, and strains existing social networks as immigrants and long-term residents sort out their relationships. When a community is relatively large and already established, such as the historic city of Aberdeen (Scotland) on the verge of North Sea oil development, the effects are less severe, mediated mostly by prices, slower to develop, and easier to manage. When the community is small and initially lacking in infrastructure and amenities, such as Williston, North Dakota, or Rock Springs, Wyoming, in a previous era, the effects are sudden, disrupting, create severe chokepoints in infrastructure and supply, and can have systemic effects.

Many residents welcome boom times as the first, transient phase of economic expansion, and see it as "the big chance" to build a sustainable economy on a new foundation and to bring new opportunity to a previously stagnant local economy. Others see it as a permanent, unwelcome disruption that disturbs the status quo, especially if they have deep roots in the community. How people view this depends importantly on their own situation, as young people and those who stand to benefit economically tend to welcome the boom times; long-term residents with deep roots but no economic stake in the economic development, together with recent arrivals who came because they liked the town (and who tend to be even more conservative about change than the local families), tend to be pessimistic.

Many boom towns were associated historically with gold and silver rushes, such as Virginia City (previously mentioned). Perhaps the most famous boom town in history was Ballarat, which began as a sheep station northwest of Melbourne, Australia. Almost overnight after gold was discovered Ballarat became a global destination, receiving an estimated 20,000 migrants in a few months in 1851. (Ballarat continued to produce gold for decades and eventually matured into a stable, thriving community.)

Sociologists began to study the boom-town phenomenon in depth starting in the 1960s, and many early studies focused on Rock Springs and Green Valley, Wyoming. Similarly, many communities in the northern Great Plains, particularly Williston, North Dakota, have become boom towns as a result of servicing of becoming regional centers for the exploitation of shale gas, coal, and enhanced recovery of oil.

Today, the leading example of a boom town that eventually achieved some semblance of stability may be Fort McMurray. Fort McMurry is a city in northern Alberta (Canada) that is the center for development of the oilsands, which is an unconventional hydrocarbon resource. When it was smaller in scale, the oilsands industry experienced wild booms and busts. The population of the city, which is now about 62,000, fluctuated noticeably with the economic cycle, with large in-migration when oil prices were high. This placed a large strain on local services such as hotels, restaurants, medical care, and distributors. Since the oilsands industry has stabilized (almost entirely due to export of synthetic crude oil to the United States) over time, the economic cycles have become less severe. Even before the last major oil bust in the early 1990s, the city was becoming more "normal" in its civic life, with more services for families and permanent housing for long-term residents. In the case of Fort McMurray, migrants came disproportionately from Newfoundland, which was on the other side of the country. This only partially helped mitigate the economic disaster of the cod fishery collapse, however, because the fishermen and outport residents of Newfoundland who were displaced by the fisheries collapse usually did not have the skills to work in the oilfields and oilsands of Alberta. Eventually, Newfoundland developed its own oil industry, supported in part by returning workers who had gained their experience in Alberta.

The typical boom town already exists as a community that has a small settled population (Fort McMurray had about 2,500 residents in 1960), often with deep family roots in the area. Services, such as health care and schools, and facilities, such as housing and stores, are initially scaled modestly, to the original local population and tax base. When a commodity-led economic expansion, or "boom" occurs, the established community faces profound and uncomfortable changes, complicated by the reality that local residents usually have little or no stake in the boom (except for oil and gas, in which more opportunities exist for local landowners to profit).

The vanguard of new residents tend to be young unattached men who are well paid if they are employed by a company. The prospect of work also attracts less fortunate, unemployed men who may be rootless, lacking in marketable skills, and unattractive to employers (who usually have secured desirable workers on contract long before they come to town). One effect of infrastructure shortage and soaring rent is to trap these job-seekers into substandard or illegal housing or homelessness. Some unemployed migrants, perjoratively called "drifters," are diverted into criminal activities (particularly drug dealing). The pattern is highly gender-specific. Women may be attracted to boom towns by the availability of legitimate, mostly service jobs, and some, usually with prior experience and often a criminal history that makes them otherwise unemployable, to easy money made by prostitution. However, women are much less likely than men to migrate to a boom town and this creates a huge imbalance in the gender ratio, which in turn aggravates the prostitution problem.

The stereotype, which has often been true in the past, is that early migrants are poorly educated or poorly socialized men with no roots and little conscience about what disruption they cause and the consequences they leave behind. They may be perceived by current residents as louts and drifters but are tolerated because they bring in money and their labor supports economic development. This vanguard of younger, temporary workers who are needed to build the infrastructure rarely integrate or interact favorably with the local residents. The temporary workers may entertain themselves boisterously with alcohol and often with illegal recreational drugs, which causes a new (or newly perceived) drug problem. Extremely high wages and limited opportunities to spend it fuels drug and alcohol consumption and prostitution. Crime becomes a concern and may outstrip the local policing capacity. The original local residents find themselves feeling threatened by crime and perceived menace from people they regard as strangers. Drifters and rough-looking but law-abiding workers may be difficult to tell apart by appearance. Small-town natives may feel that they are experiencing urban levels of stress and anxiety, particularly with incessant media coverage of how things are changing. These feelings are exaggerated when the incoming migrants are perceived as being different in culture, ethnicity, and attitudes. This often leads to negative attitudes toward immigration in general, nostalgia for how the town used to be (and undoubtedly will never be again), and a political backlash as residents try to regain or keep control.

When boom towns do not have the discipline of control by companies, squatters and masses of unemployed adventurers may indeed flock to the town. Increasingly, however, in modern boom towns such as Fort McMurray, the migrants are often skilled workers who came from compatible cultures, which eases the transition. Fort McMurray has done better than most but still must cope with a disproportionately high crime rate.

The original residents experience a loss of control over their lives. Their own businesses and social concerns are sidetracked, and the town seems overrun by outsiders. Divisions arise among the original community residents over local issues, particularly land use, and between those who see the boom as a "once in a lifetime" opportunity for economic growth and those who blame it for changing the quality of life and see it as an intrusion. Decisions about the business, land use, and labor recruitment are usually made by corporate officers in distant cities: at best, this happens with nonbinding consultation, and at worst, with no local consultation or regard for local negative impact. Their own sons and daughters, surrounded by the new activity and interacting more frequently with the newly arrived workers, pick up new ideas and habits and relationships that may concern their parents and that weaken ties to the traditional role models in the community. There may be strong temptation for teenagers and young adults to drop out of school because inflated wages are so high.

Housing, whether decent or substandard, is bid up to levels that locals find astronomical. Health services may be strained to cope with the health-care needs of transient workers, and local residents perceive a reduced level of service from caregivers who used to know them by name. Schools are not a problem while the new workforce is temporary, but become so when workers in the more permanent workforce arrive with their families.

As the boom progresses, investment in infrastructure cannot keep up with demand. A severe labor shortage may develop because pay is better and opportunities are greater in the commodity-led industry than other jobs could possibly offer in the region. The existing industries in the region decline because they cannot compete for labor and investment (a phenomenon known as the "commodity trap"). Commercial outlets in the area lose employees and struggle to service increasing demand and changing tastes.

Booms may be followed by a "bust." If commodity prices abruptly collapse (as has happened many times in Fort McMurray), the boom town may quickly find itself in trouble with respect to its tax base and with absorbing employment. And there is an unexpected mismatch between its newly built-up infrastructure and a reduced population. At the extreme, if the commodity price never returns, the town may become economically nonviable unless there are alternative opportunities. Former boom towns gone bust mostly survive on a small scale with a mixed economy. In the case of the former copper-mining towns of Arizona, that economy depends on tourism and often retirees taking advantage of low-cost housing.

If the town is then deserted, it becomes a "ghost town." Although this was common in the eighteenth century in the American Southwest, few communities in modern times are ever completely depopulated, so there are very few modern ghost towns. (One modern exception is Uranium City, Saskatchewan, an isolated northern town with a peak population of 5000 in 1980 that was subsequently closed and then turned over to the local aboriginal community, which now numbers about 200.) In the case of Virginia City, Nevada, on the other hand, the town came back, as a major regional tourist attraction.

To prevent the boom-town phenomenon, and the consequences of a subsequent bust if the boom abruptly fails, many resource companies now provide temporary housing for their transient workers, and limited amenities, in sites near but not inside the original communities. When the initial phase of resource development is over, these temporary settlements can be dismantled and moved. Sooner or later, however, permanent workers and the original residents must interact and find common ground.

The "DPSEEA+C" Model

One of the most successful conceptual models used to guide thinking about sustainability, human health, economic development, and social development is the "DPSEEA Model" (for "drivers, pressures, state/conditions, exposure, effects," and pronounced "dip-see") developed at the World Health Organization (WHO). This model brings together many interrelated issues involving underlying driving forces and immediate "pressures" that are influenced by conditions such as economic forces and technology. The DPSEEA model is especially useful for analyzing sustainability problems.

The DPSEEA model was based on an earlier conceptualization called the "Pressure-State-Response" model that was adapted by the Organization for Economic Cooperation and Development (an organization and economic research forum for rich countries) from a still earlier model used by the Canadian government. Before it achieved its present form, it was known as DPSIR (for "Driver, Pressure, Impact, Response"), which is a more limited formulation still in use in some organizations and for some purposes. The version of the model presented here adds an additional element called "consequences" and so will be called "DPSEEA+C," with the "C" added to close the loop in the basic model.

The DPSEEA framework envisions all processes, environmental and social, as being driven by a set of powerful "driving forces" that may be more succinctly called "drivers." These drivers may include basic social needs (such as the need for security, water, and food), the introduction of new and transformative technology (such as mobile phones), or social conditions (such as the level of economic development or a population growth). Most drivers reflect unresolved social problems such as poverty. There can be multiple driving forces at any one time influencing subsequent events, and any given driving force can influence several outcomes. DPSEEA is a structured way of looking at these linkages.

Figure 3.1, in chapter 3, presented the example of drivers and forcers of atmospheric change. This example will be carried further here as an illustration of how several potent driving forces can result in social processes that affected air quality on many levels and in many settings that resulted in a variety of health outcomes.

In the DPSEEA framework, these fundamental drivers exert *pressure* on the system as a whole. Examples include shortages of essential commodities, lack of clean water, and production of waste that must be disposed of somehow, and transportation bottlenecks. This pressure may be felt on a material, economic, political, or social level, and by individuals or populations. When the pressure is conveyed to a market in the form of demand or leads to political debate to advocate a response, this pressure can be constructive. However, if the system does not move and the pressure leads to frustration and anxiety, it can foster political instability. At the extreme, these pressures may even lead to violence and interstate conflict. The inclusion of a term for "pressure" makes the DPSEEA model explicitly political, because it acknowledges that the pressure is social or economic and that it has a political expression.

The pressures change conditions in the environment and in society and therefore change its *state*. Elevated pollution levels may result in worse health, compromised living standards, and perceptions of environmental injustice. However, pressure to reduce pollution may result in legislation, tighter regulation, investment in pollution control, and (for environmental issues) monitoring programs that in turn draw more attention to the problem. (A small drawback to the DPSEEA model is that the term "state" implies a relatively static condition, whether for the environment, the economy, or society. In reality, conditions may be fluid and the so-called state may actually be a dynamic trend.)

Table 7.2 provides some selected examples of major early twenty-first-century air quality issues in the United States and the pressures that led to interventions that changed the state, or conditions of air quality. Although somewhat simplistic, the point of this table is that major advances in air quality were pushed forward through the political process by pressure, motivated by social concern that was raised by perceptions shaped by an underlying driving force. Taken together, the condition created by these pressures became a trend toward cleaner air and a state of improving air quality. (Of course, there were other trends in air quality developing at the same time; these are illustrative.)

The state, or set of conditions as defined by DPSEEA, determines the *exposure* experienced by people individually and the overall or average and distribution

Table 7.2 Important social drivers that led to recent major changes in air quality in the United States

Time period	"Driving Force" in DPSEEA Model	"Pressure" in DPSEEA Model	Change in Air Quality Management	Change in "State" in DPSEEA Model
1986–1995	Children's health, concern for lead exposure	Parental and social concern	Removal of lead from gasoline	Lead banned from gasoline, atmospheric levels fell to negligible
1990s	Emission of acidifying emissions from power plants and other stationary sources	"Acid rain" became an issue for highly populated northeastern states	Regulation of acidifying emissions	Highly successful; major reduction in acid deposition
1999 (revisited 2008, 2012)	Technology: shift to diesel and secondary formation of ozone	Relative risk → attributable risk; excess mortality and morbidity visible and urgent	Fine particulate air pollution, $PM_{2.5}$ and O_3 standards	Regulation of diesel and other emissions led to reductions in $PM_{2.5}$ and stabilization of O_3 formation
2006	Emission of greenhouse gases	Climate change interpreted by Al Gore in movie *An Inconvenient Truth*	Climate science more widely accepted and movement toward regulation	Much voluntary activity but no Congressional action
2009	Children's health, concern for toxic exposure	Community risk perception of air quality around schools	Air toxics release and monitoring	Major EPA monitoring program, interest in toxics that could affect children

of exposure in the population. Here the DPSEEA model converges with the "Source-Effect Model." (See chapter 5.) A change in state (as a consequences of driving forces exerting pressures), leads to a change in *exposure*. A new "exposure regime" (a technical term for aggregate exposures to many contaminants and the timing of exposure to each, more often used in environmental sciences), and, by extension, a set of risks or factors, exerts its own effects such as health risk from air pollution or effects on utility prices and energy supply. When the state changes, the exposure regime changes. If the exposure remains the same, then the state has not changed, and the pressure has not changed the current status despite whatever political or economic movement may have occurred.

At each state, *actions* may be taken (or not) changing each of the elements. These may be technological changes, social or behavioral changes, education and awareness, technical intervention to correct a problem (such as health-care or ecosystem remediation) or, importantly, political changes, each of which may be a response to some aspect of the circle. Therefore actions are represented as arising from inside the circle and pointing to but apart from the influence the individual elements have on one another. (One of the criticisms of the DPSEEA model is that the relationship of the actions to the primary elements is usually not spelled out; and when it is, the relationships are confusing to diagram. Of course, in real life such actions frequently have many motivations and are frequently only loosely linked with real needs.) In figure 7.1, the usual DPSEEA diagram (as originally developed by WHO) has been modified to run clockwise, which is more intuitive and shows that actions adjust to circumstances and feedback. It is also

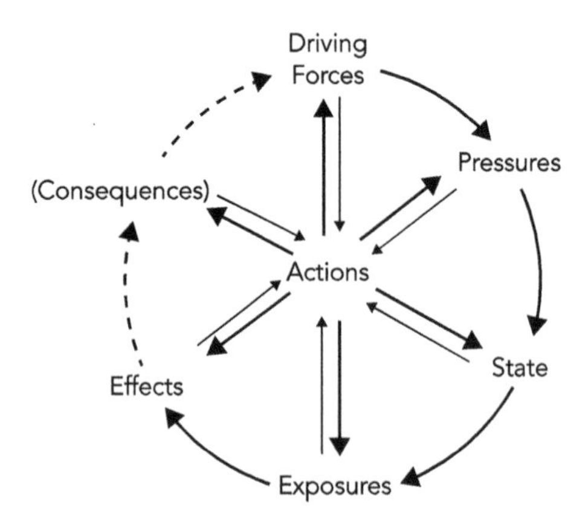

FIGURE 7.1 Graphic representation of the DPSEEA+C model, showing effect of actions on drivers, pressures, states, exposures, and effects. The conventional DPSEEA model, as developed by the World Health Organization, stops there. In the DPSEEA+C adaptation, the cycle continues (dotted line) with resulting consequences, which ultimately results in a new set of driving forces. This adaptation also takes into account feedback required for relatively small operational adjustment to the actions taken (small, inward-pointing arrows).

shown as a closed cycle because social issues tend to be repeated or be iterative and rarely do trends go to completion or do social processes reach equilibrium.

In the air pollution example, a driving force may be increasing income and a greater market for buying automobiles; the action in response to growing traffic accessing limited highway space may be a tax on purchases or gasoline, which goes into a fund to control air pollution and build roads. The pressure may be an increasing number of cars on the road, for which an action in response to traffic congestion might be registration to control the number of cars and a vehicle inspection to determine how well they are maintained. The state that results would be an increasing trend in photochemical air pollution (associated with vehicular traffic); the action in response might be standard setting by the national regulatory agency. The trend in air pollution results in increasing exposure of the population and individual residents of the community, for which an appropriate action might be to publish or display an air quality index, which both allows the individual to protect himself or herself and children on a bad air day and informs citizens whether their elected or appointed leaders are managing air quality effectively. The effects of exposure to photochemical air pollution might, for the example of ozone, be an increase in the frequency of asthma episodes and emergency room visits; an appropriate action for this might be to improve treatment and patient education for people with asthma but also to advocate for air quality management as a means of improving health and preventing lung disease.

The traditional DPSEEA ends at effect. (Only a few users of DPSEEA draw it as a circle, as in this book.) However, in the real world, the cycle is closed and continuous, never arriving at a constant, sustained effect. Even in slow-changing agrarian societies, driving forces change from time to time because of draught, population changes, market conditions, and external factors such as warfare. Also, the effects may accumulate or be so profound in their consequences that they change the driving forces, completing the cycle. Conditions get better or worse and are affected by external factors, and so the social and environmental movement described by DPSEEA is really continuous, a concept better conveyed by drawing the cycle as closed like an *O*, rather than open in WHO's conventional *C* shape.

Effect will have consequences, intended and unintended, and these will shape the driving forces that keep the cycle moving. In this chapter, the idea of "DPSEEA+C" is introduced to accommodate this idea that social movement and environmental conditions do not stop at one point and remain static.

In the example given above for air pollution, the consequence may be that these trends lead to political response and action in the form of regulations imposed by government regulatory agency, which then becomes a permanent feature of the system, together with organizations, lobbyists, and advocates whose purpose is to influence the policies of the regulatory agency and who therefore contribute a new force to the driving forces. The perception of value in clean air and the ongoing debate in society about the proper place of regulation in society and whether in a liberal democracy there is a right to a clean environment that triumphs a right to exploit a free good (i.e., the air) has contributed to driving forces and changed social fundamentals about how pollution is perceived. One might think that improvement in air quality would reduce pressure to improve the state of air pollution, but historically it has not. Instead, a more sophisticated understanding of the health effects of air pollution and appreciation for the benefits of clean air has if anything increased pressure to improve in rich and poor countries. One might also expect that lowering exposure to air pollution, and consequently to reduce the frequency and severity of health effects, would lead to better health and that would be the case. However, the perception of the degree to which air pollution affects health has, if anything, become more acute. People become even more concerned about their health and those of their children as the situation improves. The perception is itself an effect. (In this example, only the direct health effects of ground-level pollution are considered, although the response to emissions resulting in climate change is inadequate and still evolving.)

Because the DPSEEA+C model is really a model of social influence integrated with the Source-Effect Model, elements such as exposure do not have to be literal exposure to a hazard, measured in concentration or exposure level. Perception of state or exposure can be a profound pressure and motivating factor for change whether or not exposure has really changed and is adverse or improving. Risk can treated the same as actual exposure in the DPSEEA model. Exposure can be an exposure opportunity or probability, a risk or liability. In a DPSEEA+C model of crime, for example, exposure may also be to a behavior such as violence or the probability of a criminal act. Its ability to accommodate perception, risk, and exposure opportunity in addition to exposure levels makes DPSEEA and DPSEEA+C very versatile and easily applied to social and behavioral problems as well as environmental health problems, narrowly defined.

The DPSEEA+C model can obviously be used effectively to explain and analyze complicated issues in sustainability and environmental health. It also has other uses. The formulation of:

[Driving Forces → Pressures → State → Exposures → Effects → Consequences → (Repeat)] ← Actions

can also be used to generate indicators of progress. In the example of air pollution, DPSEEA+C suggests a variety of behavioral indicators in addition to the usual indicators of pollution measurement and distribution. For example, how responsive is the system to a perceived problem (i.e., pressure)? To what degree are regulatory agencies responsive to state and changing conditions? Are solutions proposed to manage effects, only, or do they address pressures and states? These behavioral indicators may be as important as monitoring actual progress in controlling pollution if they identify problems or bottlenecks that could fail or stall further improvement.

DPSEEA+C is a highly versatile analytical tool for understanding sustainability issues. It is comprehensive, flexible, scalable, and covers everything from local problems to national and regional issues. It should be part of the toolkit of sustainability professionals.

8

Ecosystem Services and Health

ECOSYSTEM SERVICES ARE functions performed by the natural environment that are of economic value—meaning that they support human communities—and that are performed by ecosystems without human intervention (other than to protect the ecosystem). One immensely important example is the collecting, oxygenating, filtering, and (within limits) purifying drinking water through natural watersheds. Problems with continuity or capacity of ecosystem services have enormous implications for sustainability and for health (see table 8.1).

These services cannot be easily replaced by technological means, especially on a regional or global scale, and so conservation is the only logical mode of management. The capacity of each ecosystem service is ultimately limited but usually huge—and often hugely abused. These resources are constantly under pressure wherever they are accessible to diversion or exploitation and from interests that desire to appropriate the resource for alternate uses. Reduction of the capacity of ecosystem services may even result in catastrophic failure (see chapter 3). Box 8.1 describes one of the most dramatic local examples in history of the loss of ecological services: the Great Stink of London, 1858. This is an example of steadily diminished capacity that resulted in an acute crisis for the government of the most powerful country in the world.

"Ecosystem services" should not be confused with institutional "environmental services." In practice, "environmental services" has come to refer to housekeeping, the environmental management of facilities, and maintenance of the indoor environment, particularly as the name of departments or service units in large institutions. Chapter 13 briefly mentions environmental services in this context. This chapter is devoted to ecosystem services as defined in the opening paragraph.

Unfortunately, when the United Nations Millennium Ecosystem Assessment documented the status of twenty-four essential ecosystem services (described later), fifteen were evaluated as being degraded or in decline, a situation likely to have a profound negative effect on human health and likely to have gotten worse.

Table 8.1 Table of sustainability values: Ecosystem services

	Sustainability Value	Sustainability Element
I.	Long-term continuity	Ecosystem services are essential for unlimited (at least within the foreseeable future), long-term continuity, because technology cannot replace them.
II.	Do no/minimal harm	Saturation or diversion of ecosystem services to other purposes can cause substantial and often devastating harm to the environment.
III.	Conservation of resources	Conservation of resources for the future requires the protection of ecosystem services.
IV.	Preserve social structures	Because ecosystem services such as water collection are essential to social structures and social stability, conserving them is essential to avoid harm to or to enhance social structures.
V.	Maintain health, quality of life	Ecosystem services are essential to maintaining health and the quality of life.
VI.	Performance optimization	Because of the value they add, and the enormous expense of providing similar services any other way, conservation of ecosystem services is essential for optimization or maximization of economic, social, and environmental performance.
VII.	Avoid catastrophic disruption	Ecosystem services tend to be much more stable than any technological replacement, and so present a lower risk of catastrophic disruption than artificial alternatives.
VIII.	Compliant with regulation	Protecting the ecosystems that provide ecosystem services is compliant with regulation and best practices.
IX.	Stewardship	Protecting the ecosystems that provide ecosystem services is consistent with stewardship and is a responsibility that goes beyond ethics in dealing with other people to include other species, life, and some notion of nature.
X.	Momentum	Ecosystem services are quite capable of supporting continued progress under its own momentum and therefore promote stability in the short term.

BOX 8.1

Denial of Ecosystem Services: "Great Stink"

Perhaps the most dramatic example of denial of ecosystem services may have been the Great Stink of 1858, the culminating event of a long period in which the Thames River was overloaded with sewage. Fundamentally, it was a local, even hyperlocal, municipal problem. However, it just happened to occur outside the windows of the building housing the governing body of the most powerful country in the world at the time.

Historically, London was built on a number of small streams and tributaries of the Thames, some of which bore names such as the Fleet (covered over in 1666 but the origin of "Fleet Street"), the Tyburn, the Effra, and the Strand (now a boulevard, then the shoreline of one of the streams). These streams were filled in, channeled, and covered over from medieval times into the twentieth century. The result was that the natural streambeds with their associated plant life, biodiversity, and natural flow characteristics were lost. They are now popularly called London's "Lost Rivers."

For centuries, the principal function of these degraded streams was to channel storm water, so that the streets would not flood. After 1815, an expedient but imperfect solution to London's increasingly severe waste disposal problem, consisting of simply dumping waste into the storm channels, turned these streams into closed, running sewers.

At the time, most Londoners depended on cesspits to receive waste. These shallow pits were scattered around the city and gave it a fecal odor in warm weather. The invention of the underground sanitary sewer in the 1850s by Edwin Chadwick (b. 1800–d. 1890) and the construction of a drainage system diverted human waste from the pits into what became an underground combined sanitary and storm sewer system. Worse, there was much cross-connection and intermingling with London's water supply, which was often drawn from nearby wells fed by these same polluted streams.

The volume of sewage soon exceeded the much-reduced natural capacity of the streams to dilute and mitigate microbial and, increasingly with industrialization, chemical pollution. London experienced frequent cholera outbreaks during this period, one of which was associated with the seminal event of modern epidemiology: British physician John Snow's analysis identifying the Broad Street pump as a point source of transmission of cholera in 1854. This was a triumph of analysis of a disease outbreak in time and space and established the fundamental methodology that became epidemiology.

However, in those times even cholera was just one among many serious health issues and not necessarily the priority. Typhoid was also rampant.

Tuberculosis was emerging. Recorded mortality in large English cities in Victorian times was higher than at any time since the Black Death (i.e., plague).

In 1858 London experienced an unusually hot summer and a long dry period. The flow of the Thames and its tributaries slowed considerably. Raw sewage spilled into the river barely diluted and often got deposited on the river banks. The stench was unbearable. Most important of all, however, it directly affected Parliament and the Law Courts. Finding the situation intolerable but stuck in London for the summer session, it took the members of Parliament only eighteen days to draft and pass a bill for the reconstruction of London's sewer system.

The failure of the sanitary sewer system, the overflow problem from combined sewer outfalls, and the unreliability of the water supply for London resulted in a comprehensive civil engineering plan, undertaken at enormous expense for the time. Led by Sir Joseph Bazalgette (b. 1819–d. 1891), this massive engineering feat took well over twenty years. Fortunately, and unusually for the time, much of the system was intentionally over-built, with capacity beyond minimum requirements. This allowed it to continue long past its expected lifespan and to accommodate much larger population growth than ever anticipated. The system is still carrying London's sewage, but there are now plans for a massive new receiving tunnel that, if built, would ensure service for at least another century.

This case is also a good example of how such problems are not neatly classifiable in mutually exclusive classes: this is because the problem involves natural ecosystem services, artificial ecosystem services (discussed in the next chapter), pollution, ecological change favoring transmission of disease (the epidemiological triad), and socially mediated effects (public outrage over the odor and fear of the resurgence of cholera).

Contemporary examples of important ecosystem services that have failed in the past or are failing now include the following:

- The High Plains Aquifer system, which is one of the world's largest collections of groundwater formations, ranging from Texas and New Mexico to South Dakota, where it is the source of drinking water for 82 percent of the population (2.3 million). Because extraction of groundwater for agriculture has been massive and rapid while the recharge rate of rainwater percolating into the formation is almost negligible, the underground water level is dropping quickly. The Oglalla Aquifer, which is part of the system, is currently failing, compromising irrigation and placing agriculture at serious risk in states such as Kansas.

- The collapse of the Grand Banks cod fishery, described in detail in chapter 7. This is an unusually straightforward example of the abrupt loss of an ecosystem service and the industry that depended on it.
- Hurricane Katrina, a devastating storm in 2005 that came close to destroying New Orleans. It was much more destructive than it would have been had the wetlands of southern Louisiana been preserved and water flows in the region not been diverted.
- The boreal (northern) forest of Canada and Russia, which provides climate mitigation, carbon sequestration, renewable resources, and numerous other services from the local to a global scale but are threatened by climate change, unsustainable use, a pest infestation, and habitat loss.

The Scope of Ecological Services

Ecosystem services include the supply of renewable resources such as fish or forest products (sometimes called "provisioning"), water supply really, water diversion and flow regulation (through wetlands and mitigating run-off into bodies of water), regulation of climate and local microclimates, disease prevention, pollution mitigation (through the dynamics of airsheds, water flow, biological remediation, and adsorption), nutrient cycles (such as nitrate uptake), and pollination. Ecosystem services also include economic uses such as recreation (including boating, hunting, fishing, hiking, and the esthetic experience).

Specific Ecological Services

Ecological services can be classified in different ways: by medium, by function, or by end use. The Millennium Ecosystem Assessment, referred to above, was a massive collaborative project to describe and inventory the world's ecosystems and the functions they serve for human activity. (This classification system is slightly different from that of the Assessment, in the interest of didactic clarity.)

Support for Human and Natural Existence

Human existence would be impossible without the ambient environment and natural ecosystem services. Human life today depends on natural cycles, sources of supply, and renewable resources. If, at some point in the future, it becomes possible to dispense with natural ecosystems, it is highly unlikely that this will be a voluntary development. More likely, it would represent a descent into desperation as humankind runs out of options in the face of catastrophe.

- *Biodiversity.* Biodiversity is critical to the stability of ecosystems. Ecosystems derive their stability from redundancy and efficiency. Some species occupy a particular niche that human beings find useful, such as those that prey on economically important pests; those that degrade biological waste material, such as molds; and those others that maintain the flow of energy and nutrients in the trophic levels of the ecosystem, such as menhaden in the ocean. Typically there may be several species occupying a similar niche (such as herbivores grazing on a particular plant), but each has a somewhat different ecological niche so the most intense competition between species is not direct. (For example, many terrestrial species eat leaves, but giraffes eat leaves other animals cannot reach.) When an ecosystem loses key species, it simplifies: the pathways of redundancy become fewer, and the ecosystem becomes less resilient to stresses, whether natural or human in origin. Biodiversity is therefore fundamental to maintaining the natural environment and the existence of a wide range of ecosystem services.

- *Primary production.* Primary production is the capture of energy and synthesis of food in the first trophic level of nutrients. The captured energy supports the whole ecosystem and therefore the entire food chain above it. Since all life in the ecosystem depends on primary production, denial of this service means the collapse of the natural biological community and all ecosystem services that it provides. Over-harvesting of primary production, for example the capture of krill in the ocean, could remove a pillar of ecosystem stability and risk collapse and endangerment of numerous species that depend on it.
- *Evolution.* All living beings today are the product of an evolutionary process, and so the abundance of life forms that participate today in ecosystem stability, that provide economically useful ecological services as described below, and that were available for domestication or contemporary genetic harvesting are the result of an evolutionary ecological service that occurred in the past. Although the future of contemporary humankind and civilization may not depend on continued vitality of natural selection (there is not sufficient time for that), whatever (and whomever, if human beings are allowed to continue to evolve) comes next will depend on it. If evolution by selection of what is best adapted to the natural environment is replaced by human-directed selection, genetic manipulation, and forced evolution, it will not mean the end of natural selection. Natural selection will simply shift to favor species that can survive and thrive in whatever niche is available regardless of human intent and preference (and probably in niches undesirable to human health).

Evolution is not usually thought of as an ecosystem service: this is because changes that result in stable adaptation and new species require a time frame normally beyond human accounting, although natural selection itself can and does occur over a very brief in short-lived species such as bacteria. The most famous example of rapid natural selection involved the predominant color of peppered moths in and near Birmingham, England, which changed from speckled light cream to dark brown when the city industrialized and became polluted and then went back to light as the city lost its smoky factories. Although there have been attempts to refute this iconic work (principally by those who deny evolution), the actual evidence strongly supports that natural selection occurred by the mechanism of birds spotting and eating moths that were easy to see because of the contrast to the color of their background surface.

Production and Supply ("Provisioning")

The literature on ecosystem services usually refers to the provision of essential supplies as "provisioning," which is a somewhat awkward term. "Production and supply" seems more felicitous. This function refers to the supply of commodities that are useful to human life. They include, but are not limited to, the following:

- *Food.* Many people in developed as well as developing countries depend on "country foods" harvested from natural sources (such as berries and game); the natural ecosystem also supports and interacts with the artificial agricultural ecosystem (see chapter 9), for example by buffering the effects of pesticide resistance, holding water, and supporting pollination of plants (although this is also done by commercial beekeeping operations, which are now in jeopardy).
- *Bulk water.* Water is an enormously valuable bulk commodity even if rarely appreciated as such and almost always under-priced. (See below.) The supply of water that comes from rivers and lakes is used as a chemical input (water is not usually thought of as a chemical, but it is one), cooling medium, cleansing agent, solvent, fire-extinguishing medium, and many additional uses discussed in greater detail in the next section.
- *Potable water.* Drinking water is, of course, necessary for life. Water from natural sources is not usually potable, requiring treatment before it can be safely consumed; however, the supply of water that can be rendered potable is limited. Collection of water in a natural watershed is an immensely valuable service because surface water is easily contaminated, and protection of that watershed prevents contamination and may even make some treatment steps (such as filtration) unnecessary.
- *Natural materials.* Here, "materials" is used in the sense that they are used to construct or make things. Wood from forests, wild animal furs (in an appropriate manner without cruelty), and plant materials are renewable resources available through harvesting natural growth, which is an ecosystem service.
- *Natural substances.* Here the term "substances" is used to mean "constituents" or "commodities with an end use." Medicines are often initially identified from the action of natural products, such as taxol (not to be confused with

tamoxifen), a drug from the yew tree that is used in treating some types of cancer. Until there is a convenient way of synthesizing plant products like taxol that have a highly challenging chemical structure, the supply of such natural pharmaceuticals will be severely constrained. Natural plant dyes are popular and still in use alongside synthetic dyes, which were developed in 1865. Perhaps the most exotic natural substance is ambergris, the biliary secretion of the sperm whale that is used in making fine perfumes. Until 2013, there was no perfectly suitable replacement for ambergris. A bioengineered ambergris analogue can now be produced in quantity. Natural ecosystems are both storehouses of natural substances that can be utilized until alternative production methods are available and libraries of diverse chemicals that can be studied for active compounds with particular effects.

- *Energy and fuels.* Biomass fuels such as wood and wood pellets are produced from renewable natural products and represent another commodity from natural ecosystems. By convention, wind, geothermal, and other renewable energy sources are not considered ecosystem services, although they are provided by the natural environment. That is because they are the physical result of solar energy (which is an open, not a closed cycle) and geological processes, not ecosystems. On the other hand, hydropower is often counted as an ecological service, since it is heavily influenced in the hydrologic cycle by ecosystems and is a closed cycle. Although fossil fuels are the direct result of the biological productivity of ecosystems of the past, they are never considered as ecosystem services because they are not only nonrenewable but were also accumulated in the distant past and require extraction and consumption.

Regulation and Homeostasis

Natural ecosystems and resources have the capacity to reduce extremes and changes in the environment that human beings find undesirable. The result is a world that generally stays within bounds of human habitability. A regulatory ecosystem service is one that limits the magnitude and disruption of change and so tends to regulate and dampen the consequences of a perturbation. It can be thought of as a feedback loop in which a change provokes a countervailing correction. The result is a mechanism that keeps the environment similar or within limits, at least in the short term. This is similar to self-regulation in body temperature, body chemistry, and intracellular conditions (called the "intracellular milieu") observed in living organisms. This tendency to self-correct and to keep internal changes within limits is called "homeostasis" in biology and medicine. These same mechanisms have some excess capacity to deal with human activity, as long as the perturbations occur on a small scale relative to the variation in natural systems. Thus, they are able to mitigate some forms of human ecosystem disturbance, and this gives rise to the "mitigational" category described next.

Because homeostasis is a fundamental concept of physiology within human beings and other animal species, the notion has been a powerful idea in ecology and gave rise to the immensely popular "Gaia hypothesis" of the planet as a living organism on a massive scale. However, it should be clear why ecosystems have this built-in resilience in the first place: they evolved, just as species evolve (and often over a longer time period), to accommodate and exploit these same environmental conditions and to recover from (and even take advantage of) extremes that occur from time to time (e.g., temperature, water availability, food supply, and so forth). (A striking example is the Australian central desert, where plant species are superbly adapted to long dry periods, flash floods, and fire, such that they often require these "disturbances" to regenerate.) The ecosystem therefore depends for its integrity and productivity on occasionally reaching its limits. Within these limits, the ecosystem is highly adaptable and maintains what appears to be a type of biological homeostasis.

Specific examples of regulation and homeostasis include:

- *Carbon fixing, natural carbon cycle*. Natural ecosystems, especially the tropical forest of the Amazon and the boreal forests of Russia and Canada, fix a substantial amount of carbon in biomass, removing carbon dioxide from the atmosphere and preventing natural accumulation that would lead to temperatures intolerable to humans. This ecosystem service operates in addition to the natural buffering capacity of the oceans. This function has served humankind well since the last Ice Age, through volcanic eruptions, fires, and the first few thousand years of human activity. It is now stressed beyond natural limits by atmospheric change, with greenhouse gas emissions and atmospheric loading exceeding its capacity and thus resulting in climate change. This is an example of a natural ecosystem function that is naturally regulatory and sufficiently homeostatic to cope well with natural variations but that is only partially "mitigational," as described in the next section (meaning that it cannot satisfactorily cope with the magnitude of human activity).
- *Temperature modulation*. Natural ecosystems reduce local temperature in many ways: providing ground cover, absorbing light energy, transpiring water and so raising humidity, conserving soil water, and, of course, providing shade.
- *Water cleansing and decomposition of organic matter*. Flowing water is richer in oxygen than standing water and therefore has more capacity to degrade and decompose organic contaminants chemically. Flowing water also has more capacity to dilute both chemical and microbial contamination, reducing the concentration and risk. This "self-cleansing" capacity of flowing water has been known since early in civilization, but water's capacity to cleanse itself has been exaggerated for almost as long. This has led to over-reliance on rivers and other bodies of water for waste disposal.

- *Chemical decomposition and biological decontamination.* Related to biological remediation (see below), natural ecosystems perform the same services of decontamination and sequestration to prevent excessive or toxic levels of chemicals that have a natural origin. Organic chemicals from plant and animal species and metals from natural geological sources (including mercury and arsenic) are generally dealt with easily by natural ecosystems, in part because they have evolved to deal with them. The ecosystem has considerable but ultimately limited "reserve" capacity to cope with an additional load of contamination, in which case this ecosystem service then becomes "biological remediation."
- *Water retention.* Wetlands (even if they happen to be dry most of the time), forests, and grasslands along rivers and lakes are critical to absorbing water and holding water back to slow flow during times of high precipitation. The result is mitigation of flood risk, slower and less destructive flow during flooding, and more time for surface water to recharge groundwater. As mentioned earlier, the destruction of wetlands in southern Louisiana were a major contributing factor to the destructiveness of storms such as Hurricane Katrina, which had devastating effects on communities.

Mitigation

Mitigation is used here to mean the correction or reduction of harm from human activity that affects the larger ecosystem. Mitigational services balance the effects of perturbation, contamination, and ecosystem disruption, and are discussed in previous chapters. They represent one aspect of ecosystem resiliency. Intentional engineering of ecosystem services for this purpose moves the discussion in the direction of "artificial ecosystems," which is the topic of the next chapter.

Some examples of mitigation by ecosystem services include:

- *Airshed management.* Natural terrestrial ecosystems are important regulators of air quality. Open spaces in which emissions to air are restricted by accessibility or land-use controls create the volume for diluting air pollution and so reduce its concentration and impact. Where open space is limited and the pollution load is high, air quality can deteriorate rapidly. For example, Griffith Park, Los Angeles' only large municipal park, is an inadequate airshed buffer for the San Fernando Valley, where air quality has historically been exceptionally poor; moreover, sometimes the park contributes to poor air quality when there is a wildfire. Many forms of vegetation have the capacity to absorb and remove volatile organic hydrocarbons from air, principally benzene, toluene, xylenes, formaldehyde, and some organochlorines such as trichloroethylene. (Some plants release volatile organic compounds, principally terpenes, that may either scavenge and remove ozone, at low concentrations of air pollution or contribute to photochemical air pollution, at high concentrations.) Forests appear to be particularly effective in absorbing and removing photochemical air pollution.

- *Watershed protection.* Natural ecosystems collect water, and through the mechanisms described above for water cleansing have a limited capacity to prevent water pollution. When sources of water contamination are rigorously prevented, as in New York's Catskill/Delaware Watershed Protection Program, surface water quality can be so high and reliable that drinking water treatment requirements can be made less intense, maintaining safety at a much lower cost.
- *Biological remediation.* Just as metabolically active organs (particularly the liver) in human beings and animals have specialized biochemical pathways that metabolize and decompose toxic chemicals (to a limit) or sequester them as storage depots, so natural ecosystems have pathways that degrade chemicals by metabolism or that store them and make them unavailable for circulation. Much of this metabolic capacity exists in the rich diversity of microorganisms in lakes and rivers (usually ones with a slower flow), which can degrade contaminants biologically through enzymatic pathways. However, these same pathways sometimes activate organic chemicals that were initially relatively harmless or may convert chemicals (usually metals) into forms that are more toxic than the original contaminant (such as the conversion of metal mercury into organic forms of mercury). These same chemical pathways may also accumulate metals, especially, to toxic levels in plants, invertebrates, fish, and other aquatic species. Also the chemical, metal, or organic may be toxic at some level to the organism that is metabolizing it, so that higher concentrations kill off the very means of decontaminating the contaminant. Therefore biological decomposition of chemical contaminants tends to take place in a range of concentrations and falls off at high levels, depending on the contaminant and the composition of the ecosystem.
- *Soil quality.* Natural ecosystems have a limited capacity to regenerate diversity and maintain productivity, which was the origin of the practice of "shifting cultivation," keeping land fallow after a certain number of crop rotations. During the fallow period, organic matter re-accumulates, nitrogen and phosphorus are conserved, natural fauna returns, and physical characteristics of soil return to their pre-planting features. Intensive modern agriculture has substituted artificial replacement of nutrients for this natural regeneration, called "authentic regeneration" by some. This service is primarily mitigational, since it restores qualities lost from overuse. However, it would be considered a "supportive" service (as described in the next subsection) in historical terms, because the practice of shifting cultivation supports agriculture.

Support

Supportive ecological services are those that support economically important activities from a human perspective and that allow other ecological services to maximize their benefit. These ecosystem services include, but are not limited to the following:

- *Nutrient cycling.* Cycling and recycling of nutrients, both the major nutrients carbon, nitrogen, and phosphorus and the micronutrients that are important for metabolism and enzyme activity, is a basic function of ecosystems. Together with photosynthesis, it is the essence of food production, for example.
- *Pest control.* Natural ecosystems keep economically important pests in check in several ways. One is to prevent or reduce the emergence of pesticide-resistant strains by sheltering a reserve population that is under less selection pressure but that freely intermingles with the pest population exposed to pesticide. They evolve with predators to prey on important insect pest species or with plant species to compete with "weed" species for a similar ecological niche. Natural ecosystems rarely, if ever, eradicate insect pests because the viability of the predator species depends on survival of a sufficient population of prey as a food source. As a result, the level of spoilage or consumption that may occur in a crop with no pest control intervention is usually unacceptable to farmers and gardeners. The use of sustainable natural predator populations to keep pest destruction to acceptable levels and pesticide use to a minimum or nothing is the objective of "integrated pest management." Removal of a key predator in the natural environment may lead to an explosion in the population of pest species. Introduction of an "invasive species," not indigenous to the area capable of exploiting an ecological niche, may create a new pest problem, in the absence of natural predators or competitors.
- *Fertility support.* Natural ecosystems play an important role in supporting the biological processes that support fertility and dissemination in plant species. Seed dispersal may take place by water, for example in mangrove swamps, and by airborne dispersion. Seed dispersal by birds and animals (technically called "zoochory") is an important example of an ecosystem service because it involves many fruit species. Pollination is another means of supporting fertility of plant species, with a crisis occurring at the time of this writing because of the collapse of bee populations, for reasons not entirely clear but suspected to be an indirect result of toxicity form neonicotinoids, a relatively new type of pesticide.

Cultural Uses

Ecosystem services also include cultural and spiritual functions that are meaningful in human life and civilization. Their value is hard to quantify, but it is very real. As an ecological service, more narrowly defined, natural ecosystems can be recognized as providing the following:

- *Aesthetic value.* Natural beauty and form have strong aesthetic appeal and provoke the same reactions of astonishment and awe as fine works of art. Even natural phenomena that are not pleasant to watch evoke a sense of wonder and appreciation. Unfortunately, this argument tends to put a premium on accessible and conventionally beautiful landscapes such as alpine meadows and discounts the value of ecologically important but often less spectacular sites such as deserts, grasslands, and mud flats (each of which can be beautiful in its own way). Aesthetics was the founding principle of the conservation movement in the United States, resulting in the preservation of numerous parks and national monuments before there was a complete understanding of ecosystem viability. However, this may not have been as arbitrary as often alleged, since for many ecosystems (such as rivers and streams) an attractive appearance correlates with ecosystem integrity. However, for some others appearances can be misleading, as in the clear water columns of unproductive seas and lakes and the intimidating, even grotesque (to humans) appearance of swamplands that are highly productive biologically. It may also be, and has often been postulated, that human beings are most attracted aesthetically to ecosystems that resemble the grasslands and forests in which the species evolved. This theory fails to explain the near-universal admiration for ocean scenery and mountains and the common perception by those who have not lived there that prairie landscapes are boring.
- *Recreation.* Without question, natural ecosystems provide opportunities to walk, hike, run, jog, climb, swim, boat, and enjoy beauty. Some nature experiences provide serenity. Others provide risk and the opportunity for extreme sports. One might postulate that recreation in nature or a nature like setting (such as a park) is a return to a state in which humankind lived in and worked within nature using many of the same skills (such as running) important in chasing, hunting, fishing, and food gathering. By providing a similar experience, recreation outdoors exercises and stresses those parts of the body and mind that became best adapted in humans to the natural environment. In the nineteenth century, there was a strong belief that time spent "taking the air" (especially in cold weather) kept children healthy, and this was prescribed as a cure for diseases such as tuberculosis and some other diseases. Another point of view might be that spending time out of doors usually means healthy exercise and exposure to sunlight, which resets circadian rhythms (sleep/wake cycles) drives vitamin D synthesis in skin, and lifts mood. A common point of view is that parks should be accessible in every neighborhood so that residents, particularly children, can experience being out of doors. Another point of view is that strenuous outdoor recreation is largely an elite activity, requiring access,

transportation, and expensive equipment (climbing equipment, special boots or shoes) that put it out of reach of the poor. The recreational value of being outdoors is limited or not easily accessible for many people who live in cities, especially if they are not affluent or live in a crime-ridden neighborhood. Thus, opportunities for outdoor and especially wilderness recreation has historically presented important issues in environmental justice.

- *Scientific discovery.* Curiosity and scientific interest often go along with time spent in nature. Nature hikes, birding, beachcombing, and more strenuous wilderness adventures are opportunities to learn, observe, and internalize concepts of biology and ecology, even when unrelated to the formal conduct of research or systematic education. Thus, there is great value in" field trips" to see biological phenomena firsthand, and "ecotourism" to observe first-hand the qualities and complexity of an ecosystem or habitat, which leads to greater understanding and appreciation of the reality. It then becomes much easier to visualize relationships, appreciate the magnitude of change, remember important principles, and understand the implications of change. Aldo Leopold (b. 1887–d. 1948), the premier ideologist of conservation, wrote that "recreational development is a job not of building roads into lovely country, but of building receptivity into the still unlovely human mind."
- *Existence value.* "Existence value" is the concept that some ecological and cultural phenomena are of such profound importance that they confer benefit by simply existing; knowledge of their existence alone has such value that people will and should incur a cost to protect and maintain them even if they never see these phenomena. The originator of the term, economist John V. Krutilla (b. 1922–d. 2003), used the term "sentimental value." It is the leading example of "non-use value," the idea that some things have economic or trading value even when they are not used. Examples of existence value might include the cultural appreciation of the iconic whales of the deep ocean (which most people will never encounter in real life, save for short "gray whale" tours), Antarctica (a small part of which can be toured, but at a high cost), the Grand Canyon (which is considerably more accessible, although the North Rim is not easy to visit), charismatic megafauna species such as elephants in Africa, and cultural monuments such as the Taj Mahal. The concept of "existence value" has been recognized by a number of governments, particularly those with exceptional natural monuments, such as New Zealand and Canada. Critics have argued that existence value is meaningless in economic terms, represents advocacy concealed as economic analysis, and is so subjective as to be theological in its complexity. They are correct insofar that the concept of existence value overlaps with the spiritual dimensions of sustainability, as discussed in chapter 12.
- *Option value.* Another nonuse value is option value, which involves the present value of a resource conserved for the distant future when sustainable

management might be possible in, for example, space, remote locations or the deep ocean. The most common nonuse value familiar in daily life is "bequest value," which is the value of saving something to provide an inheritance or legacy; the benefactor will not use the asset but the beneficiary may. This has obvious resonance for sustainability. Examples of option value might be mineral resources left in the ground, anticipating some future time when exploitation will be practical or technology to extract it will be less damaging or expensive.

- *Tourism*. Closely related to other cultural uses, but with more commercial implications, tourism is a powerful economic driver and has the advantage, in theory, of creating strong incentives to preserve the attraction. In practice, crowding, poor land use, distractions, opportunism, and traffic may degrade the experience if the attraction is not well managed.

Water

Water is so important that it requires further elaboration. Water is fundamental to many of the ecosystem services described above. Only 3 percent of the world's water is fresh water. The basic water cycle ("hydrologic cycle") of evaporation, condensation, precipitation, accumulation (as surface water, snowpack), and run-off into open water is supplemented by two side cycles: one for groundwater (infiltration, percolation, accumulation in the aquifer, and access or artesian sources) and one for plants (absorption, transpiration). Fresh water is treated as a renewable resource because it flows more or less predictably in lakes and rivers and the water cycle of evaporation, condensation, and precipitation has provided a constant, recycled supply most of the time. In time of drought and climate change, however, it becomes all too clear that, at least in relevant time frames, fresh water behaves like a nonrenewable resource. There are only so many water molecules in existence on earth, and access to the water that exists is often tenuous. Groundwater, especially, may recharge very slowly, and when that occurs extracting it is tantamount to "mining" fossil water, since the accessible resource may have accumulated over millennia (as in the case of the Oglalla Aquifer) and will not replaced in any humanly relevant time period.

Water is unequally distributed. The majority of fresh water on the planet is located in only ten countries. The largest volume of precipitation is in Brazil, the largest volume of flowing freshwater is in Russia, and the largest volume per capita is in Canada. The largest repository of freshwater in the world is the chain of Great Lakes shared by Canada and the United States. The United States and China are both among the top ten countries in the world for total renewable water resources, but both have vast arid regions. On the other hand, thirty-three countries are so deficient in water resources that they depend on other (adjacent) countries to supply them with this basic necessity.

Water is considered a common good. And because it is a necessity of life, costs in many countries are kept artificially low for reasons of social equity and stability. In most countries, there is no "abstracting fee," or price, for removing water from either surface water (rivers, lakes) or groundwater sources. However, water is immensely valuable as a necessary resource for life. When externalized costs are taken into account, the price of water is generally much lower than its usage would merit. A higher price for water would support conservation, effluent controls, groundwater recharge measures, and wiser use but a higher price would also impose a hardship on low-income families and importing countries.

Water has many important uses directly related to health, some obvious and others less so: drinking, cooking, cleaning and washing, sanitation and flushing, bathing, watering crops (including irrigation), a habitat for food, and a medium for aquaculture. Freshwater use that indirectly involves health includes recreation, hydroelectric power, cooling, industrial production, inland transportation (on boats, ferries, and barges), and groundwater recharge. Fresh water is also habitat apart from food species.

Considering these various uses and their health implications, public health recognizes classifications of water-associated diseases as follows:

- *Waterborne diseases.* These are conditions that arise from contamination of water used for drinking or in the preparation of food. They may include disease from biological agents such as viruses, bacteria, and parasites, or chemicals such as arsenic (found in groundwater in many parts of the world).
- *Water-privation diseases.* Conditions that result from inadequate personal hygiene, such as failure to wash hands or bathe. Primarily, they involve infectious causes of diarrhea spread by the fecal-oral route.
- *Water-based diseases.* Water may be not only the means of conveyance but the habitat for some pathogens, their hosts, and their vectors that directly affect human beings. The waterborne disease schistosomiasis is the prime example and is ranked second in burden of disease in the developing world after malaria.
- *Water-related disease.* Water is habitat for arthropod disease vectors, such as mosquitoes, which in addition to malaria can carry dengue, chikungunya (a denguelike disease currently establishing itself in the Caribbean), yellow fever, and various forms of encephalitis.
- *Water-dispersed infections.* These conditions occur when pathogens that enter or proliferate in fresh water are ingested, of which there are numerous examples, or inhaled in water aerosols (droplets suspended in air), such as occurs with Legionnaire's disease (usually from the cooling system of buildings).

Although water provision is an ecosystem service, effective use of the water is an engineered process. The resource must be intensively managed for safe use. It is also a commercial and agricultural management issue: when food is raised in one country or region and exported to another, it is the functional equivalent of exporting the amount of water used to grow the food including the current water content of the food.

Not surprisingly, water-source protection and treatment to prevent disease are major concerns of public health. This normally involves protection of the watershed and collection systems, intake of "raw water" into the system, flocculation and coagulation (which aggregates the particles into clumps, or "floc"), sedimentation (to remove the floc), filtration (to remove the bulk of microorganisms), disinfection (chlorination or chloramination), storage (in a reservoir or elevated tank), and distribution, to the end user.

Likewise, wastewater protection is a critical public health technology once the water is used and drained or flushed. Wastewater is carried in septic sewers to a central wastewater treatment plant where it is screened (to remove large objects), aerated to facilitate bacterial digestion of organic material, held in a settling tank (in which heavy solids fall to the bottom and light fats and oils rise to the top), and decanted. Most of the organic matter and bacteria are removed with the solids. This is called "primary treatment," and at this point the wastewater is discharged into surface water if permitted. However, some risk is involved, as primary treatment is not effective in removing pathogens on a large scale. "Secondary treatment" is standard for municipalities, in which the wastewater is again aerated and held in a digester, where a managed culture of microorganisms further degrades organic matter, or "activated sludge." The leftover solids, or "sludge," may be fermented in the absence of oxygen to generate methane for fuel or, if the content of metals is low enough, dried and used as fertilizer and artificial topsoil for agriculture. (Biosolids are often involved in local disputes over nuisance and land use, however.)

There are many variations on secondary wastewater treatment. At one extreme, biological treatment involves carefully managed flow of treated wastewater into constructed wetlands that can support habitat for aquatic species, remove nutrient chemicals, aerate water naturally, support metabolism by aquatic species and natural decontamination, and selectively take up contaminants by plants. At the other extreme of technology, membrane bioreactors are devices that use activated sludge treatment in combination with solid-liquid separation across membranes, to increase efficiency and transit speed of treated wastewater for discharge into waterways. However, membrane bioreactors tend to be expensive.

"Tertiary treatment," in which chemical contaminants are removed, may also be used to bring the effluent water as close as possible to the quality of natural surface water before it is discharged. Sometimes the final effluent is disinfected before discharge.

Clean-up of wastewater is particularly important in the near future because of climate change and dimishing water resources in many arid lands. It will eventually become essential to use treated wastewater for irrigation and as a source of drinking water and therefore, to overcome the remaining impediments (mostly related to the risk of viral contamination). The health risk implications are obvious but with appropriate technology there may also be also substantial opportunities for reducing health disparities in parts of the world where water is scarce.

Value of Ecosystem Services

The economic value of ecosystem services is immense but never tracked by traditional accounting. The benefits derived from ecosystem services are treated as "free goods," as if they carried no price. Access to ecosystem services are incorporated into utility costs only insofar as infrastructure costs and price are concerned. However, the value of ecosystem services dwarfs sectors of the human economy. A minimum estimate (calculated as a low estimate in 1997, adjusted to 2013 without taking into account any increase in value) is between US$23 to 78 trillion worldwide, compared to the 2010 US GDP of $15 trillion and the estimated global combined GDP of US$72 trillion. In other words, the value of all services provided to humankind by the ecosystem is comparable to and probably much greater than the combined domestic products of all countries in the world, but it so poorly accounted for that the uncertainties are almost as large as the minimum estimates.

Ecological services usually have a huge but not unlimited capacity. Because human society as it developed was embedded in (and depended on) these services, and because replacement is impossible, impractical, or prohibitively expensive, ecological services must be relied upon. Once their capacity is saturated, the effects can be gradual, as in diminishing quality of air and water, or catastrophic, as in flooding due to failure to conserve wetlands.

Loss of ecosystem services essential to health or supporting the health of populations can be catastrophic. Even partial or local loss of services may result in critical health consequences, particularly in the case of those services involving water.

For the most part, ecosystem services have been partially protected by legislation and regulation when they have been recognized, but the magnitude of their value and the abundance of services they provide has not been appreciated by decision makers or the public. A relatively new approach advocates having users pay to preserve and protect ecosystem services: this is an idea known as "Payment for Ecosystem Services" (PES). This has the advantage of monetizing the value of the services and making people aware of what a good deal they are getting by letting nature handle these essential services.

Payment for Ecosystem Services

Payment for Ecosystem Services (PES) is a relatively new idea that involves money and value transfers to people who are stakeholders affected by the preservation of property essential to ensure continued ecosystem services. It is critical that such payment be sustainable over the long period, scaled appropriately to the economy (so as not to cause inflation and market distortions), and appropriate to the value of the service. In concept, the owners of critical properties are paid a "rent" for the services of their properties and in return are obliged to refrain from certain activities, preserve certain land use, protect essential infrastructure, and forgo other economic activity that may interfere with the ecological service use. This is quite different from a subsidy or aid: a service is actually being performed in return.

The premier example of PES in the United States is the Catskill/Delaware Watershed of New York (referred to above). This is an enormous tract of land in the Catskill Mountains in which a compact was agreed upon in 1997 between local government and private interests and the city of New York. New York provides substantial direct monetary payment, derived from water fees, and indirect payment in the form of technical assistance to the region, which gives residents stable income, lower taxes, economic opportunity, and a high quality of life. In return, construction is limited, waterways cannot be diverted, farming (dairy and vegetable) is constrained from heavy use of pesticides or other chemicals, land use is restricted, wetlands and forests (78 percent of ground cover) must be conserved, and sanitation is required to protect surface water. Wastewater treatment plants have been installed for municipal areas in the region, and there are programs to support organic farming, mixed land use (including hunting and fishing), and regional economic development. As a result, the quality of water from this watershed is so high that water treatment can be kept to a minimum (skipping filtration altogether, although the water is of course still chlorinated), saving huge infrastructure costs. For an annual cost of the program on the order of $100 million (itself a high estimate), New York City receives benefits of about an equivalent amount (equal to the cost of operating a filtration plant), with savings in capital costs on the order of $10 billion (the cost of building a filtration plant that is not necessary because of watershed protection). The system delivers about 1.5 billion liters of water per day to the city, reliably and much less expensively than an engineered collection and purification system would.

Ecological Services and Traditional Knowledge

Another example, this one from Australia, is the joint management of significant natural sites between state and Commonwealth government and "traditional owner" Aboriginal peoples, who regained title to the land. As part of the atonement movement to redress past treatment of Aboriginal peoples in Australia and to recognize Aboriginal land rights, the Commonwealth government and some state governments negotiated a return of land title together with a novel land-use arrangement with the Aboriginal peoples of the continent. Under this agreement, the ownership of what had been national parks was transferred back to the tribes that traditionally occupied the area.

In addition to cultural sensitivity, the arrangement made practical sense. Historically (and prehistorically), Aboriginal groups have been meticulous in protecting their lands. The expectation, therefore, was that management of these lands would actually improve, at lower cost.

The first instance of joint management was in a geographically well-defined area shared by four aboriginal groups, now Gurig Gunak Barlu National Park. The most famous example has been the return of Uluru (formerly Ayers Rock) and Kata Tjuta (formerly the Olgas) to the indigenous Anangu people as Uluru—Kata Tjuta National Park, which is considered to have been an outstanding success. The federal government returned the land to the Anangu, then leased back the land for use as national parks, giving the Aboriginal community a stable source of revenue derived from park entrance and user fees.

Although Aboriginal communities in Australia have had to overcome many hardships and are alienated, to various degrees, from traditional culture, many have maintained their traditional knowledge of the land. Those communities have retained deep empirical knowledge of how to manage the ecosystem sustainably. The community (less distinct than a "tribe" might be among North American aboriginal groups) maintains rights to hunt and forage on the land and to set brush fire as they see fit with the benefit of traditional knowledge, since fire is essential to regenerating the natural ecosystem. They also have the right to declare certain areas off limits, especially during rites and ceremonies, and to tell the tradition of their people in their own way. Jobs are also created in park facilities and at interpretive centers (rangers) and through retail outlets for Aboriginal art and crafts.

Valuating Ecosystem Services

Serious studies of ecosystem services are almost always economic valuation studies. The values placed on ecosystem services are primarily theoretical, since they are almost never sold, and are useful primarily in demonstrating the importance of natural functions and comparing their enormous magnitude and value with that of human engineered alternatives. Ecosystem services are "bundled," meaning that one resource (such as surface water in flowing streams) may serve many functions and so generate a multiplicity of values.

Different economic models have been used to assess the value of these services. The details of these methods are beyond the scope of this book, but they include the following:

- *Replacement value.* How much would it cost for human beings to perform the same service using an engineered system?
- *Avoided cost.* How much cost that would otherwise be incurred is avoided because natural ecosystems are available to do the work?

- *Factor income.* How much does the ecosystem service add value at every step along the value chain (e.g., from fish habitat to consumption)?
- *Willingness to pay.* How much are people willing to pay to maintain the ecosystem service? Estimates are imputed from "revealed choice" expenditures, often called "hedonic pricing," that show how much people are willing to pay to access or use the resource. These expenditures include: the premium that consumers pay for natural products over artificial commodities (such as fresh ocean fish over aquaculture-grown fish of the same species), the average travel and hotel costs of tourists who visit a particular natural site, and the price differential between, say, a house with an ocean view compared to a similar house lacking the view.
- *Conditional pricing.* How much would people be willing to pay, in theory, for various alternatives presented to them?

Critics of ecosystem service valuation (from a development perspective) correctly point out that it is an exercise in justification for conservation, since the magnitude of estimated benefit is ordinarily huge (they would say often inflated), features huge uncertainties, and does not take into account the opportunity costs of foregone development or exploitation. However, this argument tends to assume that the costs actually are similar and are being distorted. In fact, the distortion is in the other direction. It is indeed inconceivable (but also highly unlikely judging from the numbers and the long time span of sustainable use) that a valuation study could be used to argue that an entire local ecosystem or habitat is not worth saving. However, that is not because the values are distorted but rather because they take everything into account, while conventional valuation does not. The value of the exploitative use is tangible, accrues to a group of people who are highly motivated (often for one-off gain), and infuses money into the conventional economy, which is amplified by the multiplier effect of wages, trade, sales, and transfers into jobs and infrastructure. The value of the undisturbed ecosystem is usually invisible or at least taken for granted; it benefits everyone to a large degree in the aggregate but to a degree that is perceived as proportionately small (a classic example of the "commons") and may be so inapparent as to be invisible to the community (e.g., airshed protection). The benefits also accrue indefinitely and are not subject to depreciation, thus some of the most important terms in the equation can grow very large and even approximate to infinity.

Realistically, it is hardly conceivable that the total ecosystem service contribution of an undeveloped property, properly accounted and taking into account future use, would be less over time than the sales price of a building, new housing development, golf course, or facility site—in other words, assets that will incur costs to maintain and that will depreciate. Yet such trade-offs are made all the time in the private sector; this is because the noneconomic, conservation use of the property is not properly valued and there is no counterparty in the market to put a price on conservation. Valuation is generally done only for a small number of projects in critical environments. Overall, therefore, the balance is already skewed heavily on the side of development and new projects.

Critics of ecosystem service evaluation from a conservation perspective point out, also correctly, that it is a sterile exercise. They argue that ecosystems have value far beyond their economic utility and can never be completely or satisfactorily replaced with human alternatives. They also argue that opportunity costs are short term, while ecosystem services are not only extremely long term but of unpredictable—but probably critical adaptive—value to future generations. They point out that since all life depends on ecosystem services, and therefore that all economic value depends on maintaining an intact and viable global ecosystem, parsing out the value of any given ecosystem service is naive and silly. At best, this critique would say it plays into the agenda of monetizing everything and holding nothing precious in its own right.

However, valuation of ecosystem services does have practical value. It provides an inventory of essential functions that must be understood for their role in sustainability, identifies key resources and the limits on their capacity, justifies regulatory limits, and supports the imposition of user fees, fines, penalties, and damage awards. Of particular importance, valuation may be used to calculate reasonable payment for ecosystem services, as described earlier.

Natural Threats to Sustainability

Sustainability, as it is usually understood, assumes human agency and human benefit. However, natural environmental events and exposures also affect health. The effect of the environment on human health is not limited to changes in the environment initiated by humans. The natural environment can also have adverse effects on sustainability and human health risk, and the results can occasionally be monumental or even catastrophic (see chapter 4).

Leaving aside the evolutionary past, one can identify the following classes of current threats to human health that arise from natural systems and have a connection with sustainability. They include

- natural disasters, such as earthquakes and hurricanes, which are natural in origin but the effects of which may be significantly worse or better depending on human anticipation and so have a component of human agency (natural disasters are also discussed in chapter 3)
- natural exposures, such as arsenic, which arise from entirely natural sources (such as groundwater and coal) and that become problems when human populations access the natural resource

- natural processes, such as weather, which are perturbed by human activity (such as climate change), resulting in an exaggerated impact
- natural resources, such as shale gas (natural gas in deep underground shale formations), which, when exploited, present problems of both sustainability and human health impact associated with the technology used
- undisturbed or semidisturbed ecosystems, such as the rainforest, which are increasingly penetrated by human beings, bringing communities into contact with novel pathogens (such as HIV and the Ebola virus) and other hazards

These issues are significant because their effect is substantive, the primary drivers are not necessarily the result of human intention, and they do not fall neatly into classifications of pollution or ecosystem disturbance. The community of the earth sciences (geology, geochemistry, mineralogy, hydrology, and related disciplines) has recognized this class of issues and has developed a specialized field known as "medical geology."

9

Artificial Ecosystems

ARTIFICIAL ECOSYSTEMS ARE organized physical, chemical, and biological environments that have some of the same characteristics of natural ecosystems but are created and maintained by human beings. Several types of artificial ecosystems are recognized, although only agriculture, urban landscapes, and built environments for human habitation (buildings) will be discussed in this chapter (see table 9.1).

Some examples of artificial ecosystems include the following:

- Landscape-scale biological communities residing in artificially created environments that rely on managed systems to maintain their continuity and function: examples include agriculture and managed forestry
- Landscape-scale human communities residing in artificially created environments that rely on managed systems to maintain their continuity and function (e.g., cities and towns, indeed most human communities)
- Small-scale artificially created "built environments" that rely on managed systems to maintain their continuity and function, usually buildings
- Formerly natural ecosystems that are subordinated into a larger artificial ecosystem, such as the rivers of London referred to in chapter 8, that are incapable of maintaining their biological communities in a natural state
- Ecosystem replicas, which are designed to duplicate features of natural ecosystems but are also open systems that would quickly degrade or fail without intensive human management. Examples of these replicas include zoos, parks (other than wilderness), gardens, and "biodomes," or, on a small scale, aquariums and terrariums
- Enhanced or protected natural ecosystems that are preserved and made more biologically or economically productive and subordinated to human use, such as urban parkland or the practice of disposing of scuttled ships or placing engineered concrete structures to create artificial reefs in the ocean

Table 9.1 Table of sustainability values: Artificial ecosystems

	Sustainability Value	Sustainability Element
I.	Long-term continuity	Artificial ecosystem services are essential for long-term continuity of human communities but are completely dependent on technology and intensive management.
II.	Do no/minimal harm	Mismanagement of artificial ecosystem services can and often does cause substantial and often devastating harm to the environment and direct human health threats, such as pollution of air and water.
III.	Conservation of resources	Conservation of resources for the future requires the management of artificial ecosystem services to minimize resource consumption and maximize efficiency.
IV.	Preserve social structures	Because artificial ecosystem services, such as water distribution, are essential to social structures and social stability, reasonably equitable and effective management is essential to avoid harm to or to enhance social structures.
V.	Maintain health, quality of life	Artificial ecosystem services are essential to maintaining health and the quality of human life, because that is where people live.
VI.	Performance optimization	Maintenance and management of ecosystem services is essential for optimization or maximization of economic, social, and environmental performance; however, management by human institutions is inevitably imperfect, leading to misjudgment, mistakes, neglect, and corruption, all of which threaten sustainability and health in the communities served.
VII.	Avoid catastrophic disruption	Artificial ecosystems, relying as they do on technology and engineered systems rather than self-governing feedback systems, present a real risk of catastrophic disruption.
VIII.	Compliant with regulation	Operating the infrastructure that provides artificial ecosystem services must be compliant with regulation and best practices or health may be at risk.
IX.	Stewardship	Protecting the artificial ecosystems that provide essential services also protects natural ecosystems upstream that feed them and downstream that receive waste; it is consistent with a responsibility for the health of the human population served, as well as protection of the environment and therefore environmental sustainability.
X.	Momentum	Artificial ecosystem services are not capable of supporting continued progress under their own momentum and must be continually monitored and managed.

- The "built environment" where most people live, within structures of human construction called "houses," which, when occupied and animated by a sense of personal place and family life, become "homes"
- Buildings and the indoor environment
- The workplace, a built environment of great social complexity
- Communities and human settlements, characterized by density and infrastructure
- Human life-support systems designed for protection in hostile environments or to sustain short-term exploration or travel or economic use, such as Antarctic scientific research stations, offshore oil platforms, long-running submarines, space capsules, and the space shuttle
- Demonstration or simulation projects, such as Biosphere 2, that are created for scientific or creative purposes
- Computer-generated virtual ecosystems and models that have no real analogy to ecological services

Obviously the definition becomes a bit arbitrary at the edges. A space capsule may be and is often described as an artificial ecosystem, with many features of a closed system (such as rebreathing atmospheres and water recycling) while it is aloft or in orbit; it is not designed for sustainability over the long term, or for human habitation on a longer time scale. The features of sustainability built into these life-support systems are intended to last as long as it takes to keep the human payload (an astronaut or a cosmonaut) alive for the mission and a reasonable time during which rescue can be attempted. Thus, although common usage often refers to such life-support systems as artificial ecosystems, they actually represent something altogether more modest.

Characteristics of Artificial Ecosystems

Artificial ecosystems have the following features in common:

- They are designed to maximize human benefit but not to optimize the productivity of a biological community.
- They are open systems (as defined in chapter 4), dependent on inputs (such as fertilizer in the case of agriculture and food from distant sources in the case of cities) supplied from outside the ecosystem, rather than being maintained by nutrient cycles and trophic levels and being naturally powered by solar energy as most natural ecosystems are.
- Biomass created within the artificial ecosystem (produce in the case of agriculture, waste products in the case of a city or building) are removed from the ecosystem so that it also is an open system in that it does not retain its production.

- Artificial ecosystems are dependent on human agency for their basic functions, and when management stops, entropy prevails. The result is that when the system is no longer actively supported it degrades, collapses, or reverts to a successor ecosystem, usually to a much less desirable state in terms of biological productivity or human utility.
- Artificial ecosystems lack redundancy, diversity, and stability and so are much simpler than natural ecosystems. Few artificial ecosystems can match the biological diversity and complexity of interaction of even the simplest natural ecosystems.
- They have ceased to evolve, in the sense that natural selection stops. If selection takes place within the artificial ecosystem, it is guided or protected by human agency.

Although the points made above emphasize the "un-naturalness" of artificial ecosystems, they are by far the most familiar surroundings for almost all people; the manipulation of the environment for security and comfort is the basis for all human communities, without exception. In fact, people in the contemporary world never live in natural ecosystems without artificial adaptation, even if this adaptation is limited to just constructing shelter. Notwithstanding their importance for ecological services and in the imagination, wilderness and pristine nature are hardly ever a viable habitat for human communities. On the other hand, almost everyone in the world lives in entirely artificial ecosystems such as large cities, suburbs, towns and villages, and farms, and if they do not live in human settlements they still spend most of their time in constructed shelters. These artificial ecosystems also include places where the landscape has been largely constructed and so drastically altered as to be entirely unnatural, even though they are popularly considered to be part of "nature." Examples are the hedgerows and manicured meadows of England's "green and pleasant land" (a reference to a poem by William Blake, [b. 1757–d. 1827]); the fields and rolling hills of Germany or much of North America (both previously dense forest); the farms, paddies, and terraces of Southeast Asia; the polders in the Netherlands (which were reclaimed from estuaries or the seabed by pumping and drainage); the waterside floating cities of Thailand; and the floating houses of Amsterdam and Rotterdam (built as adaptation to accommodate sea-level rise resulting from climate change).

More controversial is whether restoring a highly degraded natural ecosystem to something approximating its former natural state turns it back into a "natural" ecosystem. This question has been raised repeatedly by scholars of conservation biology following ecosystem restoration projects such as drained wetland (as with much of the Everglades in Florida), recovery of streams and bays that have been badly polluted, dam removal and aquatic habitat restoration, reintroduction of wildlife into a habitat in which a key species went extinct, or replanting of native grasses on a prairie that was dominated for a century by agriculture and introduced species. These examples of "ecosystem restoration" represent a specialization of conservation biology that has had notable successes in recent years. However, restoring a damaged ecosystem is expensive, difficult, data intensive, highly difficult with conflicting stakeholder interests, and sometimes counterproductive. It is almost never possible to completely replicate the preexisting natural state of the ecosystem or to create a new ecosystem that meets specifications for an ecosystem that previously existed in the location but was lost. The test of whether a restored ecosystem is an artificial creation or has actually been returned to a natural state logically depends on whether it sustains itself without human intervention and evolves over time, not on whether it meets defined specifications or "looks right."

Some ecosystems are self-repairing (such as farmland that returns to second-growth forests), and desirable endpoints may be a compromise reflecting what is possible. Other ecosystems pass through successions and evolve over time anyway, and the selected end point for restoration may be quite arbitrary from a biological point of view, since the ecosystem would naturally change on its own over time.

A similar question could be extended to whether a vulnerable natural ecosystem, once it is protected by human effort (such as land-use planning or conservation measures, including declaration as a park or wilderness), ceases to be "natural" and becomes "artificial" by virtue of continuing to exist only by the tolerance of human society and under the protection of human authority. Most people would probably say that having been a self-sustaining natural ecosystem, it remains a natural ecosystem even if it could no longer survive without protection.

Perhaps the most profound question, however, is to what degree human beings should be considered agents of change in the natural environment before considering the ecosystem "artificial." Even in the Paleolithic era, the capacity of human communities to change the environment was profound. Fire was used to change landscapes to a more favorable habitat and to encourage regeneration of desirable species for good hunting and gathering. Does this mean that the ancient red earth landscape of the Australian outback or the Canadian boreal forest is actually a human creation? These landscapes were shaped mostly by controlled burning to encourage regrowth of desirable species by native peoples. During the Neolithic era, mammoths and other large species (the evidence for ground sloths is circumstantial) that would have played key roles in the biological community had apparently been hunted to extinction by human beings. Since modern elephants are known to change their environment (creating clearings by trampling saplings while browsing), no doubt the extinction of the mammoth had effects on their ecosystem. By the Bronze Age (starting roughly in the fourth millennium BC), large swaths of Mesopotamia and later both Greece and China had been deforested. Was this the "natural" work of a particularly aggressive species of primate (humans), or did human disturbance eventually constitute steps in the creation of an artificial world, the one which persists today and that our forebears only assumed was "natural"? These are questions beyond the scope of this chapter but are intriguing no less in environmental studies than in archeology and ecology.

In order to keep the discussions manageable, this chapter will only consider artificial ecosystems at the landscape scale and human-scale "built environments," in other words, buildings.

Agriculture

Agriculture is perhaps the clearest and most important example of sustainability as a cultural construct and of the role of human behavior in shaping the environment; it is also an example of the peculiar combination of stark economic reality and sentimental irrationality that pervades choices about sustainability. Agriculture demonstrates the complicated interrelatedness between human population and the environment and the human exploitation of "renewable" resources. In short, agriculture is foundational to human sustainability and at the same time demonstrates inherent contradictions within sustainability and health risk.

Agriculture, together with teamwork (such as coordinated hunting) and aggregation into structured communities on at least the tribal or village level, was among the earliest and most successful human behavioral adaptations to mitigate risk. Small tribal units faced potential extinction if they could not feed themselves because of drought or a failure of forage or lack of wild game. Managing this existential risk of small tribal units through agriculture was an essential step in the history of human sustainability. Without agriculture, continuous settlement would probably not have been possible for populations larger than an extended family and would always have been tenuous. It is inconceivable that the early villages that expanded and led to the first Neolithic proto-cities could have developed without agriculture. Jericho (Levant, 9000 BC), Çatalhöyük (Anatolia, 7500 BC), Uruk (Mesopotamia, 4000 BC), and then the many ancient cities that followed would scarcely have been imaginable, and their population would probably have reached the limit of subsistence agriculture, not to mention water supply and waste disposal. Agriculture also allowed a surplus of food and raw materials to be accumulated, which could be used to even out supply in lean years or used in trade.

About 10,000 years ago agriculture on a sustainable and subsistence level created such a change in human life that this era is called the "Agricultural Revolution" by anthropologists. From hunting and gathering, human communities (thought to be small nomadic clans at the time) acquired the ability to produce larger and more predictable quantities of food in locations more conveniently situated close to home. This reduced their dependence on the variable abundance of wild game and natural stands of edible plants and was therefore sparing of human energy. This newfound convenience also raised the possibility of growing a surplus and opened a door to new technologies for the more efficient growing and harvesting of crops and the eventual genetic manipulation (initially through selection and cross breeding) of species or strains that would be most useful.

The technology of cooking had already existed for tens of thousands of years and had increased the benefit of hunting by improving the extraction of nutrients and energy from meat by chemically altering it to make it more digestible than raw meat and less apt to spoil quickly when stored. Similarly, production of a larger quantity of edible plant materials made it practical to process foods in new ways. In the same way that cooking had created a more efficient protein source, more advanced cooking technologies such as baking made starchy foods more easily digestible and nutritious. The process of acquiring and consuming food became much more efficient, and it was possible to produce a surplus for storage and trade. Much more human energy derived from nutrition was now at the disposal of human communities to be applied in inventing civilization and making war.

At least, that is what anthropologists think happened. What is clear from the evidence is that the onset of agriculture appears just before a then-unprecedented increase in human population around 8000 BC and the emergence of the first cities, such as Uruk and Jericho. This increase in population, the great migrations of peoples, and the emergence of the first recorded traces of abstract reasoning and language has been called, in the aggregate, the "Neolithic Revolution." The technological innovations that emerged during this period led to the so-called Bronze Age and then the Iron Age (beginning *circa* 1200 BC), which depended on a sustainable food supply achieved at a low enough energy cost that part of the population could specialize in technical work and creating art, while an elite could indulge in leisure activities, such as writing epic poetry (chapter 12).

Today, of course, agriculture is essential to sustain the human population, and productivity must grow at least commensurate with population if standards of living are to be maintained. On the whole (see the discussion on famines in chapter 3) it has kept pace despite predictions of catastrophe. (See chapter 3.) The latest set of challenges to agricultural production, related to global climate change, will play out over an extended period of time and in ways that are still unclear.

Agriculture, it could be argued, is a human construct and fundamentally unsustainable beyond a subsistence level. And yet it has "sustained" itself and civilization for millennia. Agriculture has also transformed the natural landscape so that landforms perceived today are regarded as "nature." Agriculture also cannot yet be replaced with an alternative technology, although agriculture itself has and will change over time. (Whether tissue culture for edible products is agriculture or something else is debatable.)

Agriculture as an Artificial Landscape

As agriculture spread, it also changed the landscape. Forests were cleared (often by fire), marshes were drained, and paths were blazed into the formerly intact landscape in order to bring produce back to where it would be consumed and (presumably not much later) used for trade. This transformation must have begun at a time when human beings still had little ecological impact and probably before they developed the technology to survive the Ice Age and drive megafauna such as mammoths to extinction. The role of human beings in shaping the landscape then developed gradually and inexorably. The process of deforestation, agricultural transformation, landscape appropriation for human purposes, and urbanization is basically the ecological history of every place on earth that has been densely settled by human beings: Mesopotamia, the Indus Valley, the river valleys of China, the primeval forests of southern and then northern and central Europe, and, much more recently, most of the Americas and now Africa and Amazonia.

Agriculture may also be related to another form of cultural landscape, namely the maintenance of habitats conducive to hunting and to transitional ecosystems conducive to harvesting (such as the Australian outback where Aboriginal clans intentionally burn areas to promote regrowth and diversity). Indeed, it is possible that intentional manipulation of the environment by fire to increase yield from hunting and the gathering of grasses with edible seeds was as much at the root of the origin of agriculture as the conventional speculative explanation, which is the accidental discovery of a stand of grain or other food plant where seeds had been spilled or discarded. This is simply not known and probably unknowable. It is clear that if agriculture arose from sustainable harvesting practices then a transition to sustainable subsistence agriculture would not have been a huge leap. Human communities were tiny and probably had little impact on the larger-scale ecosystem for all but the very end of human pre-history. (See box 9.1 for more on cultural landscapes.)

Drivers and Pressures

Agriculture today supports a world of 7 billion people, and this is a world with a declining rate of poverty (at least for now). Therefore, the population in general has more disposable income to buy better-quality food (not necessarily tastier but less often damaged or visibly contaminated) that is more nutritious (more protein and energy content, with sufficient micronutrients), and lasts longer (processed to prevent spoilage).

BOX 9.1

Cultural Landscapes and What is "Natural."

Are agricultural landscapes "natural"? It could be argued (and usually is by environmentalists) that agriculture creates an artificial "built environment" and that human beings have been shaping the environment for millennia through technology, collective behavior, and shared conscious intent. It may also be argued (and frequently is by anthropologists) that the early influence of human beings and human communities on the ecosystem was perfectly "natural," in the usual meaning of the word.

The argument for a landscape as artificial as a field to be considered "natural" derives from assumptions about the relationship between human communities and the landscape. Human beings are not unusual among species in being social and in forming communities that change the geological landscape. Coral reefs, for example, are the products of numerous organisms organized in colonies and building structures. Human communities coevolved with other species and had the same predator-prey relationships. Human beings clearly evolved, and the nature of their communities changed along with natural changes of the landscape over time. Eventually, human beings learned to change the landscape by intent, particularly by using fire. (Aboriginal communities around the world burned small clearings in the forest or in brush to increase biological diversity and to regenerate desirable plants. Some surviving aboriginal peoples still do.) In this view, the effect human beings and communities had on the earth until modern times was not unlike that of other animals that change the physical environment, ultimately to their own benefit.

For example, the vast boreal (northern) forests in Russia and Canada that sit on the globe like a crown appears at first glance to be untouched wilderness. However, although it may be primeval in the sense that it is ancient and parts of it are true wilderness, it is not untouched or undisturbed by human activity. There is abundant evidence that from the earliest times people have set fire to clear land and have shaped the contours and geography of the forest for human benefit: in doing so, people have affected its species mix by ecological succession and by creating ecological margins (like the transitions between forest and meadow) that support biodiversity. Humans have certainly hunted and used the rich resources of the forest freely, with an impact limited until very recently, in the larger scheme of things, by small numbers and Neolithic technology.

At some point (the inflection point selected varies with the author), technology introduced a new level of control and scale of change that has resulted in human beings becoming the dominant shaping force in the world, even for surface-level geologic and planetary change. Those who are persuaded by this argument often refer to the current geological era as the "Anthropocene" and consider it to have replaced the Holocene, which is the era since the last Ice Age (10,000 years ago), as a meaningful geological period. Proposals for the beginning of the Anthropocene depends on the preference of the author, but the hunting of large animals ("megafauna") like mammoths to extinction, the beginnings of large-scale agriculture (perhaps 8,000 years ago), the acceleration of upward trends in trace elements indicative of mining activity in the world during the Roman Empire (2,000 years ago), and the onset of the Industrial Revolution (300 years ago) have all been proposed—but these points are arbitrary because the influence of human activity on the environment developed gradually at first, but more or less exponentially, over thousands of years.

This discussion is in part an argument of semantics (what is "natural"?) and in part an implicit argument over ecological values, since the implication is that because agriculture is a human construct, an agricultural landscape is not part of the "natural world" and therefore should not be given the same respect and priority for conservation as a pristine environment. Sentiment clouds this argument, since the evocative and seemingly eternal gentle hills of rural England, the "amber waves of grain" of America, the vast dry steppes of central Asia, and the well-tilled fields of the valley of the Nile are every bit as artificial and human-created as the deforested valleys of Lebanon (where no cedar forests are left), the seemingly natural but created tranquility of West Lake in Hangzhou, the meticulously terraced rice paddies of south Asia, and ancient irrigation works in Sri Lanka. Rather than engage in an endless circular argument about whether humans belong to nature or stand apart, these human-created landforms are termed "cultural landscapes" by geographers and anthropologists and accepted for what they are: human inventions on a geological scale.

Figure 9.1 is a natural-appearing but thoroughly unnatural installation: an artificial lily pond. Its significance of for sustainability is explained in the caption.

These "drivers" (referring to the DPSEEA+C model introduced in chapter 7), together with culture-specific food preferences, create "demand," which is a form of social "pressure" that encourages farmers to produce more but at some risk of overproducing and driving the price down. Demand may also be segmented by preference, which is reflected in price (such as the premium paid in developed countries for organic produce), crop selection (hence the dramatic and relatively sudden rise in popularity of the endive from French specialty to

FIGURE 9.1 A tranquil lily pond at the National Museum of Australia in Canberra. This image is laden with cultural significance. Its visible surface is half-covered with lilies. Thus it reflects perfectly a puzzle or riddle that was popular in the 1960s: "There is a lily pond in which each lily buds into two new lilies every night. After a month, half the surface of the pond is covered with lilies. If you want to prevent the entire pond from being choked off, when should you act?" The answer, of course, is on that very day, because the next day the pond will be completely covered. This is a Malthusian analogy for the exponential growth of population and how quickly exponential growth reaches external limits. (See chapter 3.)

Photograph by © Tee L. Guidotti,

American food trend), and harvest practices (the global increase in fish consumption having led producers into aquaculture), among other adaptations. Above all, the most intense pressure is to produce more food more productively and at a lower cost. These pressures change the "state" of the existing industry so that in the case of intensive modern agriculture it has become dependent on hydrocarbon fuel inputs, synthetic chemical fertilizer, intensive pest control (predominantly chemical), and (for some crops) practices such as forced ripening. The state of the food industry also has changed to become globalized, low margin and so dependent on cheap labor, and in many countries selectively dependent on subsidies (e.g., sugar cane in the United States) and protected markets (rice in Japan), reflecting policy objectives. The state of the technology and organization of agriculture as a sector results in opportunities for "exposure" to various local and potential determinants of health: historically by foods supply and shortages, and in modern times literally in the case of pesticide residues, or figuratively in the case of exposure to price fluctuations. These exposures in turn produce "consequences," which may be understood as direct health effects mediated by food supply, food safety, and indirect health effects mediated by food selection in response to price. The direct and indirect health effects both affect different communities in different ways. Agricultural workers, particularly seasonal ones, generally experience low wages, poverty, lack of access to health care, and unstable community support. However, the urban poor may benefit from improved nutrition brought about by increased productivity—and thus lower-priced and higher-quality food.

Contradictions of Agriculture

Agriculture on a commercial (not necessarily subsistence) scale, as currently practiced, is the clearest example of renewable resources that lack sustainability, at least in the long term. The reasons are familiar and frequently cited but are worth listing here:

- By definition, agriculture involves the selection and cultivation of a particular species, resulting in a *monoculture*, which is a uniform biological community (whether plant [crop] or animal [herd]) with a minimum of biological diversity that would support it in the way that a natural ecosystem would have relationships and stability through interaction.
- Characteristics that would be unimportant in a natural biological community because of diversity are potentially catastrophic in large populations; for example, if a planted strain of grain is susceptible to a particular pest, the entire field is susceptible and can be devastated when the pest appears.
- Because the monoculture has uniform vulnerabilities, pest populations may grow exponentially if not artificially constrained; in practice this requires use of pesticides and the unintended consequences of contamination and ecotoxicity and the constant race that comes from selection for resistance.
- Inputs of resources, especially water, can be and usually are enormous, and supply is often limited or nonrenewable. For example, the Ogallala Aquifer (a huge part of the High Plains aquifer system, an enormous but shallow natural reservoir of groundwater drawn upon by farmers in eight states) is now tapped to below existing well depths and contaminated by pesticides (primarily atrazine), as a direct result of corn production; this is a classic case of the "tragedy of the commons" (see chapter 3).
- Because the monoculture has uniform requirements, a large and continuing investment of the energy is required for transport, cultivation, and

harvest; the energy investment, principally from diesel fuel and the industrial Haber-Bosch process for producing the ammonia used in fertilizer, for example, approaches 5 percent of the energy yield for some crops.

- Because of energy inefficiencies in biological systems, the energy investment for animals such as cattle does not come close to the energy content of the feed they consume.
- To ensure market acceptability (in terms of freshness, color, lack of visible defects or blemishes, ripeness, and specifications for consumer or industrial processing), food species have been bred for so many characteristics that compromises have had to be made, with the common complaint that taste has been sacrificed, especially for tomatoes.
- In animal populations, concentration of livestock inevitably leads to an acute waste disposal problem, which has become a water quality problem in much of the world: for example, run-off from chicken operations into the Chesapeake Bay.
- The economics of supply and demand put great pressure on producers to increase yields at ecological costs; the use of veterinary antibiotics and supplements (including arsenic) presents a contamination problem and human health risk.
- Selection and breeding for desirable characteristics for agriculture has led to the near and even actual extinction of wild strains and their genetic variation, losing valuable properties such as disease resistance and resilience to stress such as drought; there is now a rebounding interest in "legacy" varieties and in seed banks to preserve genetic variation for the future.
- Many crops are highly labor intensive in cultivating and especially harvesting and therefore require access to a reliable but low-paid labor force; this has resulted in many social problems that, depending on the national situation, are related to immigration policy, to the creation of a permanent impoverished rural underclass, or even to perpetuation of slavery or indentured servitude.
- Food security is tenuous in many countries and thus has created controversial ways of securing food supplies. Some private companies and investors in the Organisation for Economic Co-operation and Development (OECD) countries and China and a few countries that have the means, particularly from oil wealth, have purchased or leased large tracts of land in other countries (principally in Africa and Brazil) dedicated exclusively for future production reserved for their populations. (Some of this land is expected to be diverted to energy production and away from food, if biofuels production expands.)

Each of these topics has spawned many books, and elaboration is outside the scope of this one.

Notwithstanding the contradictions listed above, agriculture as an enterprise is fundamental to the sustainability of the current and future human population and of all but a diminishing few nomadic societies. It is difficult to imagine feeding the current or even a much reduced population of the earth without contemporary methods of food production despite all the problems listed above. Agriculture therefore presents a paradox: it is a sector with many unsustainable features that is nonetheless the basis for human sustainability.

How can this seeming paradox be resolved? Here are some possibilities:

- Agriculture, as noted, is not wholly unsustainable in all places. Once the initial investment is made in clearing, tilling, and planting, sustainable agricultural practices can continue indefinitely when scaled appropriately to the carrying capacity of the ecosystem.
- The adverse consequences of agriculture can be and are mitigated in practice so that the cumulative effects of unsustainable practices do not produce catastrophic results.
- Inputs into agriculture are priced at a relatively low level compared to the same inputs for industrial production. This is most obvious for wages, possibly also true for diesel fuel, and probably most significant for water. For example, in California the price of an "acre-foot" of water (the amount required to cover an area about the size of a soccer field to a depth of 30.5 cm) has been $100 to $1,000 for farmers, depending on the water district, but it has been as high as $5,840 for municipal systems, industrial users, and private homes. This has led most resource economists to conclude that water is significantly undervalued in the state. The recent epic drought in the state may change this.
- Agriculture renews itself, with different crops, new technology, and selectively sustainable practices (for example, crop rotation) because growers have a powerful motivation to increase or maintain production and not to allow failure.
- Agriculture has been present on the planet for so long and has been such an important process shaping the landscape, that the planet is already permanently changed by it. Agriculture has become an integral part of sustainability—for the ecosystem, for species conservation, and for human welfare.

Agriculture and Culture

Not surprisingly for the provision of necessities of life, agriculture is freighted with many implications for sustainability besides food survival. Food agriculture reflects tradition, family history (for many), sentimentality and aesthetics, land-use history, taste preferences and culture, and the spirit of independence, which is linked to a decidedly nostalgic view of traditional agriculture. Images and metaphors of agriculture (such as an abundant harvest, "good breeding," "you shall reap what you sow," etc.) pervade language and reflect how close contemporary cultural roots are to the land, even for urban dwellers.

The small family farm and the small freehold farmer seem to have achieved iconic cultural status in every society that farms, as a symbol of security, wholesomeness, family cohesiveness, and cooperation. The reality, of course, is that in most countries considered to be major agricultural producers the family farm long ago gave way to large industrialized agricultural operations in order to achieve not only greater efficiency but also greater economies of scale and production. This trend has been most obvious in the United States and Brazil and is accelerating in Africa. The concentration of small farms into big production units was forced by "collectivization" in the former Soviet Union, with calamitous results on productivity under Communism. However, aggregation into larger operations in a market economy has been a consistent trend associated with increased output.

Countries go to great lengths to preserve the family farm as an image for cultural reasons, regardless of its efficiency. This is illustrated by US popular culture and by protectionist agricultural policies such as those in France and especially Japan. In Japan, the domestic production of rice has long been much more expensive than importing the same supply; however, rice farmers are protected for reasons of tradition, culture, and political clout. The primary means of protection are subsidies and tariffs, both of which introduce inefficiencies into agricultural economies and lend themselves to preferential treatment. In the current era of fiscal constraint, some of these subsidies are coming under increasing scrutiny, especially in the case of crops that are profitable.

A reaction has set in against the current agricultural status quo in the form of organic, locally sourced, and sustainable agriculture; however, these closely related trends depend heavily on the image of the family farm and the implication that the sustainable practitioner is an underdog fighting against an industrial behemoth selling tasteless, artificial food. It is not clear that the productive capacity of these alternative food production regimes could match demand worldwide at an acceptable cost if, for example, every large food-producing country were to convert to organic farming. It is also not clear that local sourcing of food is always beneficial because it undermines the sustainable economies of distant and more rural food-growing areas. On the other hand, the popularity of organic food products in developed countries and the affluent population of developing countries has become mainstream in just a few decades and has heavily influenced conventional (nonorganic) food production methods.

Another attitudinal dimension of food acceptability is perceived purity. Historically but also in contemporary China (and elsewhere in the developing world), food adulteration is a serious and occasionally lethal problem. Notwithstanding the benefit of additives that prevent spoilage, estimates of risk/benefit, and scientific judgments regarding safety, the presence in food of preservatives and other chemicals disturbs many people who would prefer their food to be free of additives.

Food purity, the absence of additives, and assurance of taste and quality are deeply rooted in the European tradition. A touchstone of this point of view is the body of legislated food purity laws in Europe, of which the most famous is the sixteenth-century German "Reinheitsgebot," or "purity law," which was an ordinance adopted in Ingolstadt (Bavaria) in 1516 that mandated that beer should be made only from its essential ingredients (water, barley, hops). Beer makers invoke the Reinheitsgebot even today and claim to comply with it as an assurance of quality, care in brewing, and purity. Likewise, organic farming makes a virtue of refraining from the use of synthetic pesticides. Together with this tradition of food "purity" is the tradition of *appellation*, the European custom of naming the region and provenance of fine foods as a sign of quality, an indication of unique taste, and a mark of pride in production. By comparison, food in North America is generally considered to be a graded commodity, with food types and sources of similar quality being more or less interchangeable. This is changing, as producers in the United States and Canada begin to brand goods by location and producer beyond a select few products, such as wine and cheeses.

Genetically Modified Organism (GMO) Foods

Against this tradition of food purity and localization, one of the most dramatic and revealing events in modern agricultural history occurred in 1996 when the Monsanto corporation introduced a genetically modified seed that would be resistant to glyphosate, its herbicide product, which was going off patent. Farmers could plant the seeds and then spray herbicide at relatively lower concentrations to control weeds without reduced crop yields, whether for food or cotton. The result was a powerful backlash.

For millennia, people selected the most robust, productive, or desirable strains among native plants and animals, crossbred them until strains would breed true, and controlled fertilization to control the genetic fidelity of the strains in order to achieve a monoculture. Direct manipulation of the animal or plant genome had not been possible until advances in genetic science. Its arrival appeared to cross a threshold of acceptability because it was perceived to be qualitatively different from manipulating the genetics of a plant or animal by breeding. The controversy that erupted in the 1980s over genetically modified (GM) foods, however, was about much more than whether genetic engineering was an unprecedented technology or merely the latest advance in genetic manipulation. It had to do fundamentally with who makes decisions about food and by what right.

The essential difference is that breeding for desirable properties results in a reassortment of existing genes in the species. Genetic engineering to produce a "genetically modified organism" (GMO) involves altering the genetic component of the species, currently by introducing a gene from an unrelated species so that the resulting strain has properties not present in any strain in nature. The technology for doing this, "biolistics" (for "biological ballistics") or the "gene gun," is remarkably simple and involves coating a tiny metal particle with plasmid DNA and literally shooting it into the plant cell with an compressed-air gun.

Monsanto was founded in St. Louis, Missouri, in 1901 as a chemical company. It initially specialized in producing the artificial sweetener, saccharine; so the company has a very long history in the manufacture of food additives and food-related products. For most of its history, however, Monsanto was a diversified chemical products company. In recent years, the company has redirected its business to focus on the biotechnology of agriculture, where it believed that it had a competitive advantage. Part of that advantage arose from its invention, in 1970, of the aforementioned herbicide, glyphosate, which was marketed under the name Roundup®. Glyphosate was effective against broadleaf plants, which included many locally unwanted species (otherwise known as "weeds") and could be applied before the weed plant broke the surface (i.e., it was "preemergent"). Most herbicides in use at the time were highly toxic or indiscriminate in the weeds they targeted. They were often difficult to apply. Some had short half-lives, and others were biopersistent. Glyphosate had a low toxicity for human beings and was easy to handle. It quickly became the most heavily used pesticide in countries where it was available. However, glyphosate is not entirely selective and affects some crop species as well.

In 1983 Monsanto achieved a technical breakthrough with the introduction of a foreign gene for glyphosate resistance into the DNA of corn. Testing for safety done in 1987 was very controversial in principle at the time, but the technology was deemed safe by the US Environmental Protection Agency and was approved. Users buy the seeds for one-time use but farmers are not permitted (by a 2013 US Supreme Court decision) to use seed from the subsequent crop for new plantings. An original plan to make the plants that grew from GM infertile, (so-called terminator seeds) was abandoned after fierce opposition. Either way, the company effectively had a monopoly on the supply for the duration of the patent, which expired in 2014. Using this system, entire fields could be planted with Monsanto-produced GMO seeds. Glyphosate could then be applied over the entire field to prevent the emergence of weeds without interfering with crop yields. This also allowed farmers who planted the glyphosate-resistant seeds to grow their crops in much more densely packed rows, which greatly reduced the need for tillage and maintenance in the field, which in turn, greatly reduced costs and increased productivity. Within a few years, most of the corn produced in the United States was Monsanto's engineered variety.

The company appeared to have been completely unprepared for the outrage and calumny it attracted in the 1990s and that continues to be directed at the company. It had prepared for a debate over the safety of ingesting GMO foods by sponsoring studies that produced ample scientific evidence that there was no apparent toxicity risk. The backlash came in the form of outrage that a single entity such as Monsanto could decide to tamper with something as fundamental as food and could effectively impose its will and control the world's supply of certain crops by making it competitively untenable to produce them any other way. There was also a concern that Monsanto had more or less ignored the problem of pollen drift and was unconcerned that GMO strains would "contaminate" conventional strains in adjacent fields (which appears to have happened with corn in Mexico), removing meaningful choice as to whether to avoid Monsanto GMO strains.

On the level of international trade, most GMO foods for human consumption have been banned from the European Union because they contravened the traditional rules and cultural preferences on keeping foods pure and natural. On the level of grassroots activism, Monsanto became the perfect villain because of its behavior and because its products threatened the viability of small-scale European and organic farming that depended on marketing products as pure and traditional.

Monsanto initially seemed to have underestimated the depth of this feeling and treated the concern as antiscientific and naive, which infuriated opponents who had behind them centuries of tradition and food science based on maintaining purity. Monsanto envisioned this technology as a solution to many problems of agriculture in poor countries, with boosted yields providing food security at reasonable cost and the possibility of establishing new industries based on food production. Because the EU would not approve GMO food imports, countries that were expected to benefit, such as Malawi, were constrained from adopting Monsanto products.

As of this writing, however, there is no substantiated evidence that GMO foods carry health risks any greater than conventional agriculture. From a technical point of view, therefore, the controversy may appear to be counterproductive, since it denies people the benefit of affordable and plentiful food. From the cultural point of view, however, the Monsanto incident was perceived as intolerable coercion and arrogance, with one company presuming to change the direction of food and to arrogate to itself the right to make decisions for the entire world. Seen in that light, it is not clear whether anti-GMO activists or Monsanto itself bears greater responsibility for delaying the introduction of GMO as an acceptable food technology.

On the horizon are several new food-production technologies, all of which are practical and raise no obvious purity issues. These include tissue culture of plant and animal foods, "additive printing" of edible products, and epigenetic control of gene expression to express desired traits that are already in the plant genome. Another is to perform selective breeding, as humankind has done for millennia, but to use gene expression as an indicator of desirable traits rather than waiting for the "phenotype," which is the actual expression of the trait.

Urban Ecosystems

Urban ecosystems combine artificial, built landscapes and elements (such as buildings) with remnant or present and natural elements on a regional scale. Cities, settlements, and the systems that support them form a complex structure that mimics the functions of natural ecosystems. Artificial ecosystems such as parks and waterways also subsume or incorporate the remnants of natural systems within their boundaries. They require constant monitoring and maintenance and inputs of energy to operate, so they are not self-sustainable. Because they are always governed by some sociopolitical structure (even private property), normally collective decisions are made about their construction, operation, and the allocation of services and resources.

Cities can be seen as artificial ecosystems, built by people but preserving remnants of natural ecosystems such as parks and waterways as part of their structure. Urban systems are complicated artificial ecosystems, the viability of which depends on natural and agricultural ecosystems both in the surrounding area and at distant venues. Urban concentrations are therefore both complicated ecosystems in their own right and major determinants of stability or instability in natural ecosystems. These have substantial direct and indirect effects on the their regions and ecosystems and are significant determinants of health for the populations, human and nonhuman, residing in the urban area. Urban systems are complicated blends of artificial and natural ecosystem services, such as water supply, distribution and transportation, and air- and watershed management.

The viability of urban ecosystems depends on natural and agricultural ecosystems, both in the surrounding area and at distant venues. This has been called (by Rees) the ecological "footprint" of the city (also used for countries). These have large direct and indirect effects on the districts in which they are located and are significant determinants of health for the populations, human and nonhuman, residing in the urban area.

Historically, cities have been centers of social change, culture, learning, innovation, and opportunity that have been engines for changing attitudes and cultural development. However, they have also been crowded, polluted, materialistic, and often corrupt. If a "healthy" ecosystem is a sustainable ecosystem in which the health status of all participants is optimized, urban ecosystems may contribute constructively to ecosystem viability or can be strongly negative in their effects.

Much of the attraction of cities depends on density. The potential for interpersonal interactions leads to cultural and economic activity but also increased possibilities for person-to-person disease transmission and for concentration of waste and therefore pollution. For this reason the incidence of many diseases is highly dependent on population density, influenza being a prime example. The density of urbanization concentrates environmental impacts such as pollution into a small area characterized, among other physical factors, by impermeable surfaces (streets, parking lots, buildings) and therefore heavy water run-off, and heat generation creating "heat islands" that are warmer than the surrounding countryside.

The social implications of density have been very controversial over the years. The alarming findings of some studies performed in the 1960s on rats suggested to many scientists that dense concentration of the human population not only aggravates but may cause social problems. It now appears that many of these findings were spurious and that the association is not at all straightforward. (See box 9.2.)

Density of urbanization has many benefits because it permits efficiencies in resource utilization, especially energy. It also facilitates the management of risk and many social problems (such as access to health care), allows more efficient distribution of goods, and provides a more efficient labor market for economic development. Density also supports more expensive and efficient infrastructure. The optimum density clearly changes with culture, technology, and affluence. The largest cities of medieval Europe were never populated with more than 150,000 people, in part, it has been suggested, because of practical difficulty disposing of human waste. And given the frequency of outbreaks of epidemic disease in Europe at the time, even this was probably too dense. (Chinese cities of the same era were much larger.)

Sustainability is just as important for these urban ecosystems as for natural ecosystems, both for the stability of the urban system and for managing its impact on the "hinterland" that serves it (and that it, in turn, serves, with trade, culture, industry, and financial services). Urban concentrations are therefore complicated ecosystems in their own right and major determinants of stability or instability in natural ecosystems. Indeed, sustainability for urban ecosystems is really an ideal rather than an achievable goal because urban ecosystems are open systems, often drawing resources and energy from long-distance sources. So what does sustainability mean for urban ecosystems?

A sustainable urban ecosystem, conceptually, has reduced impact on the environment (footprint) as a whole compared to a conventional city; it is a more pleasant and creative place to live so that residents will wish to preserve its features rather than overbuild; and it is an efficient structure that concentrates human impact and spares the negative consequences on less dense parts of its region. This approach to enhancing the urban environment and making it compatible with (rather than antagonistic to) environmental protection is consistent with the principles of sustainable development. It is a fundamental issue in further progress and an unresolved problem in urban and regional planning. Indicators of the relationship between ecosystem and human health are required.

BOX 9.2

Rethinking Urban Density

Density to the point of crowding, which is the subjective experience of having "too many" people surrounding a person, is usually considered to be one of the worst features of city life and a source of some of the worst urban pathologies. However, the evidence that crowding in itself has a detrimental influence on people is actually quite weak. The popular idea that crowding necessarily leads to aberrant and dysfunctional behavior was largely based on a series of now-discredited studies.

In the 1960s and 1970s, at a time when overpopulation was widely discussed (see chapter 3), attention in popular debate turned to crowding and its presumed effects in promoting crime, inducing "anomie" (a sense of detachment and apathy, with a negative attitude toward life) and creating stressful conditions on city residents. Concepts of anxiety and stress (especially in the work of Hans Selye [b. 1907–d. 1982], Swiss-Canadian) were coming together and seemed to suggest that modern living was pathological. It was also a time when there was general alarm at the high rate of crime in major American cities (long before the crime rate tumbled in the 2000s) and policy makers were preoccupied with the "urban crisis" of inner-city decay and with the apparent creation of a permanent underclass of impoverished minority residents. This discussion tended to focus on high density apart from the confounding factors that go along with it such as urban blight and social disorganization.

Against this backdrop, John B. Calhoun (b. 1917–d. 1995) and others at the US National Institute of Mental Health carried out a series of studies in the 1960s in which they attempted to reproduce crowded conditions using rats on a farm in Montgomery County, Maryland. A small enclosure with four compartments was created, with a glass roof for observation. The structure of the artificial habitat was thought to have the same characteristics as impacted inner cities, with high density and congested traffic patterns that impeded but did not block access to food and movement around the enclosure. The animals were given unlimited water, food, and nesting materials in order to remove these constraints on population. The most important of these artificial rat habitats was called "Universe 133," which briefly entered popular culture as a byword for a dystopian environment.

These studies showed an apparent deterioration of social structure among rats in a densely crowded environment, resulting in behavior that included aggression (including cannibalism), food competition, "homosexuality" (inferred from same-sex coupling behavior), "promiscuous" coupling by females (inferred from frequency of partners), reproductive failure, and many other so-called deviant behaviors. A minority of animals showed passivity (supposedly resembling human anomie), but the dominant ones, who were able to control and defend space, were able to maintain relatively normal behavior. In one famous study ("Universe 25"), the rat colony experienced a massive die-off around day 600, which was attributed to intolerable stress.

Comparisons with the inner city were inevitable and encouraged by Calhoun, who took an apocalyptic, Malthusian view of the world. (See chapter 3.) Calhoun called this deterioration in the social structure the "behavioral sink" in rodents and "spiritual death" in human communities, and he compared it to the biblical Book of Revelations. The work became broadly known and entered popular discussion after an article on the work appeared in *Scientific American*. These studies are still frequently cited as evidence for the profound distress that occurs when individuals, whether rats or humans, are crowded together into dense communities.

However, subsequent efforts to replicate Calhoun's observations showed inconsistent results, until it was noticed that density seemed to be less important among the rodents as a behavioral driver than unavoidable, constant social interaction. When the animals were given access to their own space, no matter how small, the population did not show the same effects. Likewise, efforts to demonstrate analogous effects in human populations failed because density alone seems to play little role in human discomfort. In these studies, the rate and quality of social interactions seemed to far outweigh any effect of density alone in conditioning behavior, as long as sanitation and health were maintained.

The sociological evidence for density being the controlling factor was always weak for human populations, since cities such as Hong Kong (which was much more densely populated than New York) did not show the predicted adverse behaviors attributed to crowding. Unlike experimental rat populations, human communities are highly structured with mechanisms for individuals to regulate or adapt to the frequency of their encounters with others. Increased density is sought by many as an opportunity for stimulation and interaction and tends, all other things being equal, to increase the number of possible opportunities for trade or personal gain for every individual, although not everyone may be in a position to benefit.

The original studies on crowding depended heavily on the earlier work of Selye on stress physiology to explain the changes that were observed. Later, stress pathways were better analyzed in both human and rodent models, and it became apparent that not all frequent social interactions provoked negative stress in human beings, which was a tacit underlying assumption of the density studies. In reaction to simplistic studies of crowding, there has developed a considerable literature that runs counter to the original conclusions of Calhoun and others and suggests that it is meaningless to isolate density from other social factors.

In retrospect, it is clear that this literature was laden with bias based on assumptions about what inner city life was like and that the investigators ignored social outcomes that did not support analogies to their thinking (e.g., evidence of healthy dense communities and evidence of aggression and deviant behavior in the middle class and tribal violence in low-density environments). Although it is clear that increased density increases the number of potential interactions for any individual, it is far from clear that the frequency of these interactions alone are responsible for intolerable stress that leads to "deviant" behavior, much of which may only be "normal" individual behavior released from social inhibitions.

Historically, cities have been centers of social change and opportunity that have been engines for changing attitudes and cultural development. If a "healthy" ecosystem is a sustainable ecosystem in which the health status of all participants is optimized, urban ecosystems may contribute constructively to ecosystem viability or can be strongly negative in their effects.

Monitoring the viability of a city, in particular, and the integrity of its essential systems is not simple. Indicators of the relationship between ecosystem and human health are required just as they are in natural ecosystems. A set of reliable indicators for urban ecosystems can be used to determine needs, to guide interventions, and to monitor progress. These indicators must reflect trends and not status at just one point in time. In urban ecosystems, one indicator in the set of indicators should be behavioral; this is because human beings control the system. Examples of such indicators include the responsiveness of governance to change, to the protection of environmental services, and to ecosystem protection. A hypothetical scorecard measuring the viability of a city might take into account the capacity of an urban ecosystem to meet the needs of its residents, its adaptability to new challenges, the opportunity it provides to its residents (especially the young and innovative), the security it offers to residents, and (less obviously) the effectiveness of self-correcting mechanisms to solve problems.

The Built Environment (Buildings)

Although, strictly speaking, "the built environment" covers anything constructed by humans, in practice the term is mostly used for individual buildings. These may be houses, apartment buildings, schools, hospitals, shops, production workplaces, or any other form of building. However, there is one type of building on which sustainability and health issues have been focused and that have been central to thinking about sustainability issues in general: the modern office building. This has become the prototype for studying the principles and practices of sustainability as well. For this reason, following a short introduction to building sciences, this section will emphasize office buildings, after which residential uses will be described briefly. Schools will not be discussed aside from this brief mention, as the issues surrounding them are complicated, specialized, and very much affected by high foot traffic and the propensity of children to bring with them and share communicable disease pathogens. Workplaces that involve more specialized occupational hazards open an even broader range of work-related sustainability and health issues.

Buildings and Health

Architecture is often highly visionary but has to be grounded in the materials that are available and construction techniques that contractors can reasonably provide. The construction industry has massive inertia and the industry does not quickly adopt innovations or new materials. Individual contractors are focused on their current work, and the market does not readily accept early adoption. This is in part because of concerns for liability if the innovation does not work well and the need to ensure that the final structure will satisfy the owner and developer and will meet specifications, such as water impermeability. On the other hand, the operational systems of new buildings, such as heating, ventilation, and air conditioning (HVAC), have been revolutionized by advances in information and control technology. Building methods certainly do change but not nearly as rapidly as methods and processes in other sectors and not without extensive trial and demonstration to ensure that buildings using new technology will remain as acceptable to the owner as buildings using conventional methods.

Aside from "showcase" buildings and small-scale demonstration projects (e.g., to show innovations in sustainability) construction techniques and methods change only slowly. The last big revolution in construction methods came with the "frame and skin" construction method, which is made possible with steel structures that bear the weight and provide what is in effect an armature, or a skeletal structure, on which a façade is hung. The method avoids having weight-bearing walls and pillars and allows taller buildings with more floor space. This innovation took hold before the Second World War.

Residential housing is even more conservative. Aside from a wave of enthusiasm for "manufactured," or prefabricated buildings in the early twentieth century, most houses are constructed much as they have been for years, with only incremental improvements in materials and design.

Like all sectors in mature industries, the construction sector is characterized by a relatively few leading thinkers who innovate and monitor trends and a larger group of technical experts who receive and disseminate innovation and new ideas. This dissemination group consists of architects, consultants, designers, and engineers. In the construction sector this dissemination group is very small, relatively conservative, and fragmented. However, it gets a great deal of media publicity for innovation and forward thinking, particularly architects. Yet for most construction work, such high levels of technical sophistication are not necessary and are not supported because it is costly. Most buildings are built for functional purposes and not to advance or explore the limits of technology. A utilitarian building with a faÇade that is interesting or artful is more than sufficient for most purposes.

This may be changing. Potentially disruptive innovative new construction methods are on the horizon, such as new materials with useful properties (e.g., surfaces that resist contamination or catalyze the breakdown of air pollutants), methods borrowed from manufacturing (e.g., three-dimensional printing on a large scale), and building in sustainable features built into the building (e.g., solar panels in the skin) rather than relying on systems installed separately into or onto the structure. There are also new incentives for innovation in the form of certification for sustainability (see "LEED" certification, discussed later in this section), cost cutting to improve margins while supporting acceptable quality, reversal of the historic migration from city to suburb and revitalization of urban districts with small but high-value lots, and increased interest on the part of owners and developers, largely driven by sustainability issues and shortages of affordable housing.

Building-Related Health Issues

Buildings exist to protect people from the elements and so would be expected to insulate people from factors detrimental to health. However, buildings can potentially keep pollutants trapped inside, and sources within the building can contribute additional exposures. There has long been interest in building-related health issues, mostly focused on such ventilation-associated problems. This interest increased after the 1970s, when many buildings constructed and operated to be energy efficient with unsatisfactory air exchanges were implicated in low-grade ill-health and discomfort among residents. These problems should not exist in a well-constructed and well-managed building; however, even in well-built buildings reports of similar symptoms are not uncommon. Investigating such issues has become a major issue in public health and occupational health, and there are implications for sustainability and especially for property values and building occupancy.

Buildings are more than containers for people. They are machines that create an inside environment that is conducive to human life, security, and health, as well as for the protection of essential possessions and the conduct of essential or culturally important activities. Buildings protect people from environmental influences on health and effectively insulate them from the outdoor and especially the "natural" environment.

Buildings exist to protect people from the environment, rather than to place them at risk. However, the inside of the building is its own environment, and is controlled, whether passively by shade or cover in a crude shelter or elaborately with HVAC (heating, ventilation, and air conditioning) in modern buildings. When the interior of the building is contaminated from the outside or traps emissions arising from sources inside, human health can be affected.

Health problems affecting the occupants of buildings can be classified in a few categories, some of which overlap:

- *Community health problems that are attributed to the building.* These include outbreaks of acute upper respiratory tract disease (colds and influenza), cancer among occupants (sometimes occurring in apparent "clusters," which are usually not true outbreaks), or common allergies. Communicable disease in the community obviously will also occur in occupants of a building who live in that community, and sometimes this is misinterpreted as a building-related problem.
- *Building-related diseases.* Diseases that arise from the characteristics of the building itself or from contamination within the building are rare but occasionally occur and can be very serious; examples include Legionnaire's disease (caused by a bacterium growing in hot water, usually in the humidification system), mold-related disorders (conventional allergies and a particular type of lung disease called "hypersensitivity pneumonitis"), and fungal infections (cryptococcosis and histoplasmosis) sometimes resulting from exposure to bird feces.
- *Building-related exposure to irritant chemicals.* Dusts and airborne chemicals can irritate mucous membranes, causing mild but annoying eye, nose, and throat irritation in people who are predisposed, usually because of allergies, and occasionally triggering an asthmatic reaction or sinusitis. This usually occurs when there is construction and involves high dust levels, solvents, and high concentrations in air of *bioaerosols*, which is particulate matter consisting mainly of bacteria, mold spores, and mold fragments, or formaldehyde in the past (rarely a problem anymore since urea-formaldehyde insulation was banned in 1982). But it can also occur when occupants of the building are wearing fragrances (perfumes, contain aldehydes and sometimes ketones).
- *Point-source emissions within the building.* (See chapter 5). It is uncommon for there to be a source of chemical exposure inside typical office buildings, but such sources are expected in production workplaces such as shops and

factories; the most common such problems affecting office buildings and houses involve combustion products from kitchens, sewer gas (usually at low concentrations) from drains when the water in the drain trap (the J-shaped pipe that holds just enough water to serve as a barrier) dries out, carbon monoxide and diesel emissions from a loading dock, nail polish (acetone) and fragrances (aldehydes), breakdown products from disintegrating plastic when directions on cleaning or maintenance are not followed correctly, use of cleaning agents at full strength when they should be diluted, furnishings and carpet brought into the building, and, very rarely, chemicals entering the building from outside sources.

- *Asbestos-related problems.* These problems occur in older buildings in which asbestos insulation and asbestos-containing products were used (and discontinued in the United States after 1972 but still occasionally encountered during rehabilitation and removal). The principal risks of exposure to asbestos are cancer (both lung cancer and a highly aggressive thoracic cancer called mesothelioma) and, with heavy exposure, a serious lung disease called "asbestosis," any of which take years to develop.
- *"Sick building syndrome."* This is a particular cluster of symptoms described in box 9.5, sometimes associated with adverse building conditions (such as inadequate ventilation) but often not. Concern over this problem and its considerable cost, when it occurs, has spawned a whole industry of building experts. The term is often misused to describe any building-associated health issue. (See box 9.3.)
- *Psychogenic building-related outbreaks.* These outbreaks are a behavioral response to the psychogenic stress that occurs when many occupants are persuaded that there is a health risk in a building, even when none is present; they are often associated with an odor or with deep anxiety and are observed more often in schools, where the results can be impressive. This explanation is too often invoked when other problems have not been ruled out and should always be based on evidence.
- *Odors.* Odors, even when they are not consciously perceived, can contribute to psychogenic effects and may play a primary role in a building-related problem and can make people feel uneasy, threatened, and even nauseous. Some odors (such as solvents, acrid burning plastic, earthy smells, and especially that of sewer gas) suggest building problems and raise anxiety, while others (such as feces, putrid food, and body odor) stimulate a physiological response appropriate to disgust and manifest themselves as anxiety, nausea, and shallow breathing.
- *Comfort and productivity.* Indoor environmental quality plays an important role in the occupants' comfort, in the prevention of minor illness or aggravation of medical conditions, and in the ability to perform their work productively and without distraction.

BOX 9.3

Building-Related Outbreaks

One of the most common problems in occupational health, especially in affluent countries, is nonspecific building-related illness, often but misleadingly called the "sick-building syndrome," which properly refers to a set of nonspecific symptoms that includes fatigue, inability to concentrate, eye and throat irritation, and lightheadedness, which are almost always the leading complaints in this situation. This problem usually occurs in modern office buildings, seldom in older structures, and in what are traditionally considered "safe" jobs without unusual or toxic occupational exposures. Although more than one person in the environment usually experiences similar symptoms, the problem is experienced by a minority of occupants. (When the majority of occupants complain, that is usually indicative of a single diffused or widespread source of an airborne hazard.) Symptoms improve when the patient is away from work and return upon reentry or reexposure. Symptoms occur first in people with allergy and asthma. Investigation of such cases is time consuming and usually involves industrial hygienists, specialized occupational health professionals equipped to perform exposure assessment. Some cases may reveal a specific hazard, such as very dry air; in some cases there is a clear psychogenic component, and often the building is poorly ventilated. Improvement of air exchange in the building will often result in symptomatic improvement for most workers. However, for some of these cases, no cause is ever found, and exhaustive evaluation fails to uncover a factor.

When a building develops a problem with occupants reporting symptoms consistent with the "sick building syndrome," the usual response is to evaluate indoor air quality in the building to determine if the space is contaminated or under-ventilated. Beyond a preliminary survey and inspection, extensive measurements are often (not always) unnecessary and they can be very expensive.

A basic evaluation of a space in which occupants complain of "sick building" symptoms includes

- visual inspection (emphasizing mold and other sources of allergen, as well as housekeeping)
- checking for odor ("sewer gas" or hydrogen sulfide has a distinct foul odor)
- HVAC assessment and air exchange
- conditions within the workspaces (temperature, humidity, airflow)
- testing for carbon dioxide level (an indication of ventilation)

- testing for carbon monoxide (testing for toxic levels, more than once because sources are sometimes variable)
- testing for volatile organic compounds and small particles (potential irritants but normally present at low levels and always higher in kitchen areas)
- determining whether recent construction has taken place or new furnishings have been introduced

Expensive testing of either the patient or the work environment in such cases is rarely justified and almost never productive; however, it is almost always undertaken in order to demonstrate to employees, occupants, and renters that nothing has been overlooked. The occupants of a building expect a thorough search and often demand it. Efforts to avoid conducting an expensive survey are usually criticized as evidence that the owner does not care about the health of occupants and is not willing to spend money to protect their health.

An alternative approach, used successfully but only available to licensed health professionals (who can promise confidentiality of medical information), is to conduct symptom and health history surveys and then interview occupants who complain most often and heavily of symptoms. These people are usually women (which in years past has siometimes led managers to dismiss their concerns as "hysterical") and usually have allergies (and sometimes asthma) making them hypersensitive to changes in their environment. They are often among the 5 percent who are not comfortable in indoor environments even when they comply with the ASHRAE guidelines (see text). Often they have dry eye (sicca syndrome, an eye condition in which there are not enough tears) and sometimes dry skin, which is aggravated by lower humidity. Sometimes the apparent building problem is the public face of a deeper issue, such as fear that there is a cancer-causing chemical in the building, job dissatisfaction (although this should not be used as an excuse to dismiss the problem), or anxiety over an impending or recent move.

The fact remains that for many (and perhaps most) buildings with complaints of the "sick building syndrome," a source often cannot be found, and considerable time and effort is expended on exhaustive, comprehensive environmental monitoring (occupational hygiene surveys) without a satisfactory result or explanation. Endless investigations for hidden or low-level airborne hazards have been unproductive in these cases. This has given rise to the working theory that the "sick building syndrome" is, in many cases, a phenomenon of personal susceptibility in which a person is exquisitely sensitive to variations in the environment within the range that would be considered "normal" (that is, within ASHRAE guidelines that are based on a consensus).

Perceptions of a "sick building" may also arise when the building is the common experience for occupants in the building. They may transmit germs to one another, especially respiratory infections, and perceive that what is actually a community-based outbreak of disease (usually an upper-respiratory infection such as a bad cold) has come from conditions in the building.

Sometimes symptoms of stress occur or unrelated symptoms are exaggerated in perception when occupants have a heightened awareness of health because of some event, knowledge, or rumor. For example, they may be aware that someone they know developed cancer and are concerned whether it might have been caused by something in the building. Odors may convince some occupants that there is a toxic chemical present or that the building is dangerous. Sometimes the building has nothing to do with risk of illness, but there is a perception that the building is uncomfortable, chaotic, or inhospitable. If housekeeping is inadequate, the building may be perceived as dirty and even dangerous. These situations are particularly difficult to manage because of the factor of fear. Occupants may be genuinely fearful that something in the building will affect their life and health and difficult to persuade otherwise.

Thinking about the functions and operations of buildings and beyond their structure and architecture has led to a new understanding of what has been called "building science" or "building dynamics." Although it applies equally to all buildings (including residences, office buildings, schools, and workplaces for production), the concepts of building dynamics have been worked out primarily for office buildings, in which no unusual occupational hazard is assumed.

The three distinct waves of concern over health issues and buildings are as follows:

- *First wave of concern.* The first generation or wave of concern focused on building safety and resulted in fire codes and provisions for the codification and control of safety hazards that exists today.
- *Second wave of concern.* The second generation of building codes was a response to modern building methods and problems of ventilation, as well as awareness of the "sick building syndrome," which at the time assumed that most or all such problems were due to "tight" (airtight) buildings.
- *Third wave.* The third and current generation of thought emphasizes opportunities to the enhance and improve health through the relationship between buildings and occupants. This school of thought is still evolving.

Most existing building code provisions were created for reasons of fire safety. Great fires that occurred in London in 1666, Chicago in 1871, and Baltimore in

1904, among other cities, had catastrophic results in the short term, although the cities quickly rebounded. Disastrous events, such as the San Francisco Earthquake of 1906, had a tendency to end in even more destructive fires. The ability to escape from fire became a major concern, particularly after the shocking Triangle Shirtwaist Factory fire in New York in 1911, in which 186 young immigrant women working in a low-wage "sweatshop" perished by flames or were forced to jump to their deaths because exits from the building had been blocked by management to keep them from leaving work unobserved. The result was the adoption of a much more stringent fire code in New York, which was a major improvement for the time. Over time, fire codes evolved into more comprehensive building codes.

The early risk of typhoid and cholera led to health regulations being initially directed toward water and sanitation. Regulations on ventilation and allowable occupancy largely reflected efforts to reduce the risk of tuberculosis.

The second wave of concern resulted in a second generation of building codes and was a response to modern building methods and problems of ventilation. It began in 1922 with the promulgation of guidelines for heating, ventilation, and later air conditioning. This movement gave rise to a set of consensus standards developed by what become the American Society of Heating, Refrigeration, Ventilation, and Air-Conditioning Engineers (ASHRAE) and centered on establishing standards for adequate ventilation and the most comfortable temperature and humidity range for most people in a workplace without special requirements, such as an office. Ideally, after experimental tests, each guideline would be considered comfortable by 95 percent of occupants in a building, which still leaves 5 percent of occupants dissatisfied: too hot, too cold, too dry, or too humid. ASHRAE standards are set largely for comfort but also for an environment conducive to productivity; this is on the assumption that health is adequately protected within this range. For the most part it is but comfort is more individual.

Building design and construction methods changed dramatically in the 1950s and after and came to emphasize load-bearing pillars and frames rather than load-bearing walls. Also emphasized were exterior building "skins" that were "hung" rather than constructed of masonry and centralized ventilation that depended on the HVAC (heating, ventilation, and air conditioning) system rather than passive ventilation through windows. One effect of these changes was that a large number of buildings from that era developed similar problems with inadequate ventilation, water intrusion and retention (sometimes with mold growth), and insulation with urea-formaldehyde foam, which released formaldehyde (an irritant gas) into the indoor air. Combined with this problem was the presence of airborne irritants in many manufactured products of the time, such as formaldehyde outgassed from fabric sizing to prevent stains and volatile organic compounds from new carpets and carpet pads. Because these buildings were completely dependent on their HVAC systems for climate control, and because running the HVAC system was expensive (especially under cold or very hot conditions) the buildings were often under-ventilated. The accumulation of irritating chemicals in the indoor atmosphere provoked discomfort in many occupants, especially those who had asthma or allergies, conditions that are well known to increase sensitivity to irritation nonspecifically. Often, these building-related problems could be eliminated by increasing the air exchange. These problems led to the term "tight building syndrome," because the building was thought to be too tight (i.e., hermetically sealed with inadequate air exchange).

A major driver resulting in under-ventilation and changes in building design was the effort to achieve energy efficiency, which became an increasing priority over time. The first generation of energy-efficient buildings, in the late 1970s and early 1980s, often had problems with indoor air quality and with water retention precisely because they were very tight.

A new but closely related problem emerged at the same time as "tight building syndrome" and overlapped with it. That was the recognition of the diversity of "sick buildings," which are buildings in which occupants have a high frequency of complaints, regardless of ventilation and the presence or absence of emissions sources. Many building occupants were observed to complain of a highly stereotyped cluster of symptoms: eye irritation, shortness of breath, fatigue, inability to concentrate, irritability, and more variably headache and dry skin. This became known (and accepted terminology by the Environmental Protection Agency) as the "sick building syndrome." The complaints rarely varied or included other major symptoms. It was observed that ventilation did not reduce complaints in more than a fraction of these cases. Again, the symptoms were reported more often by people with asthma and allergies and by women (who are in general more sensitive to irritation and odors than men). Sometimes a problem could be found in the building that explained the symptoms, such as secondhand smoke, mold growth, airborne dust, irritating emissions from a kitchen or other source, construction work, or a plant. Not infrequently, it would be found that complaints were reported more often by individuals who were dissatisfied with the building or their jobs or that the complaint had a psychological or intentional motivation, which led to considerable confusion and gave some building owners and managers an excuse to dismiss the problem. However, it is a general phenomenon in occupational and environmental medicine that persons who are depressed, dissatisfied, or in conflict with their supervisors or coworkers perceive symptoms more acutely (including pain) and report more complaints, so by itself this observation did not explain much. Unfortunately, these problems are often difficult to solve.

The third generation of thought on the relationship between buildings and health perceives buildings as an opportunity to enhance health and to redefine health goals for occupants of buildings today, to support physical and mental health in the people who work in them, and even to enhance health through a building experience that leads to greater satisfaction. The thought leader in this area is the US Green Buildings Council, which administers the LEED (Leadership in Energy and Environmental Design) rating system, which has become a benchmark for new construction and major renovation. The current drive toward energy efficiency (for example LEED certification) and "green buildings" does not ensure that a building is healthy—just that it is environmentally sustainable. Health and sustainability should go together, but this is not necessarily the case. Energy conservation measures, construction materials, cleaning products, and ventilation practices have sometimes resulted in symptoms in susceptible persons. New criteria for LEED certification are expected to address this gap.

The creation and maintenance of healthy and healthful buildings may be an opportunity to make gains in the health of the occupants. To unlock these potential health and productivity gains requires identification of opportunities to influence health and establish standards and norms for managing buildings for the health and productivity of the people who live and work in them.

Office Buildings

Office buildings have been the focus of modern sustainability management and the fusion of sustainability and health concerns. This is probably in equal part because office buildings present a generic template for innovation, and there are limited potential sources of unexpected or severe hazard compared to other workplaces. Also, expenses are closely tracked and can be evaluated, and there is fear of the "sick building" label that has plagued modern construction and office life (see box 9.3).

Much attention has been given to office buildings without special occupational hazards, where people work with modern information, documents, and communications technology. Office buildings are generally assumed to be free of significant occupational hazards—but this is not the case. Even office buildings have some hazards peculiar to a particular workplace in the building, such as might arise in particular places within a building: a kitchen, maintenance, health-care provider (such as a medical or dental office), or a ground-floor service such as dry cleaning. The protection of workers and visitors in a workplace with particular hazards is the proper domain of the field of "occupational health" and, while critically important, is outside the scope of this chapter. Even offices without special hazardous exposures are sometimes associated with occupational health issues and special problems such as solvent fumes, dampness, mold, and dust.

Housing

The founder of modern nursing, Florence Nightingale (b. 1820–d. 1910), the British social reformer and pioneer of the nursing profession, once said "The connection between health and dwelling is one of the most important that exists." In her era (the mid-nineteenth century) that was undoubtedly true. It is less true today with modern housing, but when houses and apartments go wrong the consequences can still be serious, particularly for children.

The challenge of designing an esthetically satisfying, energy-conserving, and environmentally sustainable home has attracted the world's best architects. Sustainability and health for new residential buildings must avoid recreating the safety and health hazards of the past while preventing irritant and allergic responses to new construction materials and achieving adequate ventilation with energy efficiency. This, of course, is much easier when the structure is designed to optimize health and sustainability. Existing housing stock, on the other hand, presents a myriad of problems that require new approaches to retrofitting and rehabilitation, which is generally more difficult. Most difficult of all is overcoming property management issues in maintaining the older dwelling for sustainability, enhanced efficiency, low health risk, and for fitness for health.

Sustainability and health risk are less often achieved together for houses than for office and commercial buildings. This is largely because a home is usually a private dwelling that is subject to individual choice and privacy. Property rights preclude social control of homes but office buildings with public access and workplaces are subject to public health and occupational health regulations.

Homes are outside the usual environmental regulatory framework. Unless there is an acute public health problem, such as a lead hazard or a dangerous condition, homes are not subject to regulation by public health authorities and do not fall under regulations of, for example, the US Environmental Protection Agency. They are assumed to be the responsibility of the owner, who (if the owner is not the principal occupant of the dwelling) has a contractual relationship with the occupant (e.g., tenant). Since 2009 the EPA has offered a voluntary certification program for home construction called Indoor airPLUS through which builders can demonstrate compliance with guidelines. The National Association of Home Builders also has a National Green Building Standard, similar to LEED but for homes.

Owner-occupied properties are controlled by the owner who, within limits of imminent hazard to others, can do whatever he or she wants to the property. An owner cannot allow an unsafe condition to exist that could harm other people off the property or that could harm someone who is on the property with permission or for a legitimate purpose. The law does not protect trespassers to the degree that it does people who are invited or allowed onto the property (e.g., visitors, guests, service representatives, emergency response personnel). Children who trespass on a property are treated differently, on the assumption that they do not comprehend risks, and property owners are generally held responsible for unsafe conditions that could harm a child wandering onto the property, especially if the hazard is an "attractive nuisance" (something that may lure or interest a child, such as a swimming pool or a broken and unsupervised playground set).

Landlords may have right of access and inspection of rental properties, but most owners are not interested in constant supervision of their property, and tenants would not appreciate this intrusion if they were. Thus, for rental property the landlord is dependent on the responsibility of the tenant to protect the property, and the tenant is dependent on the responsibility of the landlord to keep the property in a safe condition and invest in it. Sometimes either or both are disappointed.

Houses and apartments are where most people live, and so conditions are largely determined by their habits, lifestyles, choices (e.g., pets), tolerances, and the money available for maintenance and cleaning. Houses and apartments are usually cleaned less frequently and often less thoroughly than public spaces, and maintenance is usually more casual. As a rule, they tend to be more damp, especially in bathroom areas. As a result, environmental conditions within houses tend to be much more variable than environmental conditions in offices.

Some generalizations can be made linking sustainability and health in housing. Historically, poor housing was a major factor in the transmission of communicable diseases, either airborne, by sanitation failures (such as inadequate plumbing), or by personal contact. Tenements were breeding grounds for tuberculosis and other diseases, because of the high density and forced close proximity to infected occupants. Housing was often damp and cold, and lead was everywhere. Such poor housing was the only shelter available for the poor and for immigrants (e.g., on the Lower East Side of New York City at the turn of the twentieth century), as is still the case in many developing countries and, despite much better enforcement of housing codes, in some disadvantaged neighborhoods in North America.

Although adults may certainly be at risk, particularly the elderly and disabled, health risks in the home are primarily a threat to children. For otherwise adequate housing, the following factors are usually recognized as most frequent and critical with respect to serious health risk:

- safety hazards (worn electrical cords, ungrounded electrical equipment, fire hazards, broken steps, sharp or hot objects within reach, hot water heaters with the temperature set at too high at >49°C, and other physical hazards)
- secondhand smoke
- allergens (such as dust mites in bedding, cockroaches, mold, cat and dog antigens for those allergic to them)
- household cleaning and maintenance products, automotive products (especially antifreeze), and solvents
- carbon monoxide (from incomplete combustion in heaters, fireplaces, stoves)
- interior lead paint

- damp indoor spaces (aggravating respiratory symptoms and disorders)
- mold growth (allergen and irritant)
- pests (especially rodents and cockroaches)
- pesticides (toxic or allergen)
- radon (a radioactive gas, highly variable depending on geography)
- unsecured guns in the home
- unsecured seats and restraining devices for small children
- drowning hazards (swimming pools, partially-filled bathtubs)
- unsecured animals, usually dogs

Considering this link between housing and child health, it is often easier for landlords to exclude families with children rather than correct deficiencies that would place children at risk.

One unique problem of homes is the issue of the "hygiene hypothesis." This is a theory, for which there is abundant evidence, that children who grow up in environments exposed to a wider variety of "antigens" (substances that provoke an immune response) from more and many different plants, bacteria, and even fungi are less likely to develop asthma and other forms of allergy (on the cell-reaction side of the immune response) and are more likely to have stronger immune systems (on the antibody-forming side of the immune system) than children who grow up in highly sanitized environments. This is not an argument for a poor home environment. The hygiene hypothesis does not imply that dirty and unhygienic conditions promote good health. It only suggests that when children grow up exposed to a wide range of antigens from infancy, they become tolerant to them early and are less likely to become sensitized to them than when their home environment is restricted and they encounter the same antigens, as they inevitably will, later in childhood.

Once a child develops asthma, however, the hygiene hypothesis no longer holds. For a child who already has asthma, the antigens that triggers the asthma must be identified and avoided. These are usually antigens associated with dust mites, cockroaches, mold, and proteins from animals (cats, dogs). Of course, there are other triggers of asthma, such as secondhand smoke, respiratory tract infections (colds and influenza), cold air and exercise (triggers that tend to go together). Many of these triggers also relate to housing. Every asthmatic child's triggers need to be identified and avoided.

Interior lead paint is an almost uniquely American housing problem, concentrated in older American cities and prevalent in old homes in rural areas. These houses still have surface coats or older underlying coats of white paint containing lead. "Lead white" paint was the standard white paint and undercoat for homes in the United States until 1978, when it was banned for use in interior paints in the US in the face of massive and prolonged industry opposition. By comparison, several countries in Europe had already banned lead paint by 1909, on the basis of the scientific evidence already available at that time. When the paint peels, lead paint chips from that can be ingested by a very young child, who naturally puts things into his or her mouth. The child is attracted to lead paint chips a second time because they taste sweet. A few paint chips can dramatically elevate the level of lead in a child's blood and result in toxicity and acute poisoning (see chapter 5), with brain damage. When lead paint is pulverized or broken up, it contaminates house dust. Remediating homes with lead paint must be done carefully with containment to avoid spreading it around even further. In those communities where housing is old and has not been remediated, lead paint is usually the leading childhood environmental health risk.

Management of this problem reflects the unique problem of private houses. Children are required, in most jurisdictions, to be screened at a certain age for elevated blood lead. Instead of inspecting residences (because they are private property) or compelling remediation (because landlords can choose not to rent to families with children) the most common means of identifying homes with lead paint has therefore been to find children with elevated blood lead levels, by which time there is already some degree of toxicity. The simple solution would appear to be to require comprehensive lead abatement of all community housing, or abatement when the property is turned over (sold to a new owner). Some cities, particularly New York, have taken a strong, proactive stance and have effectively eliminated the problem. Other local governments have faced concerted opposition from real estate interests and owners based on the principles of property rights. The refractory nature of this seemingly simple problem demonstrates the difficulty of achieving sustainability and health where private property is involved and motivation to eliminate the problem is lacking.

The Workplace

The workplace is any location where work is performed: it could be in a building or a specialized structure such as a factory; or it could be in a vehicle, outdoors, or even in a home. It is simply the setting in which economic value is produced. The workplace is about wealth creation, sustainable productivity, workers, workers' families, and the community. Health is not the primary purpose of the workplace as such (health care and health products are provided by some workplaces, of course), but health issues related to the workplace affect productivity and both the health of both individuals, mostly workers, and the health of populations, mostly communities, are directly affected by health issues in the workplace. Workplace health and safety is therefore a critical component of sustainability, but it is rarely thought of in those terms. Instead, issues of workplace health tend to become bound up in the employment relationship, regulation, and expediency and little consideration is given to its role in public health, productivity, and supporting sustainability.

The workplace—whether physically an office, factory, shop, tent, construction site, logging camp, underground mine, military base, ship at sea, coastal base, submarine, home office, corner of a kitchen, mountaintop monitoring station, or whatever—is also, fundamentally, a social environment in which relationships exist defined by work, status, and role, and in which a shared (and sometimes clashing) culture, lines of communication, and often complex interactions among actors are featured. The workplace is a community and social system in addition to a venue where work gets done (or not) in pursuit of a common purpose (at least in the eyes of an employer). This feature of being a community of opportunity also gives the workplace the potential to convey messages, attitudes, and knowledge about health more effectively than in almost any other setting, including schools, which is why the workplace has been of such interest for health education and disease prevention. Also, because productivity is key to economic sustainability and both the economy and the individual enterprise stand to gain from advances in productivity that come from health gains, health promotion and wellness programs have proven immensely popular among employers.

The workplace is therefore a "built environment" of central importance for both health and sustainability. Its design, safety, efficiency, support for personal health, and process organization have a profound effect on adults of working age, who constitute the economically productive sector of the population and who therefore create the wealth and support the provision of social services for the rest of the population. Many health issues arise in the workplace and affect the well-being of workers and the productivity of the enterprise. They include

- occupational health and safety protection
- "health promotion in the workplace" and wellness programs in the workplace, which are designed to enhance the health and well-being of workers for their own benefit and to protect productivity
- productivity and health management, because workers who feel better do more and better work
- disability prevention and management of work capacity
- economic development and health, because poor health status can be a major drag on the economy
- controlling health-care costs, a uniquely American problem because health insurance is so closely tied to employment in the United States

Occupational health and safety (OHS) is a well-established area of practice requiring specialized training and offering a variety of jobs at many different levels: from inspecting workplaces for safety and health hazards and managing problems to professional specialization in occupational medicine and nursing. OHS is usually divided into occupational safety (prevention of injuries) and occupational health (prevention of diseases arising out of work and management of medical care and rehabilitation). There is also a dedicated no-fault insurance system for occupational injuries and illnesses called "workers' compensation," which provides for medical and rehabilitation expenses and for income replacement while the worker cannot do his or her job.

"Health promotion in the workplace" (a common phrase) and wellness programs use the incentives and behavioral approaches of health promotion (see chapter 1) to motivate workers to take up healthy habits. This can be as simple as health education and adding healthy alternatives to the menu offerings in the cafeteria to well-designed programs that reward workers for reducing their individual health risks. The objective is for the healthy worker to stay healthy and for the worker with a chronic illness (such as diabetes) to achieve better control of the condition for a better quality of life.

When the worker is healthier and has fewer risk factors for disease, productivity is higher and health-care costs are less. A specialized field of practice and research has developed under the name "health and productivity management." Managing common health problems can substantially improve productivity in working units. So too can managing risk factors for future health problems, which tend to be correlated with current health status and current productivity.

Disability rates are a combination of chronic diseases with complications, non-work-related injuries from which the person did not fully recover, work-related (occupational) injuries, and some (occupational) diseases related to work. Some countries, including the United States, have high rates of disability. Disability that renders a person unemployable or that interferes with getting a well-paying job prevents people from earning what they could and reduces productivity in the economy as a whole. The costs of lost productivity tend to be much greater than the cost of income support payments incurred in paying compensation to people who cannot work; compensation may be provided through private or public insurance (including workers' compensation for injuries that took place in the workplace, in countries with such a system). There is much concern among economists that in some countries (e.g., the United Kingdom) disability rates are high enough to impede economic growth and progress. Disability prevention, rehabilitation services, and accommodating disabled workers in the workplace is therefore an important aspect of economic sustainability.

As noted in chapter 6, which used malaria as an example, poor health status in the population at large can be a major drag on economic development on a national level. So can chronic (long-term) disability, which results when large numbers of workers are impaired due to complications of an injury or of chronic diseases (such as diabetes). The effect on the economy is particularly severe in countries with demographic characteristics such that the population of working-age adults is small and the workforce is truncated by early age of retirement either by choice or mandated—a set of circumstances that characterizes most developed countries. Disability prevention and management is an important response to this problem.

The workplace is a critical venue for health gains just as it is critical to economic sustainability. The strategies described above to enhance health, prevent disability, and protect productivity are all highly cost effective, with a high return on investment. If widely applied, they also have the net effect of boosting the health of the working population. They also protect and improve the lives and income of individual workers and their families. In fact, there are few venues as important to sustainability as the workplace, and yet it is often overlooked.

10

Energy

ABOVE ALL OTHER issues, energy is the pivotal issue for sustainability. It is the single most important driver of climate change, resource depletion, urban land-use planning and air pollution. It is also the most basic resource of all, as well as the economic commodity without which nothing else gets done. (See table 10.1.)

In order to operate the systems that keep modern society functioning, far more efficient, concentrated, and manageable energy sources have been needed since animal power became insufficient to operate the technology. The predominant energy technology since the invention of fire has been the generation through combustion of carbonaceous fuels, in the form of biomass and fossil fuels. The consequences of accessing these concentrated forms of energy have changed the world, both socially and physically, particularly through changes in the local and global atmosphere.

Energy-based economies are particularly susceptible to an economic distortion called the "commodity trap," which is often also called "the oil trap" because it is almost universal in petroleum-dependent economies. This is the economic principle that when there is commodity-led economic growth it can be difficult and sometimes impossible to diversify the local economy despite the influx of capital and the creation of wealth. This is because investment will seek the highest return and there would rarely be any business activity that could provide the high returns that can be had from the dominant commodity industry, particularly oil. As a consequence, the commodity venture attracts almost all of the available investment capital, extracts skilled workers from the local labor force and causes boom town–related issues that may scare away entrepreneurial activity in other businesses. Although some entrepreneurs do well by providing goods and services to workers in the industry instead of investing directly, they often have problems finding and keeping reliable employees at wages they can afford to pay, especially when they are competing with oil companies. The commodity trap is important to understand in the context of sustainability and energy because it explains many

Table 10.1 Table of sustainability values: Energy

	Sustainability Value	Sustainability Element
I.	Long-term continuity	Sustainable energy resources are reliable in the long term and renewable. Reducing energy demand makes continuity more easily achieved at lower cost and with fewer adverse consequences.
II.	Do no/minimal harm	Sustainable energy resources minimize or avoid both health consequences and adverse social consequences. Reducing energy demand implies achieving a level at which society has what it needs to avoid social and economic damage.
III.	Conservation of resources	Sustainable energy resources, being all or predominantly renewable, leave abundant opportunity for future generations. Reducing energy demand conserves resources directly.
IV.	Preserve social structures	Sustainable energy resources would minimize competition for energy resources, the potential for disruption in energy supplies, and distortion of society and the economy through energy pricing and supply. Reducing energy demand tends to minimize distortions introduced by manipulation and pricing.
V.	Maintain health, quality of life	Sustainable energy resources should not create or impose hazards to health but should support a higher quality of life and should be as convenient as possible. Reducing energy demand makes this easier and simpler.
VI.	Performance optimization	Sustainable energy resources should operate efficiently and responsively to provide a sustainable society with the energy it requires. Reducing energy demand to what is optimal, with a reasonable reserve for emergencies, enhances efficiency.
VII.	Avoid catastrophic disruption	Energy resource may cause disruption by their effects, as with climate change, or by failure, as in the depletion of energy resources. Sustainable energy resources would carry no risk of either and would protect rather than predispose society to catastrophic disruption. Reducing energy demand makes society more resilient in response to disruptive events.
VIII.	Compliant with regulation	Energy resources of all kinds require compliance with health and safety regulation and environmental standards. Reducing energy demand makes this esier to achieve.
IX.	Stewardship	Sustainable energy resources preserve and ideally protect ecosystems. Reducing energy demand supports this stewardship.
X.	Momentum	Sustainable energy resources support the progress of sustainability and protect it from disruption and interference. Reducing energy demand makes momentum easier to maintain.

of the issues that beset the economies of oil-producing countries and provinces and small economies that are based on a mineral resource such as coal in the Appalachian states. These economies may have had large capital inflows but they hardly ever succeed in satisfactorily diversifying their economic base and so risk becoming progressively more (rather than less) dependent on the resource.

Energy Sources

For natural systems, there are only two basic energy sources: solar (originating from the radiant energy from the sun) and planetary energy (originating from the earth's residual core heat, radioactivity, and chemical composition). Energy from muscle work and metabolic energy supplied from nutrition ultimately depend on solar energy as the original source. Wind energy derives from solar heating and air currents. Fossil fuels represent the accumulated legacy of sunlight eons ago. Planetary energy is of course far less significant than solar energy and is released in many forms, particularly as heat but also including including radioactivity, tidal energy, and as the substrate for anaerobic life forms, through geothermal vents, springs, and volcanic activity. It is possible that in the distant past, when life was first emerging on the planet, planetary energy played a more important role; however, this is not the case in today's world. Planetary energy will be revisited in nuclear technologies and briefly mentioned again as geothermal energy but otherwise will not be discussed further.

Another way of categorizing energy sources is by identifying them as carbon free or carbon based (i.e., based on combustion). Combustion of organic fuels necessarily emits carbon dioxide, and so this way of classifying fuel sources is directly relevant to climate change mitigation. Seen in this way, the primary challenge in energy sources and technology can be summarized as follows:

- To the extent possible, energy production should be "decarbonized" by phasing out fossil fuel combustion for high-volume uses (such as power generation).
- Save carbon-based fuels for essential uses and situations where there is a clear efficiency advantage (which may include motor vehicles).
- Replace centralized power generation, to the extent possible, with efficient, diverse, and distributed energy sources to increase flexibility, increase resilience to disaster or adverse events, and reduce carbon emissions.
- Whenever possible, implement carbon offsets for high-volume emission (ways of taking away as much or more carbon as is emitted).
- Identify energy storage opportunities, and build resilient distribution systems (not unlike the Internet) so that the electricity generation and distribution system is restructured into a low-carbon-emission (as it is unlikely to

be carbon neutral in this century), high-efficiency, rapidly responding, and resilient system.

- Reduce demand by thoroughgoing conservation measures (not just individual acts and interventions but with market incentives and system changes) and turnover in technology with innovation.
- Identify and quantify externalities, such as the cost imposed on society by air pollution or by climate change risks, and build these externalities into a pricing mechanism so that social and environmental costs are accounted for and charged to those who benefit.
- Raise the price of energy to match the cost of providing it, and by internalizing the externalities, provide short-term incentive to conserve and innovate in energy use.

Although it focuses attention on climate change, classifying energy sources by "carbonization" tends to oversimplify the broad range of energy resources; this is because the real issue with climate change is the combustion of fossil fuels, which results in the net addition of greenhouse gases to the atmosphere. In effect, the accumulated carbon stores of the past are brought forward into the present and carried into the future by extraction and combustion, and these emissions are added to emissions from carbon sources in the present. Biomass fuels, on the other hand, in theory recycles carbon that is in the contemporary atmosphere but in practice takes the stored carbon that is locked into the biomass fuel and dumps it into the atmosphere all at once, which aggravates the problem. Carbon emissions are much reduced in fuel cells compared to other technologies but are not negligible; fuel cell technology does not depend on combustion. Hydrogen is often considered a carbon-free fuel, but in fact it has to be generated and the two means of producing hydrogen, hydrolysis and hydrogen stripping, require electrical energy that is mainly produced by fossil fuel combustion and does so at low efficiency, so the net savings in carbon emissions may be poor.

Biomass

Discussion of energy supply and demand usually is generally focused on rich countries with developed economies. However, before discussing these essential topics it is important to point out that there is also an energy crisis in the developing world, which disproportionately relies on biomass.

Biomass fuels, such as wood and dung, constitute a large proportion of the fuel used in poor countries for cooking and heating. In the home, smoke levels, carbon monoxide, and residual indoor air pollution, called "household air pollution," can achieve levels acutely dangerous to human health and can be associated with chronic diseases such as bronchitis and lung cancer. On the village level there may be no practical alternative. Slightly better in terms of health (but disastrously unsustainable) is charcoal production, which is a particularly destructive form of energy conversion. On the other hand, enhanced biomass energy conversion, using appropriate technology, is highly sustainable. This includes the recovery of methane from biosolid waste biodigesters, which is already done on a large scale in sewage treatment plants, and the more speculative but feasible generation on a small scale of electricity from liquid (urine).

In some parts of the world, such as southern Africa, entire trees may be cut down to produce a relatively small amount of charcoal by burying branches and charring them with fires. The charcoal is then cut into sticks and packaged in bundles, which are taken to villages for sale in small quantities. The charcoal is then used in the home, mostly for cooking. Charcoal production has perhaps the poorest conversion efficiency of any form of energy, and the practice carries some of the worst consequences for the environment, health, and for sustainability. In poorer developing countries, making charcoal contributes heavily to deforestation. The entire tree is destroyed for a few branches that are carbonized but this done because, unlike small logs and branches, charcoal can be divided and traded on a scale that conforms to family use for cooking. The indoor air pollution created by charcoal may be an improvement over raw biomass. But since the charcoal is burned in open fires, the person doing the cooking (which is almost always the woman of the household) is heavily exposed to smoke, with serious consequences for health risk.

The remainder of this chapter deals mostly with larger-scale energy sources and applies primarily to developed economies and China.

Renewable and Nonrenewable

Energy resources are divided as a practical matter into "renewable" and "nonrenewable" sources. Renewable sources are based on continuous energy flux (such as photovoltaic or geothermal energy or tidal action), from energy capture and regeneration (such as biodiesel), and from natural cycles that derive from solar energy either from including resources that derive from natural cycles driven by solar energy (such as hydroelectric, which depends on the natural water cycle, and wind, which is ultimately driven by solar heating). The distinguishing feature of renewable energy sources is that within a time frame relevant to human society the source continues to produce energy with little change in recovery so the source can be sustained. Nonrenewable energy sources depend on a finite resource, such as fossil fuels, which are exhaustible over time. This simple categorization provides a useful shorthand for discussion of energy policy, but it is simplistic and somewhat misleading in practice. Many nonrenewable energy sources (e.g., coal) are so rich as to be nearly inexhaustible in practical terms. Nuclear energy in theory might be technically nonrenewable (because uranium resources are finite) but in practice are not likely to exhaust available fissionable element resources (especially if thorium technology is used) and would be virtually unlimited for all practical purposes if fusion energy becomes possible. Oil, which is certainly a nonrenewable resource, in practice has been limited much less by supply (and then mostly during wartime) than by price and refining capacity over its more than a century of energy dominance.

It is therefore perhaps more useful to think of energy resources as sustainable, or not, over a relevant time period, such as a few centuries decades to (beyond which the technology available is sure to have changed radically), regardless of whether the resource itself is finite. By this measure the fossil fuels may be unsustainable not because of likely future restrictions on supply within a relevant time period but by the adverse consequences of their use (particularly climate change). Hydroelectric energy may appear renewable, but the consequences of large dams, the risks of aging dams, and interference with fish migration make hydroelectric one of the least sustainable forms of renewable energy. Sustainability can also be limited by social concerns.

Energy sources can also be divided into (1) those that ultimately derive from planetary source (such as tidal action, geothermal extraction of heat ultimately derived from the earth's mantle and planetary deposits of elements that can be used for nuclear technologies), and (2) those that capture solar energy, either recent (in which case it is considered renewable) or distant as a legacy resource (in which case it is considered nonrenewable, as with fossil fuels). Table 10.2 presents the major sources of energy currently in use or on the horizon. Technology changes rapidly for some energy sources, and a comparison of costs, efficiencies, and usage quickly becomes obsolete. However, relative carbon emissions per unit of energy released change relatively little for combustion-based energy technologies: this is because if burning is complete it is an intrinsic property of the fuel.

The practical problems involved in transitioning to a sustainable energy future are enormous, but it is clear that the status quo cannot continue. The problem is not simply reliance on fossil fuels, although that is the major part of the problem. Shaken by Japan's Fukushima reactor disaster in 2011, and reflecting widespread distrust of nuclear power (see figure 11.1, chapter 11) Germany has declared a thorough revision of energy policy, called the *Energiewende* (or "energy transition"), designed to convert the national energy regime as much as possible to renewable energy sources and to phase out nuclear energy entirely. After considerable initial enthusiasm, the path now has become more difficult, and in the short term the *Energiewende* has led to increased dependence and consumption of coal, which is hardly a desirable outcome. In Japan itself, the initial motivation after the disaster for weaning the country off nuclear power faded in the face of power shortages, and nuclear power generation was restarted after a shutdown and without noticeable progress in moving to an alternative.

Table 10.2 An alternative classification of energy sources

Category	Contemporaneous	Legacy (including Fossil)
Solar derived. Carbon emitting	***Biological capture***	Peat
	Biomass	Coal
	Firewood, pellets, and agroforest residues	Oil and gas
	Charcoal	Conventional (petroleum, natural gas)
	Animal waste (dung)	Unconventional (shale gas, shale oil, coal bed methane, other fossil fuel)
	Biogas and landfill methane recovery	Petcoke (petroleum coke)
	Biofuels	
	Scrap, residue, and waste to energy	
	Wood and compressed biomass pellets	
	Short-term crops	
	Corn bioethanol	
	Cellulosic bioethanol	
	Algal biodiesel	
	Muscle Power	
	Animal	
	Human	
Solar derived. Not carbon emitting	***Physical capture***	*Ocean heat exchange*
	Artificial photosynthesis	
	Photovoltaic	
	Solar thermal	
	Centralized	
	Decentralized	
	Passive solar hearing	
	Natural systems capture	
	Wind (Aeolian), wind turbines	
	Hydroelectric and hydromechanical	
Planetary sources	Tidal	Geothermal
	Wave action	
	Ocean salinity gradient energy	
	Nuclear (fission)	
	Uranium	
	Thorium	
	Fusion	

Technologies in italics are in development: some demonstration projects may exist but have not been deployed.

Energy technologies have many health implications, both in direct terms of health and safety and indirect effects through energy and economic policy, urban and regional development, and influences on transportation modalities. Fossil fuel development also presents a number of direct environmental and occupational health issues that tend to be driven by source characteristics. Hydrogen sulfide in sour natural gas is much more often an occupational hazard than a community health hazard. Arsenic or mercury from burning coal is an environmental hazard, respectively, within a household (for certain types of coal, especially in China), and from emissions from power plants. Thus fossil fuels easily pose the most serious health risks as well as the most serious sustainability issues.

"Green" (sustainable) technologies for energy generation are not completely free of health issues; however, with the exception of photovoltaic technologies (which may involve exposure to some hazardous substances), their health risk profiles are mostly similar to construction projects and fixed installations of other kinds. Largely for this reason, and because of the similarity in their characteristics, they are discussed together below as "soft energy paths."

Oil and gas are usually considered together because the industry was, until recently, unified and because oil and gas coexist in many fields. There are substantial and growing differences between the fuels, however, sufficient for them to be considered separately below.

Coal

The immediate problem of fossil fuels, and carbon emissions contributing to climate change, is fundamentally reliance on coal. Carbon emissions cannot be reduced sufficiently to substantially affect climate change without reducing coal combustion. Even subtracting the global contribution from all forms of oil and gas would not make an appreciable difference in aggravating climate change if coal continued to be the primary global energy source. The problem is worsening rather than improving: although the coal industry is currently slowing in the United States, and China is moving aggressively to substitute sustainable energy sources whenever it can, aggregate demand (mainly from China and India) for coal worldwide is still increasing. (The paradox that China is both increasingly reliant on imported coal and a global leader in sustainable energy is a reflection of China's immense demand for energy wherever it can find it.)

Coal is a widespread, abundant, high-energy fuel that is most heavily relied upon for electric power production. The world consumes approximately 8 billion tons of coal per year, of which about half is consumed in China alone. The United States, India, Russia, and Germany follow, with consumption rising rapidly in India. Coal use is roughly stable in Russia and falling in the United States,

Germany, and Poland; however, coal use is rising in other high-consuming countries, including South Africa, Japan, South Korea, and Australia, which has the highest per capita consumption in the world and also exports large quantities of coal to China.

The environmental degradation resulting from surface coal-mining practices, including strip mining, mountaintop removal, and stream filling, is rarely remediated, denies future use, and imposes a huge but unaccounted cost on communities that is not recovered directly from producers. Environmental effects such as acid drainage and run-off may contaminate waterways with no feasible remedy, which causes massive ecological damage downstream. The occupational health risks of underground coal mining remain excessive in developed countries and include occasional highly visible disasters. The risks are extremely high in many developing countries, China in particular, which experiences dozens of fatal coal-mining incidents every year. Where there is weak regulation and high compensation (relative to the local economy, which is to say most coal-mining regions of the world), the full cost of coal production is simply not reflected in the price.

Coal was the first industrial fuel, delivering much greater energy density for its weight than wood or other biomass. It was probably the first fossil fuel as well, as there is evidence of it being used in prehistoric times. It was also the first obvious example of fuel switching, as coal replaced wood and biomass for heating when the forests of Europe and China were depleted. For centuries, coal has provided the cheapest and most widely available fuel alternative, and many countries, including (and especially) China, are dependent on it despite a genuine commitment to sustainable energy sources. Coal also has other specialized uses, such as chemical derivatives and coke production for making steel. But by far the major use of it today is as fuel for power plants to generate electricity.

However, coal is also a prime example of a resource for which extraction carries a steep cost and one that is heavily "externalized" (i.e., left for someone else to pay or to absorb the consequences). Furthermore, the externalized costs of coal utilization are high and derive from disproportionately higher carbon dioxide release, which contributes to climate change, conventional air pollution, release of air toxics (principally mercury), and downwind acid deposition (a problem that has been greatly improved upon through regulation). Among the common fossil fuels, emissions of carbon for the energy produced are highest with coal. These externalized costs are rarely examined, and with the exception of proposed carbon taxes and caps, they are almost never applied to the price of coal (or anything else).

By keeping the externalized costs of coal "off the books," so to speak, the price of coal in commerce stays low. If the true costs of production and the consequences of utilization (compared to other energy sources) were incorporated into the price of coal, the fuel would cost much more than it does at present, which would inevitably result in higher prices for power for consumers. That might constrain the use of coal and make alternative fuels with lower externalized costs (of which natural gas is the main substitute for coal at present) much more competitive against coal. However, the abundance and widespread availability of coal reserves at relatively low cost makes this an unattractive policy option to customers (especially utilities, which have their rates controlled), governments (who may be concerned about energy security and losing local employment), employers (who naturally find it advantageous to keep costs externalized), and consumers (who pay utility bills). The concept of a carbon tax, which will be revisited toward the end of this chapter, is in part a device to recover a small portion of these externalized costs.

However, coal has enormous drawbacks aside from carbon emissions and contribution to climate change. The occupational risks of underground coal mining are well known and historically have led to recurrent tragedy. At the time of this writing the most recent major event was an explosion and underground fire in a coal mine in Soma, western Turkey, on 12 May 2014: the explosion killed 301 miners and triggered unrest in the country over working conditions. The environmental risks are immense, ranging from local problems such as acid leaching and drainage from abandoned mines and mountaintop removal (a destructive process of strip mining mountaintops and dumping the rubble or "overburden" into valleys where it blocks streams and fouls habitats, to regional problems associated with combustion of coal such as acid-forming air pollution (and consequently "acid rain"), release of mercury and downwind deposition in water, and a high level of climate forcing due to carbon emissions. The poorer the grade (or "rank") of coal, the worse the emissions (but the cheaper the price): so the lowest grade of coal (lignite, one step above peat) is typically used when available and pollution is abated by capture at the stack. Vast quantities of low-grade coal are found in Western US states, such as Wyoming, carried by long coal trains to urban centers for use in power generation. This has provoked opposition from residents living near the right of way and in towns the trains pass through.

China consumes about half of the coal produced worldwide and supports the Australian export coal industry. However, in China, ambient air pollution has become a national crisis in many cities, small and large, not just in the metropolises such as Beijing and Shanghai or industrial cities such as Anshan, with pollution levels so severe that the national government is now taking steps (in 2014) to restrict coal-fired power generation and to close numerous mines producing low-quality coal. In November 2014, China committed itself to a bilateral agreement with the United States under which the US would continue to reduce carbon emissions in the present and China would follow suit after 2030, when its economy, level of development, and infrastructure permitted it to do so. As a practical matter, this means breaking the reliance on coal.

For all these reasons, coal has earned the reputation of the dirtiest conventional fuel alternative. In response, the coal industry has for many years proposed a series of measures that would make coal environmentally sustainable, collectively called "clean coal" technologies. Starting several decades ago, these proposed technologies included coal gasification ("syngas" production), coal liquification, coal-water slurry fuel, in situ gas production (in which coal underground is removed by heat and chemical reactions, without mining). One of them, called "fluidized bed combustion" (a method for increasing the efficiency of coal burning) was in fact a great technological success, improving the efficiency of coal combustion. Otherwise, these technologies have been disappointing in their lack of impact and together constitute a poor record for an industry that keeps promising to mitigate its considerable adverse effects.

In the early twenty-first century, the industry is hanging its hopes on carbon capture and sequestration (CCS), a technology originally developed by the oil industry in which carbon dioxide injection underground is conventionally used for well stimulation, and carbon capture is based on a technology also used for sulfur recovery. The concept is that if the carbon can be recovered at the point of combustion and then sequestered underground, emissions that influence climate change would be prevented. Of course, the many other negative effects of coal mining and combustion are not affected by this end-of-the-line mitigation step.

The coal industry is also under considerable pressure due to competition from inexpensive natural gas for power generation. The price of coal has been falling (circa 2014) in the United States, China, and on the global market, which squeezes the industry. On 23 May 2014, China signed a thirty-year agreement to import enormous quantities of natural gas from Russia, presumably with the intention of reducing reliance on coal and achieving gains in air pollution that would be impossible without conversion. This move may be a tipping point in the history of coal in China, which historically initiated its use as a fuel.

It is tempting for many advocates of sustainability to see the collapse of the industry as an unreservedly good thing. But this attitude overlooks the human cost of withdrawal from dependency on coal production. Underground coal mining is concentrated in many countries (especially in the United States and the United Kingdom, where coal mining is essentially now defunct) in regions with limited employment opportunity and poor economic diversification: for example, West Virginia, Virginia (in the southwest), eastern Kentucky, and other parts of Appalachia (which extends through Tennessee into Alabama). (The economy in Pennsylvania is more robust.) In southern Appalachia, the only nonprofessional well-paying jobs generally available in the region have been tied to the coal industry, particularly in rural areas. The demise of the coal industry, and the likely closure of mines in the future, will exact a high price in hardship and poverty in an already distressed region with many poor communities. Similarly, the northern plains of the Midwest and West are another coal-producing region where widespread poverty (particularly for native peoples) and coal overlap, but the reasons for this are very different.

The roots of dependency on coal and the history of social development in Appalachia have been the subject of thousands of studies and books, many of them brilliant but none of them charting a clear and accepted path away from dependency on coal and toward economic sufficiency. Suffice it to say that the crisis is looming, and the region lacks resilience and the diversified economy required to absorb the losses that will result from a crash of the coal industry, if and when it occurs. This lack of resilience is due in part to the "resource curse" or "resource trap," more often cited with respect to oil, in which economic activity becomes so dependent on a resource that no other enterprise can compete for investment and dependency on it stifles economic diversification.

At the time of this writing (in 2014) there have been no serious proposals for programs to support economic diversification and relief for areas that would be affected by collapse of the coal industry; this may be necessary if there were a recognized climate emergency. The region has seen such programs before (the Tennessee Valley Authority, for one). When the UK domestic coal industry was phased out (the United Kingdom still imports large amounts of coal), considerable economic dislocation occurred in coal-dependent areas, but it did not receive much focused attention because it was part of a general retrenchment of industry in the north of Britain. Despite targeted economic development programs introduced later, northern England remains the country's poorest region.

In the past, such regional development efforts have been devoted to making the disadvantaged *place* attractive for investment, on the tacit assumption that any place will thrive in a market economy if the barriers are lowered. The time may have come to question this assumption. Perhaps the time has come to consider investing in *people* first and to let them make the decision of where they will live and whether they want to leave.

One must ask if regional job creation programs are always a realistic prospect for the northern Great Plains distant from the boom atmosphere of Williston, North Dakota, or in the hollows and dirt roads of eastern Kentucky and West Virginia, where the drive over the mountain to the next small town is a travel adventure. There is just so much (or so little) economic potential in such a place, as beautiful as it may be. In remote parts of the country, perhaps geography and broadband access are destiny. People become trapped where the hills are steep, the valleys are deep, and the Internet is something special; these are places that are hard to get to and even harder to leave, and where the infrastructure supports poverty more than plenty. A mix of small local enterprises, tourism and explorations of history and culture, low-cost retirement living, telecommuting, support services spun-off from government decentralization, medical care, and public services may be the best economic anchors one can expect. In such situations, perhaps the best assistance is not to "kick start" economic growth that is more likely not to come but rather to assist individuals, particularly through education and access to technology. This is unlikely to be achieved, however, if education depends on the declining local tax base. It requires investment in education and human potential from outside the community. The objective is not necessarily to keep people in place but rather to give them opportunities to do what they want and need to do in their lives.

Some people who grow up in a certain place will stay there to keep things going, and if most of the sons and daughters of the land move away this may not be such a bad outcome if they achieve their potential in the wider world and still know that they can come back when they want. The communities that will always be home in their hearts may not be worse off if they support fewer people and more nature.

Conventional Oil

Oil, as a fossil fuel, is a distant second to coal as a contributor to carbon loading in the atmosphere. Together with transportation technology, oil plays a larger role in local ambient urban air pollution.

Oil by definition is and logically would be considered a typical nonrenewable resource of limited supply. In the distant future, oil supply could conceivably run out, as it has locally in many depleted fields. However, oil acts much more like a renewable resource in practice. Thus, oil is nonrenewable in the sense that no new oil is being created (although a Soviet-era geological theory, no longer credible, once held that it is) and existing supplies are finite. Even so, oil is highly augmentable in the sense that new supplies and alternatives are continuously coming on line.

Exploration continues to find large amounts of relatively accessible oil and the production of crude oil taps only a fraction of what is available compared to the total in the ground. Over half the oil actually present remains in a nominally depleted field. It just cannot be recovered with available technology. Part (never all) of this large residual amount of the oil reservoir is eventually tapped in future years with improved technology, which allows oil companies to book increased "reserves." (Proven reserves do not refer to the total amount of oil in the ground. They refer to control of oil reservoirs that can be feasibly and economically tapped with the technology of the time.) New technology has also made it feasible to tap known reservoirs that are currently technically difficult but becoming accessible through the technology such as three-dimensional seismic modeling.

Abundant deposits are also being discovered through exploration around the world, large virtually unexplored areas (such as offshore Africa), and "tight" reservoirs deep underground that have only become feasible to exploit over the last three decades. One consequence of having to go further into remote or inhospitable areas for oil is that the oil industry has to manage occupational and public health problems. Massive oilfields of the type seen in the Middle East are rare, as they have always been, and those that exist (and many rich smaller fields) are held as a national trust by national oil companies, which are owned by governments and have a monopoly within the country's boundaries, and so independent oil companies are having to explore in increasingly remote or inhospitable areas and often to produce in partnership arrangements. Exploration and production are becoming more expensive, but not because oil is scarce.

The prediction of "peak oil," the theory first articulated by the American geologist M. King Hubbert (b. 1903–d. 1989), held that global production rates would peak in the near future and then decline inexorably, leading to scarcer and more expensive oil. The point of peak production is often called "Hubbert's peak." His prediction turns out to have been correct but highly specific to a certain time and technology: that is, Texas in the 1960s and the United States as a whole in 1970. His predictions for the world in general, and the United States in the longer term, were incorrect due to advances in technology. Also often overlooked is that in the same paper Hubbert predicted the peak effect for a variety of other mineral commodities, none of which followed the behavior predicted by his model.

In recent decades, oil left behind in reservoirs that were thought to have been depleted has become much easier to extract due to technologies that go by the collective name of "enhanced recovery." Enhanced recovery techniques not only allow further exploitation of previously tapped reservoirs, as in Texas, but also much more efficient exploitation of smaller and deeper reservoirs and production from resources that would have been inaccessible in the past (tight oil, as from deep oil-bearing shale). Oil is therefore highly augmentable and the energy sector is (in theory) capable of readily substituting fuels for many purposes (e.g., biofuels for petroleum). Oil is also easily fungible as a commodity, since a barrel of oil can be graded and then refined into equivalent products. Contemporary petroleum economics therefore often more closely resembles that of a renewable resource than a traditional nonrenewable resource.

Similarly, in theory, fuel derived from petroleum should be easy to replace with other liquid fuels such as synthetic fuels (from coal), alcohol, and biodiesel (liquid fuel derived from renewable biological sources). In practice, it is not so easy to replace, as the energy released for the amount combusted of petroleum-derived fuels such as gasoline is very high, making it an almost ideal energy-rich portable fuel. The continued dominance of oil has much more to do with its convenience, performance as an energy source, fungibility (ease of transfer and substitution), and relatively low cost (as an energy source) than its irreplaceability.

The oil and gas industry plays a pivotal role in the world economy as a source of portable and efficient fuel that is easily transported and converted to various end uses: this economic centrality is likely to continue indefinitely with the development of a hydrogen-based energy economy. The oil and gas industry is also a major factor in shaping negative influences on the environment and is indirectly a major determinant of population health status through the means of air pollution, vehicular injuries associated with gasoline-fueled dependence on the automobile, transportation policy and urbanization, trucking, and commuter access to employment opportunity. Unlike coal, however, oil and gas do not play a major role in generating employment directly because the industry is capital-intensive and the labor force required to produce and refine oil and to produce and process gas is very small, especially compared to the value of the product.

Oil and gas shape the world to a remarkable extent, much more so than coal. Both oil and gas are fungible, which means that the characteristics of similarly graded crude or desulfurized natural gas is such that it is freely traded globally, regardless of origin (at least in theory). Oil generates about half of the carbon emissions of coal, and gas generates about one-third for the amount of energy produced. Petcoke (short for "petroleum coke") is a solid carbonaceous fuel made from the residue of refineries that can substitute for coal; it yields greater energy yield than coal for slightly higher carbon dioxide emissions.

Oil has been particularly important with respect to health and social change. Demand for oil is driven by heating and the use of petroleum products as a fuel for motor vehicles; other uses, for example as lubricants and plastic feedstocks, consume a relatively small amount compared to fuels.

Historically, countries such as the United States, where oil was abundant and could be extracted at low cost, had a competitive advantage for economic development due to relatively low prices for gasoline and other refined products. In the years before 2014 the global demand for oil sustained a high price worldwide but the price is currently (in 2015) falling precipitously due to abundant global supply. This and the competition among oil companies for control of proven reserves (reservoirs of oil in the ground that are quantified and reasonably certain to be extractable within a few years given then-current technology and costs), support expensive and aggressive exploration and acquisition of both new reserves and old fields that can be reworked with new technology. The advent of "unconventional" oil sources opened through enhanced recovery has vastly increased the available stocks closer to markets and eased threats of a global shortage. This has resulted in the current fall in prices, but global demand remains high. This is a major concern for sustainability and health because discipline in energy conservation, reduction in oil consumption, reduction in vehicular emissions, and fuel efficiency of vehicles all depend on a permanently high price of oil. Ideally, the price would be high enough to reward sustainable practices but not so high as to shock the economy into counterproductive emergency measures or to impede development in poor countries.

Oil is energy dense and portable. It is lightweight and easily transported as refined liquid fuel products (gasoline, diesel, and kerosene) and can be easily processed for removal of sulfur and blended, which has made it the fuel of choice for vehicles. The preferred use of oil for heating and vehicular fuels results in a number of consequences for health and sustainability, among them the following:

- *Ground-level air pollution.* Air pollution, especially along highways, is a direct result of the density of vehicles using petroleum-based fuels and certain types of pollution are associated with certain fuels, such as fine particulate matter (technically referred to as $PM_{2.5}$, because the particles are sized equal to or less than 2.5 μm in aerodynamic diameter), which although produced by any form of combustion, in urban air pollution is predominantly a product of diesel fuel.
- *Transportation mode.* Because cheap oil (historically) reduced the cost of driving, it favored the automobile as a personal means of transportation at the expense of mass transit systems with less flexibility for the traveler, and the truck as a means of regional distribution at the expense of the railroad. Most directly, this resulted in a serious problem of injury from motor vehicles "accidents." (Injury prevention advocates prefer to use other words, such as "events," because the term "accident" might be construed as suggesting chance or inevitability, when, in theory, all accidents are preventable.)
- *Land use.* Land-use planning and urban design then had to respond to the changing mode of transportation, which favored distributed settlement and dispersed communities instead of central nodes of transportation (such as train stations and bus depots) and much lower residential density. The modern suburb was made possible by the automobile and with it additional hazards such as the risk of injury to children, disincentives to walk (and thus increasing risk of obesity and poor fitness), communities cut off by roadways, and highways cutting through urban areas, further contributing to the pollution hazard and causing local issues of environmental justice (as discussed in chapter 11).
- *Climate change.* The contribution to climate change from oil, as noted, is much less than that of coal because of the lesser carbon emissions. However, that does not mean that the contribution from oil is insignificant.
- *Geopolitical conflict.* The commodity trap and industry concentration referred to earlier tend to favor authoritarian governments and oligarchs, with political stagnation, instability, and often rebellion against the oppression and unfair distribution of wealth, with indirect health consequences.
- *National security (defense).* Military vehicles are designed to run on a limited number of fuels so that refueling will not be restricted or complicated in the field. Among the US and NATO allies, these fuels are primarily diesel fuel and a standardized kerosene-based jet fuel called JP-8. However, military vehicles have also been very fuel inefficient. This causes many problems in deployment and combat and has led to long supply lines that are vulnerable to attack.
- *National security (civilian).* Fuel dependency on imports leads to vulnerability to diplomatic and economic pressure. The most prominent example of this

at present is the dependence of the European Union, as a whole, on imported natural gas from Russia, and the critical vulnerability of countries in central Europe, particularly Poland and the Baltic States, which are members of NATO and therefore bound in a mutual defense treaty that would require defense by the United States, Canada, Germany, and other countries. This vulnerability extends to non-NATO countries and is a pressure point that can be manipulated in a crisis, particularly for Ukraine. This is a compelling reason for these countries to look for assured energy resources within their own borders.

The so-called oil curse is a problem of economic sustainability in countries that depend on oil that reflects inadequate investment in other sectors of the economy and lack of diversification. It is a version of the "paradox of plenty" or "resource trap" that is a common problem for economies dependent on a single resource. Although it is primarily a problem of countries that are economically dependent on oil and gas, a similar phenomenon can be seen in other countries that are dependent on a single natural resource. The "oil curse" has at least three causes. First, investors are interested only in the high returns they can gain from the resource; and because no other investment offers the same returns, investment goes in one direction, other parts of the economy are neglected and remain underdeveloped, despite the oil wealth. Second, the sudden injection of huge capital investment and cash flows into an economy unprepared to receive them leads to market distortions, inflation, and excessive dependence on imports. Third, the large sums involved and the small number of people occupying critical points of control over approval and allocation of the best opportunities lead to endemic corruption and social divisions, including extreme income disparities. Oil, for all its value, does not create many jobs, because it is a capital- rather than labor-intensive industry. This leads too often to a corrupt elite gaining from payoffs (which are rarely spent within the economy) and a huge impoverished underclass with little opportunity and seething resentment. For these reasons, many oil-producing countries have poor social and health indicators. The more successful oil-producing countries (Norway being the world leader) and regions (such as the state of Alaska and the Canadian province of Alberta) have developed trust or sovereign wealth funds that manage the large accumulation of funds and keep the economy from becoming inflationary.

Conventional Natural Gas

Although a simpler resource, natural gas (consisting of methane and other short-chain alkane hydrocarbons with or without significant sulfur content) presents a more complicated management problem than oil. Gas is generally divided into "sweet gas" (with insignificant sulfur content) and "sour gas," with a high content of sulfur in the form of hydrogen sulfide (H_2S, a highly toxic gas). Sour gas is dangerous (depending on the sulfur content) and corrosive, so the sulfur must be removed in desulfurization plants before it can enter the product stream as natural gas.

The advantage of natural gas is that being mostly methane, it is more completely burned with a minimum of carbon emissions for the energy released. It is also technically relatively easy to convert power plants from coal to gas. Fuel substitution with gas dramatically reduces carbon emissions compared to coal or oil, leading many to conclude that gas represents the ideal "bridge fuel" to reduce carbon emissions in the short term while sorting out viable energy futures that require building new infrastructure and preparing for future technologies and regimes (e.g., mixed renewable energy sources, the hydrogen economy, or even fusion energy). Skeptics answer that any apparent "bridge" to a new energy future of renewable energy would actually be an "off-ramp" from energy sustainability because utilities will not cross to the other side. Once the new energy regime is entrenched, the fear is that industry would take profits and avoid further investment to finish the transition to renewable energy. In this view, industry will have every incentive to stay with known technologies that offer a predictable, lower cost and higher return and will resist phasing out fossil fuels until they are forced to, by which time valuable time will have been lost to mitigate climate change.

For much of the history of global oil dominance, natural gas was a distraction and much less valued than oil. Unlike oil, cannot be stored or transported without a pipeline and it easily catches fire and explodes. Gas requires an appropriate infrastructure is in place (pipelines or as liquefied natural gas, LNG). In the absence of such an infrastructure, gas is not only worthless on the market but also dangerous in oil production. For this reason, gas was often "flared" (burned off in tall stacks) in new oil fields, which is a wasteful practice that depressurized many reservoirs (making it harder to extract oil) and resulted in large carbon emissions for absolutely no energy benefit. This still occurs when the reservoir is distant from a pipeline connection and exploitation of the oil field occurs before a pipeline can be built for gas. Methods of liquefying natural gas and the expanding pipeline network in North America and Eurasia has made gas a global commodity.

Fuel dependency on imports leads to vulnerability to diplomatic and economic pressure, however. The most prominent example of this at present has been the dependence of the European Union, as a whole, on imported natural gas from Russia, and the critical vulnerability of countries in central Europe, particularly Poland and the Baltic states, which are members of NATO and therefore bound in a mutual defense treaty, and other allies, and of critical non-NATO countries, particularly Ukraine. This is a compelling reason for these countries to look for assured energy resources and to exploit unconventional sources within their own borders.

Gas supplies have been profoundly affected by exploitation of shale gas using modern technology. Before shale gas became a viable option, domestic use depended on pipeline access and global gas supplies were dominated by the emirate of Qatar, which supported advances in LNG technology for worldwide distribution and had the "swing production" capacity required to ramp up production to meet global demand and to affect price. Shale gas has now changed this situation, as the United States has again become a major producer and traditional markets in Europe may soon have access to their own shale gas-derived supplies.

The best modern practices are to develop oil and gas resources together so that the gas can be collected at the same time, but this is usually not done because the price of gas is often much less than that of oil, and companies do not get a return on their investment for gas development until they reach a high level of production. Gas can also be reinjected into an oilfield to maintain pressure and to preserve it, but in practice this depends on technical feasibility.

Unconventional Oil and Gas

"Unconventional" oil and gas is a broad, imprecise term used in the industry for hydrocarbons that were traditionally beyond exploitation or that require specialized techniques to extract.

Chief among these unconventional sources today is "tight" oil and gas, which are hydrocarbon resources locked in hard rock formations that have low "porosity" (the small spaces in which the oil or gas is contained). To release the oil or gas and then keep it flowing, the rock has to be fractured and sometimes etched with acid. Fracturing is done "down well" and in the past was done with explosives (mainly nitroglycerin). This became standard operating procedure in the oil and gas industry and came to be known as "fraccing" or "fracking". Other methods were actively sought because of the high risk (particularly for deep wells), and in 1947 a method of hydraulic fracking using napalm-thickened gasoline was developed—with all the safety hazards that combination implies. However, nonexplosive fracking fluids did not appear to work satisfactorily, and the conventional wisdom was that fracking had to be done with a heavy column of fluid, much like drilling mud. After a very long and costly period of trial and error during which one Texas oil magnate, George Mitchell (b. 1919–d. 2013), would not give up, a practical and much cheaper method based on water was developed and tried successfully in 1981 in the Barnett shale formation of Texas. This became the standard method used for recovery of tight oil and gas (defined in the next section) and made possible extraction from deep shale deposits. Fracking was only possible, however, using advanced methods of reservoir modeling and by directional drilling, where many drill strings can go in different directions from a single vertical borehole supported by the same pad. The next section describes current controversies surrounding "fracking."

Shale Gas and Shale Oil

Shale oil and shale gas are "tight" petroleum resources, meaning that the rock in which the oil or gas is trapped has low permeability and low porosity, so the undisturbed flow rate is very low. Tight oil and gas require special techniques to recover but have become the most important unconventional hydrocarbon resources at present. Some shale oil is close to the surface and has been mined as a minor fuel source. However, the major deposits are deep underground and have only been exploited recently because of access as a result of better reservoir characterization, directional drilling, and fracking. Shale gas resources in the United States are scattered but with large deposits conveniently located close to existing markets and infrastructure in the northeastern United States, California, and Texas. Most of the gas that is recoverable and can be transported by pipeline is sold for the production of electricity. However, large deposits also exist in the mid-continent, away from markets and pipelines, and have resulted in explosive growth of oil production, especially in North Dakota, with all the social issues associated with "boom towns" (see chapter 7). Gas prices have fallen since the boom began, however, and this removes incentives to build collection systems and pipelines for the gas in those more remote areas. Oil production has been more valuable recently with much higher prices (until very recently, in 2014), and this has led to concern over wasting gas produced in oilfields.

The global sustainability-related implications of shale gas development have to do with the much greater efficiency of converting natural gas to energy than coal combustion, as well as with the greatly reduced release of carbon dioxide for every unit of energy produced (about half). Indeed, many observers have suggested that shale gas is a "bridge fuel" that will reduce emissions of greenhouse gas in the short term and give breathing space for the development of sustainable energy technologies not yet ready for full deployment. Opponents of shale gas see this as a problem: they fear that gas will entrench fossil fuel dependency even further.

The most controversial issues associated with shale gas development have had to do with fracking, which has become shorthand for the process of horizontal drilling into a formation, hydraulic fracturing (hence the name) of the rock under extreme pressure, and subsequent recovery of free-flowing gas. The fracking process requires a few months, during which time there is considerable truck traffic, temporary facilities such as impoundment ponds, and all the short-term problems of a "boom town" (see chapter 7), as the wells are mostly in rural or remote areas. Fracking consumes enormous amounts of water over a short period and creates a disposal problem for the much smaller volume of contaminated returned water, which is usually deposited in deep injection wells. This is because release into surface waters is unacceptable and may contaminate groundwater. There have been many allegations of health effects associated with fracking, few of which can be substantiated in the absence of careful research. Other issues include seismic disturbance (currently on the order of conventional oil or gas well completion), free methane release (which does occur but at a rate similar to conventional gas wells and much less in magnitude than the urban natural gas distribution system, which clearly needs rebuilding), and methane migration into groundwater (not from the deep-fracked formation but leakage at much less depth from an improperly sealed annulus, or hole). This last problem was used to great effect by antifracking activists in a documentary movie scene showing flames shooting from a kitchen faucet, but it does not seem to be a common problem.

Leaving aside the issues involving fracking, there are many technical issues associated with shale gas and shale oil. Transport of shale oil is a major concern. Many deposits are close to major markets (especially in the northeastern United States) but some of the largest formations are a long way from markets and pipelines. This has required use of railcars to transport oil, which is an inefficient mode of transport and poses a potential hazard. Crude oil normally does not explode or catch fire during a rail accident. However, in 2013 a train consisting of seventy-four cars carrying unusually light and flammable North Dakotan crude, extracted from part of the Bakken shale formation, escaped from its handbrake and rolled into the center of Lac-Mégantic, Québec, where it derailed and exploded. The conflagration killed forty-seven people and destroyed thirty buildings, and initiated a necessary discussion on safety. Tight gas, which usually occurs in the same formation as tight oil, may be produced far from markets and far from pipelines. It is a highly flammable gas. With no way to ship or conserve the gas and the presence of a considerable explosive hazard that could blow out wells and other facilities, this gas is sometimes "flared" (burnt off), which releases large amounts of carbon dioxide and methane (an even more potent greenhouse gas) with no economic or social benefit at all.

Shale gas development is moving ahead rapidly in Pennsylvania and Wyoming, which have been the leaders in monitoring, regulation, and policy development. Shale development is also on the rise in West Virginia, where it could challenge coal (the traditional economic base for the state) for dominance. There is strong resistance to shale gas development in New York, the governor of which imposed a moratorium on new Marcellus Shale activity following a state-level review. Even if the moratorium is eventually lifted, drilling would never be allowed in or near big cities or in the watersheds of New York City (see chapter 8) or Syracuse (in the western part of the state). Moratoria on fracking have also been imposed in California, France, Québec, and New South Wales (Australia). Fracking is intensely controversial in the United Kingdom. In Poland and Ukraine, which have promising shale resources, political leaders view the technology as a path to energy independence; even so, initial results in Poland have been disappointing, and it is unclear whether this represents overestimated production potential or lack of expertise in the initial attempts. In China, initial enthusiasm for fracking has waned because the geological formations there are more complicated than initially thought and fracking requires water in quantities and places that China cannot afford at present given its serious regional water shortages.

The issues associated with fracking for shale gas have divided the environmental advocacy community. Some national conservation organizations view it as a potential benefit, allowing the world to delay irreversible climate change through a net reduction in greenhouse gas emissions and are persuaded of its value as a bridge fuel. Other environmental activists, mostly local and the local and state chapters of some of these same national organizations, are implacably opposed and seek a permanent moratorium on shale gas development.

Because of these concerns and issues both favoring and discouraging development, fracking has become a bellwether issue of the environmental movement and popular sentiment regarding environmental protection. Regardless, however, it is clear that shale oil and gas dependent on fracking will not go away and that it has permanently reshaped the energy economy of the United States.

Oil Shale

Confusingly, oil shale is not the same as shale oil, although the two can occur in the same formation. Shale oil is tight oil produced from shale deposits. Oil shale is a mineral consisting of fine-grained sedimentary rock impregnated with kerogen, the hydrocarbon substance common to oilsands and very heavy oil. Oil shale differs from oilsands in being more rocklike and harder to extract, while oilsands are crumbly and easy to break apart. Oil shale production was pioneered in Estonia and a small industry was developed in the early twentieth century; recently it has been key to that country's strategy for energy independence. In the 1980s there was great interest in mining oil shale in Colorado, particularly as oil prices were rising. There are also large deposits in Russia. However, in the end oil shale exploitation was not deemed economically feasible, and in 1982 Exxon closed the only significant pilot operation in the United States. Oil shale continues to be used on a very small, local scale in Europe and China.

Heavy Oil

Heavy oil is a more viscous, longer-chain hydrocarbon resource that requires special processing for production. It is not usually considered separately in analysis of energy policy because once it is "upgraded" it enters the product stream with conventional petroleum. Heavy oil comes in a range of weights, from close to conventional petroleum to the thick, waxlike substance called "kerogen," or bitumen, as in the oilsands. These are more difficult to extract and pump than conventional oil. Also, heavy oil tends to be high in sulfur and thus needs to be processed for sulfur removal. These additional costs make it less valuable than conventional oil. Upgrading is a process of producing a more valuable crude oil by "cracking," a process of breaking down longer-chain alkanes into shorter-chain molecules. The resulting processed crude can be easily shipped and refined and is much more valuable. Heavy oil is found worldwide. Production of heavy oil is especially important in Venezuela, Canada (separate from oilsands production), and Mexico, which export primarily to the United States, and Kuwait, which exports primarily to Asia.

Oilsands (Tarsands)

Oilsands are the most important unconventional reserve (unexploited resource), consisting of solid hydrocarbon called "bitumen" embedded in a matrix of sandstone and clay. The largest deposits are in the Canadian province of Alberta, where the largest formation is known as the Athabascan oilsands (named after the Athabasca River, which runs through it). The equivalent term, "tarsands," is often used pejoratively; however, the term was in common use when the resource was first developed. The material feels more waxy than tarlike. Extraction of oil from the oilsands has been technically difficult and much more expensive than pumping conventional oil. In 1926 a practical method of separation using water and steam was developed by the Canadian chemist Karl Clark (b. 1888–d. 1966). This technology was improved by increments until the process became economically viable and is now competitive with conventional oil at high prevailing prices. The oilsands now produce almost 75 percent of Canada's oil for export, virtually all of it to the United States.

The oilsands break the surface in several places in northern Alberta, particularly around the city of Fort McMurray (Municipality of Wood Buffalo). The largest operations mine the oilsands on the surface on an enormous scale and then reclaim the land, apply topsoil, and seed for revegetation. How successful this has been in terms of ecological restoration has been the subject of fierce controversy, but it is clear that the natural ecosystem is not completely restored. The area in question is a fragile, muskeg ecosystem, slowly growing and frozen over much of the year. A large-scale monitoring program has been proposed to assess the environmental impact comprehensively and without relying on company data alone.

Deeper deposits cannot be strip mined, and this would not be acceptable in any case, especially given the enormous area involved. Commercial production is shifting to in situ techniques, in which the hydrocarbon is separated from the oilsands matrix underground by pumped superheated steam. This liquefies the bitumen, which can then be pumped. In situ production only requires drilling and so avoids the ecological consequences of surface mining.

The resulting synthetic crude oil is "heavy" and needs to be broken down and hydrogenated (or "cracked") into lighter oil for shipping by pipeline and refining. This process requires energy input and results in carbon dioxide emissions, making oilsands upgraders major sources of greenhouse gas emissions in Canada. (However, these emissions are proportionately small compared to those derived from the eventual combustion of the same quantity of oil, as it would be from any source.)

Opposition to oilsands development is intense and based mainly on two issues: Firstly, that the oilsands are "dirty" in producing excessive carbon dioxide emissions and ecological damage from surface mining, and secondly, that the oilsands are such a vast resource that if they continue to be exploited they would perpetuate fossil fuel use for centuries and delay transition to a carbon-free energy technology. The argument for oilsands development is predicated on four main arguments: (1) much cleaner production will occur in the future, using now-proven technology; (2) liquid fossil fuels will continue to be an important energy source for decades, primarily for vehicular traffic, regardless of the economic success of sustainable energy; 3) The oilsands is a secure and reliable source of oil for the US, under the control of a close ally with shared national security interests; and 4) production from the oilsands will enter the world marketplace and be consumed one way or another and the US may as well benefit by having easy access to the energy resource. Nobody argues that the oilsands are superior or less carbon emitting than other fossil fuels, but on a life-cycle basis, greenhouse gas emissions for any fossil fuel (except methane) are more significant during combustion, which vastly outweighs release during production and distribution.

At the time of this writing, the economic future of the oilsands industry in North America is bound up with the Keystone XL pipeline, an extension of an existing transborder pipeline that would add capacity for synthetic crude oil from oilsands production. A large opposition movement has developed with the intention of halting oilsands production, blocking access to the US market, and in so doing forcing a more rapid transition to sustainable US energy resources, including biofuels. (It is not clear that this would actually happen.) The countervailing argument is that product from the oilsands will be shipped by pipeline and transported elsewhere anyway and that the United States would benefit in terms of energy security and cost by access to a secure source of oil under the control of an ally. A decision on whether the pipeline will be permitted is pending at the time of this writing and should be known by the time this book is published.

Coal Bed Methane

Coal bed methane is a gas extracted from unmined seams of coal. Coal bed methane is relatively easy to extract from seams close to the surface by simple directional drilling. Gas may be recovered from old coal mines or new mines before coal mining has begun. There have been episodes of gas intrusion into groundwater and well water in places. Fortunately methane is essentially non-toxic in water but does present a fire hazard.

Nuclear

Nuclear energy is a complicated subject in its own right, and the details of the technology will not be presented here. It is sufficient to say that all operational nuclear energy facilities at present are based on capturing energy released during fission, or "splitting" of the atomic nucleus in a controlled chain reaction of a fissionable isotope of uranium. Thorium represents an alternative fuel that is much more abundant than uranium and with technical advantages, especially in safety. However, thorium technology has not been widely deployed.

Once-promising nuclear energy has become a nightmare of apprehension, public distrust, highly visible (even if not representative) adverse events, discovery of often rather obvious design flaws (as in the Fukushima Daiichi reactors), and budgetary overruns. Nuclear energy has shown great potential for sustainability, and recent technologies are demonstrably safer; however, a series of frightening experiences with the first and second generations of technology, combined with vexing problems of nuclear waste storage and disposal, has compromised the industry and driven costs so high that even projects going forward are barely viable. Unless fusion results in a breakthrough success (and it has recently shown encouraging signs), nuclear power is not likely to expand. Even so, Japan, France, and Sweden depend on it as their primary energy source. Germany, on the other hand, has committed itself to phasing nuclear energy out entirely in favor of sustainable energy sources, even though demand has required building some new, unsustainable coal-fired power plants. It may be that the next generation of nuclear reactors will be safe and trustworthy, but it is unlikely that they will get a chance to prove their worth any time soon.

An exception to the nuclear scenario, however, may be fusion. Fusion is based on a process of capturing energy released from consolidating atomic nuclei to a more stable configuration (two atoms of hydrogen isotopes become one of helium) with the release of massive amounts of energy from the nuclear "strong force." Fusion has the potential to produce more-than-adequate amounts of electricity to meet global demand from water (as a source of hydrogen). Fusion is the same reaction that powers the sun and the hydrogen bomb. Attainment of "ignition"—the net gain of energy from a fusion reaction—has proven to be frustratingly difficult to achieve, with the essential self-sustaining fusion reaction seemingly just out of reach for forty years. However, if fusion can be achieved, the benefits would be undeniably revolutionary. The contamination and waste resulting from fusion reactions are minor compared to fusion reactions, and levels of radioactivity in reactor parts are lower than fission and have a reasonable half-life for safe and sustainable management (100 years, as opposed to millennia). In recent weeks as of the time of this writing (2014), a breakthrough of sorts was achieved in one of the two fusion technologies (inertial confinement, using lasers), and there has been steady, if undramatic, improvement in yields in the other (magnetic containment, also called the "magnetic bottle" or "torus"). Should fusion energy become viable, it would change everything and nothing. It would change everything about energy policy and supply, but it would have the enormous added advantage of being entirely compatible with existing infrastructure for electricity distribution, which means that it could go on line quickly and without disruption or the need for accommodation.

Soft Energy Paths

Energy was once thought to be in limited supply, and when energy policy was discussed the proposed solutions were almost invariably centralized and featured large infrastructure. A revolution in thinking about the problem has taken hold. Rather than considering energy to be a scarce resource for which demand was inexorably rising and that needed to be distributed from a central source, on the model of electricity, energy theorists in the 1960s began to think of energy as abundant, distributed, and easily matched to demand, which could and actually should, for the protection of other social values, be modified through conservation.

The leading thinker on these issues was Amory Lovins (b. 1947), an independent energy expert working out of the not-for-profit Rocky Mountain Institute. Lovins coined the term "soft energy paths" to suggest a pathway to energy sufficiency that relied on conservation and sustainable energy sources. This concept slowly entered mainstream thinking and is now well established in practical thinking as well as scholarship on energy policy.

Past energy planning tended to focus exclusively on centralization of power generation and the supply of conventional fuels. The assumption was that energy needed to be produced in large facilities and distributed as electricity or produced as fuels and distributed physically for local energy production. Thinking about solar energy changed that perception because photovoltaic technology is based on the observation that the flux of solar irradiation generates on average approximately 1300 watts/m^3 of energy at the surface of the earth's hemisphere that is in daylight (depending on latitude and distance from sun). That is not an inconsiderable amount of energy. Solar photovoltaic has been widely adopted, especially by local government, as a solution for freestanding street lights, metered parking, and sensors: this is a solution that would otherwise require batteries or connection to the grid. Wind is almost everywhere. Regenerative braking (and other kinetic energy recovery systems), which had been used for many years by railroads, also became available for automobiles to recover energy that would otherwise be lost. The excess light emitted from video displays can be recaptured and converted to photovoltaic energy. Likewise, the archaic idea of muscle energy from human movement has been proposed for wearable devices for civilians, for example to keep mobile phones charged, but the technological driver has been military applications. Gradually it has become apparent that energy is abundant in the world and that the true challenge is to capture, concentrate, and channel that energy. (See figure 10.1.)

Today it is clear that the role of distributed energy sources is not merely to take the load off the electricity distribution system and for living "off the grid."

FIGURE 10.1 The Bahrain World Trade Center building in Manama, Bahrain, has wind turbines built into its fifty-floor structure. This building symbolizes the transition to sustainable energy in a country previously dependent on oil and gas. (Bahrain was the first country on the Arabian side of the Persian Gulf in which oil was discovered, in 1932, but is now reaching the end of production.)

Photograph by © Tee L. Guidotti, all rights reserved.

Distributed energy can be used for local applications, for reliability, and even to support the grid, with due consideration for variability as noted above. Because of variable conditions, most technologies such as wind and solar technology are less suitable for base load than as a supplementary source of energy. The key to converting variable energy supply to a reliable source that can supply base load is the development of storage capacity. Improvements in large storage batteries and capacitors are beginning to make storage feasible on a community level, although the technology is not yet ready to serve as backup for a long period and is expensive. Further innovation may be driven by financial incentives from mandated requirements, such as those introduced in California.

Similarly, generation of fuels does not need to be confined to oil wells, mines, or farms growing crops for biodiesel. In Hamburg, Germany, one now-famous building is already completely powered by algae growing in thin flat vertical cells on the façade. In addition to generating and absorbing heat for the building, the algae can be recovered and burned as a biofuel, with properties similar to biodiesel but without competing in price with food resources.

Although technically more challenging than deriving ethanol from plant sugars, "cellulosic" or "lignocellulosic" ethanol does not compete for diversion of food crops and has a high yield. The process derives sugars from the enzymatic digestion of cellulose in woody plants and agricultural waste, and the sugars are then fermented. After a long period of development, the technology is ready for deployment at competitive cost. If cellulosic ethanol as biofuel becomes commercially viable, the preferred plant would no longer be corn but a plant called switchgrass. This plant is native to North America and found almost anywhere east of the Rocky Mountains. It was the dominant species of the tallgrass prairie ecosystem, which was mostly destroyed for agricultural production in the nineteenth century. Therefore replanting and cultivation would be tantamount to eco-restoration in the areas where switchgrass once thrived.

Soft energy paths include the following:

- *Solar energy.* The conversion of photovoltaic solar has passed quickly from a niche application to general utility for distributed, local energy, largely due to an unanticipated fall in prices for panels. The technology is now well established in the marketplace with installed capacity sufficient to easy demand on the grid and growing. The next step depends on storage technology so that excess energy generated during daylight hours is available at night.
- *Wind energy.* Wind energy is a form of indirect solar energy in which wind serves as the collecting device. Like local solar, the ultimate viability of wind depends on storage because the amount of energy generated depends on conditions that vary from moment to moment. Integration into the grid evens out local wind condition but the fluctuations in power delivered to and taken from the grid locally were initially disruptive. This problem is being solved by more advanced grid technology.
- *Renewable fuels.* Renewable fuels such as biofuels from lignocellulosic (woody) or biological culture (such as algae) have the advantage of "recycling" carbon from the atmosphere without the addition of net carbon from fossil fuels.

- *Energy conservation.* Current energy consumption is exceedingly wasteful, making conservation one of the easiest and cheapest strategies to "gain" energy resources.
- *Local energy sourcing.* In the past, the predominant design was to generate power centrally and then supply it to the user. The emerging plan is to distribute power-generating capacity and to link sources through a network for reliability and backup.

The most simple and direct approach to soft energy is to reduce demand by conservation and by reducing unnecessary consumption through design, including insulation and passive solar heating. Although technical progress has been rapid, implementation and consumer acceptance have been agonizingly slow, and in North America energy-efficient homes remain more expensive than conventional homes.

The future of soft energy is likely to be one in which energy sources are diverse, local, redundant, but still distributed through an expanded (even international) electricity grid so that fluctuations, disruptions, and short-term demand are evened out.

Mitigating Measures

Technology can offset the adverse effect of fossil fuels on health and promote sustainability. Some technologies are doing this already by promoting efficiency and energy conservation such as via environmental controls and information technology. The technologies featured in this section are likely to play critical roles in the near future.

Energy Distribution and Electricity

The main form of distribution of energy globally, of course, is electricity. Electricity is not a source of energy: it is a means of distributing energy from the source to the user. Electricity takes energy produced in one location and moves it elsewhere where it is used at a considerable inefficiency. Gasoline is a highly processed and refined product that can be conveniently shipped in bulk and divided into convenient portions (e.g., tankfuls); it can also be viewed as a means of energy distribution, as can hydrogen, which is proposed to replace it. In poor countries, sticks of charcoal often serve the same function, with disastrous environmental consequences.

The great advantage of electricity as a distribution system is that it allows many sources of energy to be pooled into a common distribution system.

Energy production can then be achieved with the best technology for the particular circumstances. Electricity therefore counts as a mitigating technology because it enables the development of other technologies that introduce flexibility and efficiency into the system even though electricity generation itself is not efficient.

Energy generation is actually a highly inefficient process and becomes even less efficient with every conversion of energy to another form. Losses through conversion are typically high, resulting in rather high inefficiencies at every step, which has its effect on pollution and inefficient capacity upstream at the point of generation. The trade-off is in utility, by providing energy at the point of use.

Energy generation can be centralized and distributed or decentralized and used locally. Historically, nuclear energy, hydroelectric generation, and coal- and gas-fired power stations have favored large centralized facilities with distribution of power through electricity. The contemporary model is a balance between centralized power generation, primarily for reliable supply to meet base requirements and predictable peak demand, and a distributed system to supplement local demand. In the future, the most likely plan will be based on storage technology and an improved grid, so that variable input from sustainable energy sources such as solar, and wind will be applied to base load without fear of insufficiency.

Electricity can be generated by any practical technology and used locally or shared through the distribution system, which is colloquially called the "grid." The production of electricity by combustion (thermal technologies), as in power plants, results in the loss of the majority of the energy as heat, at around 50 percent (for oil), 65–70 percent (nuclear or coal), or roughly 75 percent for biomass. By comparison, nonthermal solar technologies (including wind) achieve much higher conversion efficiency rates, in addition to avoiding carbon emissions. Although it is easy to store fuels, it is difficult and inefficient to store electricity once it is generated, and the search for a technically adequate system for storing it has been a story of ongoing, relatively small incremental progress for many years.

Electricity distribution requires reliable, constant voltage and almost instantaneous accommodation to peak demand. One way is to adjust generation in response, which requires close monitoring of the power grid and rapid increasing or decreasing of the power production in response in real time. (One of the advantages of natural gas for power generation is that this can be done faster than with coal.) A more satisfactory solution would be to store electricity taken from the grid and return it to the grid during periods of peak demand. Ideally this would be done electronically by some sort of battery, although mechanical means of storing energy have also been proposed (e.g., pumping water up a hill and then letting it flow back to generate power when needed, or even a giant flywheel that would store energy as kinetic energy while spinning). Electricity storage has historically been technically difficult to achieve because battery and capacitor technology had not scaled up satisfactorily to balance out the load in entire neighborhoods or towns, and safety concerns limited the uses of lithium-ion batteries. However, in 2014 new technologies were introduced that have the capacity to store suitable power overnight or longer and can even out the variable supply from solar photovoltaic or from wind turbines. California has introduced tax incentives and mandated utilities to build in storage capacity in an effort to create a larger market in order to push the technology to advance faster.

An alternative to electricity for energy distribution is hydrogen, used as an energy intermediate. Like electricity, it is an intermediate energy carrier. By separating hydrogen from oxygen by electrolysis of water or from stripping hydrogen from fossil fuels, the element can be isolated, stored, and used to generate power in a fuel cell. This is envisioned primarily for vehicles, and a demonstration hydrogen refilling station has been operating for some years in Washington, DC, despite neighborhood objections. The great advantage of hydrogen fuel is that there are no toxic emissions whatever—only water vapor. Hydrogen is essentially inexhaustible (since it is an element and returns water to the environment after oxidation), and it is lightweight and highly efficient for the energy it delivers. However, its disadvantages are considerable. It is highly explosive, so the infrastructure for handling it must be highly secure and foolproof. It is a very thin gas and is hard to handle in bulk, requiring special containment and delivery systems machined to high precision to prevent it from escaping. It cannot be delivered by pipeline. That means that hydrogen cannot be easily substituted within the existing infrastructure built for crude oil and for diesel and for gasoline refueling. Like electricity, it could be produced by the conversion of fossil fuels directly to an intermediate energy carrier but the electrolysis of water requires considerable energy, which would probably be supplied by the combustion of fossil fuels in practice.

The conventional means of recovering hydrogen (from methane in natural gas by the "steam reforming" method) generates carbon dioxide and other, more inefficient technologies may produce carbon monoxide. These drawbacks severely limit the application of hydrogen as an energy distribution system. Another problem would become apparent with large-scale hydrogen technology: free hydrogen in the atmosphere oxidizes to water vapor as it rises to the stratospheric boundary and beyond. This introduces a higher concentration of water vapor into the normally very dry stratosphere, where the water molecules then deplete ozone with great efficiency. The existence of a large hydrogen economy and industry for its production, storage, and distribution could therefore be highly threatening in terms of atmospheric change and human health risk. This observation, made in 2003, had a dampening effect on enthusiasm for hydrogen economy at the time. Atmospheric models to test this hypothesis now suggest that the effect may not be as worrisome as initially feared. Whatever the risk level (and clearly this needs to be clarified), the feasibility of a hydrogen economy is likely to depend on highly effective containment technology, control of leaks in the distribution system, and the state of the ozone layer when the technology is finally ready to be deployed.

Urine-powered generators, which work by using the urea in urine as a hydrogen source and that produce clean water as a byproduct, became global news in 2012 when four Nigerian schoolgirls designed and built a successful working prototype. Although the energy conversion is not particularly efficient, the system could be immensely practical. The technology uses an abundant resource present wherever there are people or animals, and is scaled to household use, since many houses in the developing world already rely on their own generators.

Mitigation and Source Substitution

Mitigation measures for climate change could be discussed together with its consequences (as in chapter 4), in order to emphasize the importance of mitigation. However, the topic fits as well or better with energy technology because the demand for energy and the preference for fossil fuels among energy sources are the driving forces behind climate change and mitigation must proceed in tandem with changes in energy technology and demand. Also, most of the mitigation opportunities are either price-driven or highly technological and so fit better in the context of energy pricing and technology.

Fossil fuels are not going away. Most observers within the energy industry appear to agree that coal must be phased out if carbon emission targets have any chance of being met. At the same time, there is no prospect that the world energy infrastructure and economy can diversify away from oil and gas within the foreseeable future to a degree that would make much difference. Unfortunately, total coal consumption is increasing, not declining. There is little doubt that the oil and gas industry, which supplies gasoline and diesel fuels, will continue to dominate transportation for years to come, even with the deployment of electric vehicles, which are notoriously inefficient in converting electricity generated to power. If true, the objective in the short term must therefore be to reduce the global impact of oil and gas consumption during this critical period: both to reduce the magnitude of global climate change and also to gain room to maneuver while practical sustainable energy technologies are proven and deployed.

There are, however, grounds for guarded optimism. The world as a whole has barely begun serious measures to reduce the global impact of oil and gas consumption through conservation. In the short-term, such measures may contribute far more than alternative technologies to mitigating the consequences of greenhouse gas release, even though they may not be the long-term solution to mitigate climate change.

The first great opportunity for mitigation is fuel switching. Back in 1991, for example, Shanghai, then a city of about 14 million (today a city of over 23 million and still growing rapidly through in-migration), converted from coal to gas for residential heating and cooking. Although burning oil and gas produces pollution, it produces less pollution than burning coal for the energy obtained and much less than burning biomass such as wood. So, in the short term, switching from these dirtier fuels to oil or gas is a major improvement, even though fossil fuels continue to predominate. The problem, of course, is that industrialization leads to greater consumption of fuels per capita, which offsets the fuel switching gains.

The second great opportunity for mitigation, and also for cost savings, is higher efficiency in heating application. Since it is generally easier to engineer more efficient oil-burning devices than coal-burning devices, the savings can be considerable. Cogeneration and district heating, especially in cold climates, provide further opportunity for efficiency.

In the short-term, therefore, even though alternative technologies are gradually being put on line, it seems reasonable to look at means of lowering the impact of the conventional oil and gas, since oil and gas are likely to remain a primary energy source for a long time. At this point, another factor comes into play.

What single factor simultaneously promotes alternative technologies, stimulates conservation, rewards efficiency, provides capital for economic development, and provides incentives for energy-efficient infrastructure? The answer is high oil and gas prices.

High oil and gas prices are in the common interest of sustainability in the oil and gas industry. By discouraging inefficiency, sustainable economic development is also in the common interest. This assumes, of course, that prices do not climb so high so fast that they severely damage the world economy and lead to economic regression in developing countries. There is also the potential for the revenues to be misused in oil-producing countries in furthering corruption, waste, and unsustainable infrastructure. However, as an economic signal and incentive for conservation, a high oil price is not a bad thing.

Substitution of energy sources once appeared to be a daunting challenge. Now it has become routine. In the 1990s the price of oil rose sharply, and the price of gas initially rose with it; soon after, there were predictions of a gas shortage. New technology for liquefied natural gas (LNG) followed by exploitation of "tight gas" through fracking caused the price of gas to fall sharply. At the same time, utilities came under increasing pressure in the United States to lower carbon emissions and to mitigate air pollution. This combination of developments, and the presence of abundant shale sources in the United States near large-utility customers and urban markets, made gas highly attractive as a bulk fuel for power generation (and was far preferable to coal). Power plants throughout the United States and Europe then switched to gas as their predominant fuel.

Wind energy has clearly established itself as viable in the long term through experience and viable business models. The principal problem of wind energy is that it is highly variable and so places both an intermittent load and draw on the grid, which affects all other generating units. Current methods of energy storage cannot accommodate this high degree of variation. The usual strategy followed by utilities is to rely on conventional (usually gas-powered) power plants for "base load" (the predictable demand for electricity), to use wind energy to supplement power delivered to the grid, and to ramp power production up or down depending on demand on the grid. This is technically challenging but easier to do with gas because response times are faster.

Finally, the most promising opportunity for long-term fuel shifting may be in hydrogen stripping and fuel-cell technology. Like electricity, hydrogen is fundamentally an energy distribution system rather than a form of energy. Hydrogen can be generated by electrolysis, which consumes electricity generated by some other means (and therefore is subject to compounded inefficiencies, on the order of 10 percent), using high temperatures in next-generation nuclear fission reactors (but not current technology) or from chemical stripping from methane (using natural gas or coal gas as the feedstock). Storage and transportation is difficult because hydrogen has the lowest molecular weight and diameter of any molecule (or atom) and so escapes easily; it is also explosive. Considerable innovation has made hydrogen safe for consumer use. As a demonstration project, Shell has built the world's first hydrogen filling station, located in Washington, DC. A lot more work has to be done before hydrogen replaces gasoline combustion, and perhaps fuel cells will be an intermediate; nevertheless, both technologies require oil as a feedstock while dramatically reducing emissions of conventional pollutants and greenhouse gases (to essentially zero at the site of use). This combination of preserving the "fueling station" infrastructure of gasoline and repurposing natural gas production makes hydrogen less economically disruptive and more of an incremental change than even electric vehicles. But hydrogen is getting a much later start due to the technical hurdles that had to be overcome.

Carbon Pricing

Release of carbon dioxide into the atmosphere has been the premier example of externalization of costs. The beneficiaries of fossil fuel consumption have off-loaded the ecological and health costs of the combustion products onto the community to be absorbed as consequences and subsidized by health-care costs. There are three major strategies for recovering the externalized costs, so that the "polluter pays," and the costs are figured into the cost of doing business.

Resource Depletion Allowance Phase-Out

In certain sectors, tax rules allow resource-based companies to claim tax deductions for the depletion of nonrenewable or slowly renewable resources: principally mineral resources, oil and gas, and standing timber. This means that mining, oil, and timber companies could write off the book value of their assets on an accelerated schedule at a considerable tax advantage. For many years the oil depletion allowance was set at double the depreciation schedule that other businesses would normally use to depreciate assets. This was done for two reasons: (1) because these companies were dependent on an "exhaustible" resource, their business viability was considered less than an operation that could continue indefinitely; (2) certain industrial sectors, especially oil, were considered strategically important, and it was deemed in the national interest to keep them viable and productive at low cost. Left unexplained was why it was in the national interest to provide an economic incentive to exhaust these valuable resources in the first place.

One practical issue is the timing and tapering of such phase-outs. If done too quickly or injudiciously, a phase-out on the depletion allowance for gas could encourage unwanted fuel shifting by creating an incentive for a return to coal.

Carbon Tax

The simplest way of recovering externalized costs for carbon emissions would be to estimate the total cost of the externality using an accepted model, pro rate it by the amount of "avoidable" carbon emitted, and then levy a tax on carbon emissions that would fund health care, ecological restoration, alternative energy sources and other means of correcting the externalized social cost. The technical details in achieving this, however, are daunting. The tax would be difficult to collect, except from end users as a fuel surtax, and it would be easy to game in terms of under-reporting or distorting carbon emissions data. In theory these drawbacks could be overcome, but a carbon tax is not politically popular, in part because it would be highly visible to voters. Converting the resource depletion allowance to a resource depletion penalty would achieve the same ends as a carbon tax for energy producers and so might be complementary to a direct carbon tax on users. However, anything involving the tax code (of any country but especially the United States) becomes complicated and political very quickly.

Cap and Trade

The one economic pricing mechanism in operation that seems to be thriving (after many early problems) is a market for permits in the form of unitized emissions credits for carbon emissions and to allow trading among those who hold credits. In theory, the total number of credits can be limited so that total emissions can be "capped." The market then allows the most efficient use of the permits by allowing open trading so that a company that did not need its credits could sell some of them to one that does. By steadily reducing the total number of available credits year by year, total emissions can be driven down inexorably. Companies that innovate and reduce their energy requirements would not only save fuel costs but would be financially rewarded by the revenue from selling their permits. Companies that maintain business as usual, expand their use of energy, or lag behind the industry response to energy conservation would be financially penalized by having to buy credits (but by the market, not by government sanction). Politically, the idea appeals to free-market advocates on an ideological basis and to environmental advocates on a pragmatic basis, although initial resistance had to be overcome to the idea of intentionally allowing companies to pay a fee to pollute.

Initial opposition by business interests seems to fade as the idea gains currency, perhaps because of its essential simplicity and how it operates using a familiar business model, avoiding "command and control" rigidity and encouraging innovation. Variations of this idea are now in place in the European Union, the greater Tokyo region, California, Québec, and multistate and state-provincial-level consortiums are under development for western United States and Canada and the northeastern US states. The cap and trade system in Australia, however, has just been repealed (in 2014) following a change in government, and it is not yet known whether other countries will follow that Australia's example and reexamine their carbon policies.

The primary problem appears to be how to get credits quantified and equitably distributed in the first place. Auctions with a fixed reserve (to ensure that carbon always carries a minimum price) seem to work the best in practice. The EU appears to have issued too many at too low a price. In 2013 the EU exchange was faced with a sudden and unexpected drop in demand for credits due to slowing economic activity. However, on the whole, the EU trading exchange has been successful, with a 14 percent drop in carbon emissions among participating countries since 2005. Most of this drop appears to have been due to incentives for reduction rather than a result of the 2008 recession.

Eventually, the carbon market must be globalized, but in the short term this early patchwork of cap and trade systems appears to be working satisfactorily (at least where it has been retained) despite economic miscalculations.

Carbon Capture and Sequestration

Carbon capture and sequestration (CSS) involves the physical removal of carbon dioxide from the exhaust stream after combustion, then storage, and disposal in a tight reservoir underground where the gas can be "sequestered" permanently, given the potential for a future global climate upset if the gas escaped. Proof of concept has been the presence of large quantities of naturally occurring carbon dioxide at depth, often associated with oil reservoirs, and the safe use of carbon dioxide to repressurize wells in which natural pressure has been depleted by extraction or depressurization removal of natural gas.

Capturing carbon dioxide at the source is not difficult, especially when the source is concentrated (e.g., as in a power plant). Reinjecting it deep below groundwater level is already standard practice in oil fields, and there are many abandoned sites that can be used for sequestration.

A handful of demonstration projects have been successfully operated around the world; the largest and most notable of these has been the IEAGHG Weyburn-Midale CO_2 Monitoring and Storage Project, which is a collaborative study led by the International Energy Agency for greenhouse gas mitigation (hence the acronym IEAGHG). It involves capture of carbon dioxide from a process stream from the Great Plains Coal Gasification Plant in Beulah, North Dakota, that produces syngas (synthetic gas) from lignite (a low-rank coal). The carbon dioxide is transported by pipeline across the border to the Canadian province of Saskatchewan and injected into either of two oil fields, in Weyburn and Midale, both owned by Cenovus. The carbon dioxide is used to stimulate oil production by dissolving some of the formation, after which it is again recovered and injected into deeper formations for sequestration. The experience to date has demonstrated satisfactory containment. Saskatchewan is bidding to consolidate its position as a center of the technology by following up with a large CCS project for a power generating facility operated by SaskPower.

The largest CCS projects are planned for China, in part as a way of mitigating the effects of the country's heavy reliance on coal. The first project, at the Shenhua corporation's coal-to-liquids facility in Inner Mongolia, is being closely watched as the first full-scale (rather than demonstration-scale) project.

Post-combustion carbon capture is technically straightforward, although it is expensive. After passing through a scrubber to remove mercury, particulate matter, and sulfur, the stack gas stream passes through a cool solvent (an amine solution) that absorbs the carbon dioxide to the point of saturation, a process that is exothermic (which generates heat), and so the system must be cooled. The carbon dioxide is then separated from the amine solution by heating, and then the amine solution is recycled. (Essentially the same process is used to remove sulfur from sour natural gas.) The then-pure carbon dioxide is compressed under cool temperatures, forming a liquid that is easily conveyed by pipeline or tanker, minimizing storage requirements.

Other technologies are available, generally involving gasification of coal and separation of the various components of the resulting syngas, or burning coal powder with oxygen so that all carbon is converted into carbon dioxide. Pilot projects are underway to test these alternative technologies.

"Sequestration" refers to permanent storage in places the gas cannot escape from. The liquid carbon dioxide is then injected deep underground into formations that are "capped" (i.e., in a formation with an impervious layer above). Depleted oil and gas fields are ideal, but favorable geology is widespread around the world, both on continents and undersea, and so transport to suitable sites is not likely to be a limiting factor.

The conditions of carbon capture must be controlled. Carbon dioxide normally turns into a solid under pressure, but with certain combinations of pressure and temperature it enters the liquid phase. One potential problem is that at warmer temperatures, solid carbon dioxide sublimes (i.e., passes from a solid state to gas without passing through the liquid phase) and can penetrate through pores in the rock. The geological formation and depth into which it is injected is therefore extremely important to prevent subsequent escape. One proposed passive solution to ensure sequestration is to inject the gas into magnesium-bearing rock formations, which are abundant below the surface, where a chemical reaction would sequester the carbon dioxide.

In the early days, when CCS was proposed as a viable solution to carbon emissions, there were proposals to extract carbon dioxide directly from the atmosphere and to sequester it as a geoengineering solution. That will probably never be practical because CCS extraction plants would also have to be gigantic in scale (and may require injection into the earth's mantle) to make a significant change in the ambient atmosphere.

Despite the apparent simplicity of the process, CCS at present is prohibitively expensive, in part because there are no appreciable economies of scale or standardized technology. To make this level of investment, utilities would need to be responding to a mandate for regulation of carbon emissions. This mandate was not advanced until recent renewed efforts by the Environmental Protection Agency to regulate carbon emissions, which had previously been stalled by congressional opposition and legal challenges to EPA's authority on the issue. For several years, public and political interest in climate change had dropped in the United States during the time that CCS technology was moving from pilot research to demonstration scale, and many projects were cancelled for lack of support and financing.

In one illustrative case, a successful CCS project that sequestered carbon from a coal-fired power plant in West Virginia was set to expand from a pilot to operation at scale, which covered about one-quarter of the plant's emissions, when the drive for climate change legislation was dropped in the Senate. Local regulators then advised that the cost of carbon mitigation could not be passed on to customers as a legitimate operating expense, since it was not required by law. Without passing on the expense, the projected expansion would have operated at a loss, and so the company terminated the project and dismantled the successful and well-functioning pilot unit.

Given this history, it is not surprising that there are few operating prototype or demonstration CCS plants in the United States. The largest project currently underway is in China in the coastal city of Tianjian, which borders Beijing and Hebei Province (which has extreme air pollution due to the relocation of power plants out of Beijing). Called "GreenGen," it is operated in an atmosphere of corporate secrecy.

It seems apparent that CCS, while uncertain and expensive, represents the best option to reduce carbon emissions from concentrated sources such as coal-fired power plants, refineries, and steel mills but will not be a total solution because its efficiency drops off with less concentrated sources and it does nothing to mitigate the upstream environmental damage of coal or, for that matter, oil or gas.

Geoengineering

Failure to control carbon emissions from fossil fuels will result in intolerable consequences leading to catstrophe (See chapter 3). If there is no other option, it may be necessary to take drastic action by technological intervention on a planetary level. Failure to control global emissions (particularly from coal) may ultimately force the issue.

Geoengineering is a broad term for proposed technological interventions on a planetary level to manipulate climate. As a practical matter, the only approaches under discussion are carbon removal, which is basically CCS to remove carbon dioxide directly from the atmosphere, and solar radiation management.

Carbon removal from the atmosphere can be dismissed out of hand as a viable option at present. This is because CCS has only been demonstrated for carbon dioxide-rich gas streams. The technical challenges of capturing carbon dioxide at atmospheric concentrations, even at its currently high (400 ppm) and rising concentration in the atmosphere (and doing so on a scale that would make a difference) rules out CCS as a practical option for the foreseeable future.

Solar radiation management involves the deliberate introduction of particulate matter, in the form of sulfate formed by releasing an aerosol of sulfuric acid. The resulting increase in the reflectivity (albedo) would add to the natural albedo of clouds and water particles and the existing effect of air pollution. This increased reflectivity would reduce surface temperatures on earth by reflecting back solar energy, thus preventing it from reaching the surface. Surface radiation management is technically feasible, requiring only a few hundred thousand tons of sulfuric acid per year delivered at the altitude of jet aircraft. Because the sulfate particulate would only last for about a year, it would be possible to adjust the amount delivered to avoid overshooting the target or miscalculating on the necessary amount. There are variations and refinements on this basic idea.

The effect of sulfate aerosol to cool the planet has the advantage of having already occurred in human history. Volcanic eruptions, such as that of Mount Pinatubo in 1991, provided essentially the same intervention and produced the desired consequence (a reduction in global temperature by about 0.5°C over the following year). It would not be irreversible, in theory.

Does geoengineering provide a convenient means of mitigation and an acceptable technological fix? Most people would say no. Too many things could go wrong, and success is not certain. The implications for other atmospheric processes, such as ozone depletion, would have to be modeled and closely monitored. Decisions would have to be made about who would control the project. How long the intervention would have to be continued is not clear; given the persistence of greenhouse gases in the atmosphere, the process could take centuries.

The idea of spraying acid into the earth's atmosphere to offset a problem that arose from neglect and denial is profoundly offensive in a cultural sense. If it is done, it will be because humankind is desperately trying to avert catastrophic change.

Engineering on a planetary scale holds no attraction other than as a last-ditch option for survival and mitigation to restore climate to human tolerance. To be forced to take such an extreme measure would be a failure on the part of all humankind, but it may become necessary as a last resort.

Incremental Improvement

At the opposite end of the spectrum from massive geoengineering is improvement in small increments. An ad hoc network of scientists has proposed a strategy to reduce greenhouse gas emissions in the short term using fourteen interventions, each of them feasible and yielding considerable collateral benefits. The plan was published in 2012 to virtually no publicity outside the science community except for a column in the *New York Times*. The measures concentrate on reducing black soot, which is the heat-absorbing, coarse particulate matter from combustion (especially from biomass), and methane emissions, since methane is twenty times more potent than carbon dioxide as a greenhouse gas (although it only persists for a decade in the atmosphere). The methane reduction measures would have the additional benefit of improving food security. The soot reduction measures would have the additional benefit of boosting energy efficiency, thereby reducing carbon dioxide emissions. In the aggregate, a reduction of one-half of a degree of mean temperature warming was projected, which is a highly significant mitigation.

The plan has gone nowhere.

11

Culture and Rights

CHAPTER 7 DEALT with the class of problems and challenges to sustainability that are primarily mediated by social mechanisms. Social mechanisms are, of course, embedded in a culture. The difference between a social problem, as discussed earlier, and a cultural problem is that the issues dealt with in this chapter are longer term and broader in scale. This chapter also discusses how perceptions of sustainability in general and of individual problems of sustainability are shaped by culture. In particular, the chapter will focus on the concept of rights because acceptance of rights derives from a culture of fairness and respect (see table 11.1).

Just as health professionals may endlessly debate the meaning of "health" (see chapter 1) while broadly agreeing on its general sense, so "culture" is endlessly debated by anthropologists and social scientists, even though there is broad-based agreement on what it is. A treatise in cultural anthropology is beyond the scope of this chapter, but a few salient points will help establish a common background for readers.

Since at least the 1880s, the standard definition of culture in anthropology has been that of E. B. Tyler (b. 1832–d. 1917): "Culture is that complex whole which includes knowledge, beliefs, morals, art, laws, customs, and any other capabilities and habits acquired by man [human beings] as a member of society." This and other definitions emphasize that culture is a set of values and knowledge on the cognitive level, attitudes and beliefs on the affective (emotional) level, and habits of behavior and rules for interaction. And although not immutable, culture changes more slowly than other aspects of society: for example, public opinion, political attitudes, and openness to technology.

Table 11.1 Table of sustainability values: Culture and perception

	Sustainability Value	Sustainability Element
I.	Long-term continuity	A culture that emphasizes a short-term view and that views the future as unknowable, beyond control, and insecure is unlikely to support unlimited (at least within the foreseeable future) and long-term continuity, as required for sustainability.
II.	Do no/minimal harm	A culture must have a sense of collective responsibility and must confer a sense of individual responsibility to ensure that no or minimal harm is done to the environment.
III.	Conservation of resources	A culture that does not take the long view and value the collective good is unlikely to ensure conservation of resources for the future.
IV.	Preserve social structures	A culture that sees itself as vulnerable and is rigid in its beliefs in the face of contrary information is likely to make mistakes, to do harm to its members, and to miss opportunities to enhance its social structures.
V.	Maintain health, quality of life	A culture that emphasizes the state and hierarchy is more likely to put resources into its formal structure and to fail to make investments needed to maintain health and enhance the quality of life for individuals.
VI.	Performance optimization	A culture that is satisfied with what it does and wishes collectively to maintain the status quo is unlikely to value forward progress in order to ensure optimization or maximization of economic, social, and environmental performance; it will become inefficient and stagnate.
VII.	Avoid catastrophic disruption	A culture that is not vigilant against threats will eventually face a situation where there is a high risk of catastrophic disruption and is likely to mismanage it; a culture that is consumed with anxiety about threats is unlikely to be productive because too much of its energy will be diverted into protection.
VIII.	Compliant with regulation	A culture that does not evaluate its own performance against that of others is likely to grow complacent and will not perceive a need to be compliant with regulation. This culture will be also be more likely to view regulation as an imposition and will probably fail to adopt best practices and begin to stagnate.
IX.	Stewardship	A culture that is centered entirely on an elite or on material well-being will not develop a sense of stewardship and a responsibility that goes beyond ethics in dealing with other people to include other species, life, and some notion of Nature.
X.	Momentum	A culture that requires external stimuli in order to change, or that is reliant on a small elite that is unresponsive to the needs of the majority, will eventually be incapable of sustaining progress under its own momentum and is at risk of becoming unstable in the short term.

Culture as a whole and cultural mores in the society in which an individual is born or adopts are critical to sustainability and health in many ways. Among these are the following:

- Culture is a collective way of knowing; it cannot be unique to a person or family (although families may have unique customs and beliefs) and by definition is not just an individual's own personal life choice.
- Culture is reflected in but by no means limited to the written documentation of a society; indeed, the most important aspects of culture are seldom written down because they are taken for granted.
- Culture creates a screen and a filter blocking and distorting information and messages from "outside" but also a channel for receiving messages on its own terms; if people do not recognize the context, understand the expression of the message, or trust the messenger, the message will not get through.
- Art is one way in which members of society are confronted with messages about itself and its culture, often in ways that can only be understood by mentally processing the text or image and in so doing thinking about things in a new way; these messages may reinforce cultural norms (as in religious art) or challenge them (as in transgressive art).
- Culture is shaped importantly by technology and what the society is capable of doing in the material world, but it is shaped primarily by shared values, beliefs, and rules about personal interactions: what is fair or unjust, what is polite or rude, what is tolerable or unacceptable, what is decent or obscene, what is customary for the group and what is foreign, and so forth.
- Culture defines a shared body of collective knowledge that is acceptable to be used for solving problems, enforcing laws, and making decisions; in most contemporary societies, this includes science but may also include religious beliefs, traditions, assumptions about who deserves respect and who does not, and views about other cultures and their intentions or threats.
- Culture influences language profoundly and language to some degree (but less than usually assumed) influences culture; however, it is not true that a person cannot think an original or contrary thought if he or she does not have the single word or easy language idiom to express the idea.
- Much of culture is devoted to defining relationships: kinship (fundamentally and especially in anthropology), hierarchy and status, gender roles, power relationships, amity or enmity, and whether other groups are friends and allies or enemies.
- Perceptions of what is threatening, what is desirable socially, and how much risk can be tolerated are conditioned by culture.

- In any society, there are many different strains of belief and systems of values, forming majority and minority opinion in different parts of the society and at different times.
- In any society, formal and informal rules may not coincide or may apply in different settings; for example, in many societies that nominally consider women to be inferior to men, women often exert great informal power through their role in the family and are able to get what they need despite male dominance.

Cultural values have already shifted with respect to environmental protection. The environment is far more culturally valued in American society now, following the rise of the "ecology movement" in the 1960s. Respect for and interest in the natural environment grew rapidly at that time, obviously building on existing attitudes that had been shaped by the much older conservation movement and in the beginning of the toxics movement with Rachel Carson. However, by the mid-1980s, partisan political support for environmental protection had changed dramatically in the United States, and there were strong and concerted efforts to roll back environmental standards, weaken the regulatory agencies that had been put into place under the Nixon Administration for environmental protection, and to change regulatory policy in ways that would slow and even halt response to new hazards. There was a very real risk that environmental gains would be lost and the clock would be turned back to the 1950s. That this did not happen, or at least not to the degree that it could have, was due to a deeply felt favorable cultural attitude toward environmental protection that had been translated into institutional change and had become part of the US culture and the normal way of doing things. The initial wave of alarm and concern had been replaced by the professionalization of environmental management and the assessment of what exactly needed to be done and how urgently. Environmental management was institutionalized in a myriad of federal and local agencies led by the US Environmental Protection Agency (1970). In this and many other ways, and the cultural legacy of the ecology movement of the 1960s and early 1970s kept the gains that had been made in environmental protection from being reversed with the reversal of political fortunes of its advocates.

Culture and Sustainability

There are many examples in the world of ecological changes that forced cultural adaptation and dramatic change: climate change has made many societies nonviable over the centuries. Chapter 3 discusses catastrophic change, and indeed there are many examples of societies stressed to collapse by ecological changes: several of these have been described in detail by the American scholar of geography and cultural sustainability Jared Diamond (b. 1937).

Human history and prehistory is also replete with examples of nations and civilizations that successfully adapted to dramatic changes in their environment. Australian Aborigines, in their many cultures, have for millennia adapted to changing ecosystems and have manipulated these ecosystems for their benefit (mostly by local burning) but have also carefully tended them as stewards. Archeological evidence shows that tribes adapted during prehistory, migrating or changing their culture as the environment changed around them. The ancient Egyptians, before the first pharaonic dynasty, adapted over time to the desertification of their surroundings as it dried out, and they learned to use to their advantage the annual flooding of the Nile River, which was the last remaining big river system of several that once supported human communities in North Africa.

Although there is much to justify a pessimistic view of the world's prospects, we have actually come far in the search for a new way of living. Recent decades have brought a much more detailed view of the environment, rising awareness on the part of people in the developing world, more tolerant values in accepting other cultures, and technological developments that provide tools for constructive application. In the contemporary world, the culture of sustainability has grown from a minority view and a counterpoint to the majority culture to a widely held view that requires at least demonstrable commitment.

How profound ecological changes provoke in social and cultural change is a critically important topic for these times, as diverse cultures worldwide face global climate change. Some have technological resources and wealth to minimize the disruptive effect but most do not.

Sustainability Supported by Culture

Can sustainability serve as a guiding principle in the absence of a culture that embraces its values and that appreciates stability while encouraging innovation and improvement?

The rest of this section will assume that the answer is "no," on the face of it. However, this question is not as simplistic as it may first appear. Would the need for stability become the seed of future stagnation? Will adherence to the values of sustainability grow tiresome to future generations who take it for granted and crave novelty?

The only social force powerful enough to conserve attitudes conducive to sustainability from one generation to another must be culture itself, transmitted through custom, education, family teachings, respected institutions, and (in keeping with personal and collective commitment) faith. Nothing else has the persuasive influence to force continuity over generations. Formal, school-bound education alone can transmit the knowledge between generations but only imperfectly conveys attitudes important to sustainability. Laws set limits on individual behavior, but they can be repealed, and environmental laws are unlikely to be rigorously enforced when they become unpopular. Politics changes opportunistically. Institutions may wither or grow corrupt. Only the basic values of culture do not change much over time, and when they do, they often improve.

Throughout human society, the most sustainable characteristics of society, in the sense of persistence and forward momentum, have been poverty, war, hierarchy based on birthright, and authoritarian government. However, the modern definition of sustainability (see chapter 1) rejects these negative characteristics as unsustainable because they do not lead to a future anyone would want to live in. They were "sustainable" in the narrowly literal sense in the past because the tribe, nation, or society had no recourse, no alternative social model, and only occasional opportunity to rebel effectively. Even so, numerous revolutions, peasant revolts, and upheavals happened even in ancient times.

It would be nice to think that in the modern world, adverse conditions are not sustainable. Whether this is true depends on the definition of "the *modern* world," apart from the "world as it is today." In societies that have the means to force change from below, intolerably adverse conditions are not stable in the long term because change will eventually be forced, if necessary, by revolution. More people have the means, society is aware of the ideologies, and the means of expression and resistance are more accessible in a complicated and technologically more open society, even under repression. Perhaps this is a working definition of "the modern world," but there are many places where these conditions are still unfulfilled.

Since the Brundtland report defined "sustainable development," sustainability has taken on a dimension of cultural values and political viability that allows society to envision a more or less predictable future and the continuity of society and of the resources needed to sustain it. This does not mean that the future will be completely predictable and that society will not change. It does, however, mean that the values that support sustainability, as reviewed in chapter 1, must become permanent even if their expression changes from generation to generation.

Sustainability, as elaborated in the Brundtland report, involves establishing an economic structure that ideally consumes only as much as the natural environment produces and emits only as much as the natural environment can absorb without damage. This is accomplished by reducing consumption and the scale of economic development, recycling materials, and re-using as much product as possible. The economic structure is sustainable in the sense that it can be sustained from one generation to another. It would differ from traditional peasant societies, which have been the only sustainable economic systems in the past to support large populations through many centuries, in being much more efficient in the use of resources, innovative, and based to a large extent on information management and, one might hope, happier.

There is a cultural dimension to moving toward this type of society. To accommodate and manage a sustainable economic system, the social structure would also have to change to adopt these values at all levels and as a dominant, majority culture rather than a vocal minority. Issues of equity and community control over these resources and decision making with respect to their distribution obviously arise. The concept of sustainability is closely linked, therefore, with that of community empowerment and a host of issues related to social justice and cultural expression.

Sustainability and Stagnation

Throughout history, the most sustained societies have been those with characteristics that are considered least sustainable today: impoverished peasant societies with high birth rates and high death rates, frequent warfare, marginal nutrition, and a structure based on strict hierarchy and authoritarianism. These features have been the norm, rather than the exception, throughout recorded human history.

The philosopher Thomas Hobbes (b. 1588–d. 1679) wrote that in a state of nature, human life was, "solitary, poor, nasty, brutish, and short." Then civilization happened, so to speak. As argued below, society built upon and then in most cultures largely replaced the family, clan, and tribe as the means of managing risk.

A culture that supports sustainability is likely to be one that takes the long view, values collective security, and seeks stability. This implies a respect for tradition and continuity. The danger is that such a society could become self-satisfied and lose its drive to enhance its productivity, performance, and quality of life. There could even be a risk of rigidity entering into the social structure and stultifying politics and expression. There is a question of whether the environmental heterogeneity and social pluralism of modern urban society, which many people feel tends to promote creativity (and also conflict), could be accommodated in such a system. In other words, the risk of a society devoted to sustainability may be a tendency toward stagnation.

Sustainability cannot be based on stagnation. People will not accept a view of sustainability that recreates a technologically more advanced version of a basic peasant society, especially if in countries that have only recently developed economically. For societies to accept sustainability and to continue to grow from within, the new way of living must accept cultural diversity, encourage individual expression, allow social change, offer opportunity, and examine values. There must be ways to permit opportunity and growth without ecological compromise. Constrained from concealing its flaws by unrestrained expansion because of the need to conserve resources, a sustainable society will have to improve itself qualitatively and expand selectively by creating space through efficiency and innovation. Achieving sustainability is therefore linked with policies emphasizing community, the value of information, originality in ideas, and the arts. Perhaps above all, to avoid stagnation a sustainable society will need a refined curiosity and skepticism derived from an appreciation for science as a way of knowing.

Historical European models of sustainability may suggest to some a return to something like the Middle Ages or even earlier times (an impression heightened by the Druidic tone in much of the literature of "deep ecology"), with the risk of social stagnation as a price for material stability and ecological responsibility. Umberto Eco speaks convincingly of a return to the Middle Ages in a different sense, in concepts of progress, both in attitude and cultural forms. He has written that today people are sympathetic to the idea of science but not willing to understand the world scientifically and too often defer to experts as if they were priests. The standardization of commerce and architecture, reductions in employment opportunities, and the obliteration of regional differences from popular culture that would accompany a retrogressive form of sustainability could result in a much more homogenous, simplified, and impoverished style of urban life. For all of its disadvantages, contemporary urban life and culture present great advantages in terms of social and material efficiency, individual opportunity, cultural expression, and innovation. This dynamism must be harnessed, not suppressed.

Environmentalists and sustainability advocates have been more successful in articulating a vision of harmonious coexistence with nature on a vastly reduced scale of society than in articulating an alternative vision of a dynamic society in which intellectual and information growth replace material and economic growth.

Sustainability and Dynamism

Although human beings are responsible for our present unsustainable situation, there can be no solution without accommodating human needs. "Sustainable" must include support of an evolving and opportunity-rich social environment, preferably heterogeneous in character with urban nodes as well as Arcadian hinterland and villages, or in some future generation it is likely to be cast off as restrictive and stultifying. It is not clear that in the future contemporary urban society can maintain the historic level of opportunity and cultural expression associated with urban culture, but historically cities have been the places where social interaction and the exchange of potent ideas occurred. Rural areas benefit greatly from being in the hinterland of a great city, not only economically but in cultural diffusion and access to knowledge as well.

Long-lasting change requires social and cultural change. Through collective action, governments and societies may change the presently destructive course of environmental degradation. The human economic and social systems that depend on environmental exploitation cannot merely be swept aside but must be replaced by an alternative social order. To succeed, and to be worthy of succeeding, this social order must be humane, effective in responding to social needs, equitable in the distribution of goods, and historically stable. This means answers for the perennial problems of poverty, development, and social justice.

A society that supports sustainability cannot have sustainability as its exclusive goal, because people demand more. They demand a reason for sustainability that is relevant to them and that provides for material needs, including comforts, and a sense of opportunity.

The emphasis must not be on partisan political changes but on structural change to make society responsive to these issues and to diffuse awareness of and responsibility for sustainability to the lowest possible level. The emphasis also must not be on the re-creation of a peasant society with uniform social organization and orthodox values; a new and diffused form of awareness and cosmopolitan culture, rich in human opportunities and displacing the need for material exploitation should be emphasized. The goal is to suffuse the culture with concepts of sustainability to ensure its continuity across generations but the ultimate purpose is for people to live a better life with a lighter burden on the planet.

The solution ultimately lies in seeing human society as an integral part of the planet and accepting that human communities must be accommodated in a stable world order. This implies a set of social actions that provide alternatives to both wasteful resource exploitation by industrialized societies and the often intensive resource overutilization by impoverished and less developed societies that leads, for example, to soil depletion and deforestation. Part of the answer may be sharing technological answers when they come available. Part of the answer may be a global educational and communications system in which societies have access to not only news but insight and explanation for the behavior of other people and communities. Part of the answer may be political evolution away from the nation-state and toward interdependent communities aggregated into stable regional confederations or trading blocs, while preserving local traditions and placing value on neighborhoods and roots. A large part of the answer may be in encouraging personal opportunities that are not resource-intensive and putting the energies of the economic system and the people into community development, health promotion, education, the arts, social and scientific progress, and popular culture. This also means recognizing science, just as we do the arts, as a cultural enterprise of value for its own sake in addition to serving as a driver for technology and innovation.

There must be vision in the concept of sustainability, as well as a sense of identity and purpose, for it to become a deep cultural paradigm, especially as a viable alternative to unrestrained economic growth. Sustainability must be culturally attractive, as well as technologically efficient and competent. The future will depend on human action, and there continuation of natural ecosystems because people want them to be preserved. Acceptance of sustainability by society depends on cultural values and even on spiritual notions about the relationship of humankind to the earth.

There must be a will to change. For such change to take place, it is essential that peoples of the world believe that change is possible and likely to succeed. This means that the twin enemies of global survival may be fatalism and pessimism.

Best Use or Highest Use

There is a continual conflict within society over which worldview will prevail: the use of resources for the greatest benefit of people or protection such that the natural environment can flourish without interference. The "human utility" or "best use" view is much more compatible with the presently dominant growth-oriented, economically motivated society. Even the notion of resource conservation fits well into this context because it can be translated into serving future markets. The "biocentric" or "highest use" worldview is discordant with present dominant values; one strategy for replacing the dominant values is to appeal to traditional wisdom, intuitive understanding, and quasi-religious symbolism. The present-day environmental movement borrows heavily on these traditional values in an effort to recreate a biocentric ethic as the dominant worldview.

It is not clear that the biocentric worldview will completely replace the dominant anthropocentric worldview in the move toward some form of sustainability. There is the obvious problem of inertia involved in changing all social institutions more or less simultaneously. There is also the nagging problem that alternative, stable, and biocentric social orders do not necessarily appear very attractive to most of the world's citizens, particularly those only now emerging from traditional village societies.

One possible approach to maintaining dynamism in a sustainable culture is to encourage constant change, creativity for its own sake, and experimentation. Although often viewed as hedonistic, superfluous, and nonproductive, fashion (see box 11.1) and the entertainment industry actually demonstrate how this can be done. Fashion is an unlikely but interesting model of how a sustainable society might actually engage people's interest and keep culture fresh.

Whatever worldview eventually predominates, or whether a synthesis occurs, it is fundamental to sustainability that the public believes that positive changes in the global environment are possible.

Culture and Technology

Technology, of course, plays a major role in causing many of these environmental problems. In a sense, especially if agricultural technology is considered, technology was necessary for humankind to achieve "unsustainability." It is popular to speak of problems for which there are no technical solutions and the need for a more fundamental change in society. It is not surprising that there should be sustainability

BOX 11.1

Fashion and Environmental Responsibility

Sustainability advocates may look at the fashion industry with skepticism. After all, what could be more superficial and trivial than changing styles of clothing and accessories, and what could be more wasteful than last year's fashion discards?

What if fashion is actually pointing in the direction of sustainability?

Fashion is a way of creating new interest and indulging creativity within fixed limits. People need clothes, and most clothing is prescribed by custom: dresses, skirts, shirts, pants, shoes and socks, and so forth. The scope for creativity in fashion does not lie with inventing new and innovative body coverings but in using design to enliven a product that serves a mundane purpose, through distinctive shape, color, detailing, and patterns. By creating new designs every season, the fashion industry keeps a basic necessity, clothing, fresh and interesting. In order to do this, it often experiments with outrageous innovations and concoctions that test limits but that often lead to more practical designs.

The power of fashion to make a difference in sustainability was demonstrated recently in Japan. In the aftermath of the Fukushima disaster in 2011 and the shutdown of the country's nuclear reactors, Japan embarked on a national program to adjust to reduced energy supply and the need for conservation by turning up thermostats during warm weather to reduce consumption by air conditioning. Called "SuperCoolBiz," the program was intended to promote casual wear over warmer and heavier business clothes. The objective was to lose the uniform of the "salaryman," the suit and tie, and the tailored suit of the salary woman. The ultimate goal was to allow thermostats to be raised inside office buildings and to reduce demand for electricity from air conditioning. This move was enthusiastically backed by the Japanese clothing industry, which saw sales of casual clothing triple.

The immediate motivation for this fashion intervention was the urgent need to reduce electricity consumption in the wake of the historic nuclear reactor crisis the country had experienced since the Tōhoku tsunami. This incident not only resulted in the loss of energy generated from the six Fukushima Daiichi nuclear reactors that were disabled as a result of the tsunami, but the closure of another set of vulnerable reactors at Hamaoka, closer to Tokyo, because of seismic risk. The result was loss of reserve in the nation's electricity supply, 30 percent of which is supplied by nuclear energy. Unless consumption was controlled, there were fears that unexpected demand could result in rolling blackouts.

However, SuperCoolBiz was not conceived as only a short-term solution. On 1 June 2011, Environment Minister Ryu Matsumoto, with the full weight of the government of Japan, declared that personal energy conservation and lifestyle change should become permanent in Japanese society. In other words, the starched button-down formality of the Japanese government and business sector would follow Singapore and other Pacific nations in adapting to climate rather than opposing it through technology, at least when it came to clothing.

Whole new technologies are under development for clothing, of course. New "smart" textiles, nanomaterial-treated cloth, clothes that compute or generate energy, and new generations of synthetic materials will change textile technology and therefore fashion. There may be a great deal to be gained for sustainability and energy conservation by rapid adoption of fashions that feature these new technologies. At the same time, there is already much to be gained by adopting simple cotton clothing, which has so many advantages as a fabric; however, there is one big disadvantage to cotton, which is its cultivation. Growing it requires a copious water supply and large amounts of pesticide. A technology for growing cotton sustainably would be a huge contribution.

Of course, fashion is also subject to the same principles of sustainability as other sectors. A recent popular book has traced the environmentally unsustainable path of a common T-shirt as it begins with high water- and pesticide-consuming cotton crops and goes through the production cycle of manufacturing in low-wage countries, transportation across long distances to markets in rich countries, recycling, and ultimately export and resale in bulk that often undercuts local markets in countries that once produced clothing themselves. However, fashion itself is not the root cause of unsustainability in the clothing industry. It is the way the industry creates economic value out of a commodity that has limited intrinsic value as a piece of cloth.

Textiles and design features can be made less environmentally damaging and consistent with fair trade. Fashion commodities are generally inexpensive to make and are priced to consumers for their aesthetic appeal, cultural significance, and novelty.

On a deep cultural level, therefore, there is much more to be learned from the fashion industry. One is social change. There is no area of modern human life that changes as quickly as fashion and in so many complex cycles within cycles. Yet, while hem length and width of ties may change from year to year, clothing basics follow longer-term trends, such as frocks (loose-fitting full-length garments most familiar as woman's dresses), pants (modern trousers only date from the early nineteenth century), and ties (an adaptation of the cravats worn by Serbian infantry troops, which caught the attention of the British proto-*fashionista* Beau Brummell in 1784).

Fashion is not trivial: people really care about it. Clothes serve many purposes besides concealing our less attractive bits or showing the good ones off. They are a means of communication, instantly sending a message about how the wearer wishes to be perceived by others. This reality overrides ingenuous protestations about not judging a person by their appearance, as if those aspects of appearance that are under our control (such as hair and clothing) were not a basic signal of how people wish to be sized up: threatening, fun, sexy, stolid, and so forth. Fashion and personal accents change the interpersonal environment at an intimate, face-to-face level. There is also truth in the famous soliloquy given by the Anna Wintour character played by Meryl Streep (Wintour is the formidable editor of American *Vogue*) to her new assistant (played by Anne Hathaway) in the movie *The Devil Wears Prada* (2006), to the effect that fashion is a serious, creative business that supports jobs and opportunity by giving people a chance to choose aspects of their outward appearance and that to dismiss it as shallow is a sign of callow reverse-snobbery.

White guayabera shirts, with double vertical panels, often pleated, and open collars, are widely accepted as standard business wear in Latin America, the Philippines, and elsewhere in tropical climates. They look great on both younger and older men and are well adapted to the climate. Cultural acceptance and adoption of this attire in countries where it is not already customary would be a comfortable and efficient personal adaption to lower energy use and warmer weather as climate change progresses. It is a fad that needs to happen and once it happens to become a permanent fixture of the culture.

Once established in the culture, clothing styles can become highly conserved. Certain fashions enter as fads but then do not go out of style, establishing themselves as cultural norm. Pants, for example, replaced breeches and pantaloons, neither of which ever made a comeback. Once a clothing style is established, fashion creates constant variety, an incentive for innovation, and visual interest as a form of wearable expression (if not art). Jeans, bell bottoms, leisure suits, peggers, pantsuits, and other styles that are long gone all began as fashion-driven variations of a single standard item of clothing: the union of two breech pants with buttons to become "a pair of pants." Fashion also allows some styles to be refined to near perfection and then to remain timeless, such as the "little black dress" of Coco Chanel and the jeans of Steve Jobs.

Far from being irredeemably superficial, fashion is a model for how fundamental changes in lifestyle may last for centuries but can be embellished, elaborated upon, and constantly reinvented in order to keep life fresh and interesting and to encourage innovation and creativity. Perhaps fashion ultimately has something important to say about living a sustainable life.

problems with unrestrained technology, because technology is inherently a systematic balancing of forces and interconversion of energy (mechanical, electrical, chemical, etc.) to achieve a desired goal. As such, technological processes are always subject to inefficiencies, since the conversion of energy can never be 100 percent (indeed in most technological applications it is seldom as high as 40 percent). There will always be material and energy waste and continual trade-offs between desirable and undesirable results. Forces that balance one another in one way do not necessarily balance in another desirable way: cars with more power produce more air pollution, increased production efficiency may cause unemployment, and the siting of a plant may not be appropriate ecologically in spite of economic advantages.

At best, technology only offers the range of alternatives, a toolbox of tools available to be used. The final decision on which tool to use is a decision of purpose and intention. Technology is neither panacea nor villainy but an instrument. Sustainability cannot occur without technological innovation unless human society is to devolve back to a tribal or foraging state.

Culture and Sustainable Development

The concept of sustainable development implies community empowerment, social justice, cultural expression, and ethical stewardship of resources. Once the implications of sustainable development are traced through their social ramifications, it becomes clear that nothing less than a restructuring of society would accommodate a sustainable economic structure. To accommodate and to manage a sustainable economic system, the social structure would have to change from an emphasis on increase and growth in size to an emphasis on differentiation and growth in quality. What type of community could accommodate such a transition?

Society restructures itself all the time, especially its economic structure. It simply does so within limits. A sustainable economy that treads lightly on a sustainable environment is no more of a stretch for a culture of sustainability than a market economy with a plethora of consumer choices is for a liberal (here meaning individualistic) society. People figure these things out organically, one decision at a time.

Sustainability is often described, especially in fictional utopian literature, in terms that suggest a static, technology-dependent, and culturally more homogenous regionalized society than "the modern world." It is often assumed in the utopian environmentalist literature that sustainable communities would have to be small, resource-efficient, and directly linked to a resource base in a way that would maintain some control over the exploitation of resources by the population. Traditional values found in small or rural communities that promote stability may or may not be any more supportive of sustainability than urban values emphasizing efficiency and cultural dynamism, however.

Chapter 9 discusses urban ecosystems and their characteristics, which on the whole seem to be a good fit with the characteristics of efficiency and cultural opportunity needed for an attractive sustainable future. Large-scale urban centers may not be self-sufficiency but urban nodes would have a great advantage for regional and global efficiency and technological innovation.

Contrary to the popular literary genre of dystopianism, a sustainable future may lie as much or more in cities as in dispersed small communities. Dystopian novels, projecting a degraded environmental future, are often set in the ruins of cities or vast, alienating cityscapes. A new genre set in a vision of the creative, sustainable city may now be in order.

Rights and Justice

Is there a right to health? If so, what does this mean in terms of the determinants of health (as explored in chapter 1) or access to health care? Is there a right to a safe and clean environment? If so, what does this mean in terms of "safe" and "clean," given that safety is relative and contamination is universal (see chapter 4)? Are these rights "inalienable" (meaning that nobody is free to negotiate their right away, even voluntarily), or are they relative, and if they are relative does their recognition depend on the resources available to a society?

This section begins with a discussion of health justice and health as a right and sketches in the background of rights theory. It then proceeds to discuss environmental justice. It is followed by a final section on NIMBY ("not in my backyard"), which can be perceived as a fundamentally cultural issue that is only argued on the grounds of rights and justice.

There are different types of rights, and the right to a safe and clean environment would be a "material right." Material rights are rights to something tangible such as food and clothing. These rights are usually discussed in terms of decent living conditions and protecting children. Applied to environmental rights, they would include a right to shelter, clean water, clean air, and sufficient food (including the right not be poisoned). Material rights are well established globally. The recognition of material rights has been a major theme in European social democracies but much less so in the United States, the Commonwealth, and other countries sharing a more traditionally "liberal" (in the original sense of individual-centered) political culture.

Rights

Rights, in the sense used here, are moral, legal, ethical, or economic entitlements that confer a freedom to act or to access something, either for one's own benefit or on behalf of others, such as one's family or supporters of a cause.

Strictly speaking, rights in this sense are not permissions granted by an authority, because the authority has no power to deny the right. However, not all rights are inviolate and absolute. Some rights outweigh others, such as the right to freedom of movement against the right of society to be protected by police who apprehend and detain a perpetrator. Some rights can be waived, at least temporarily, but one cannot give away a "natural right," which arises from being a human being; if one renounces it, the renunciation is not binding. John Stuart Mill articulated this well when he said that a person is not free to turn him- or herself into a slave. This is what is meant when a right is described as "inalienable."

Health rights and environmental justice cannot be separated. The concept of health rights underlies the logic of "environmental justice", both because the logic is very similar and because health protection requires environmental protection. Yet, there is broad agreement on environmental justice and little obvious agreement, at least in the United States, on health rights and justice. This is a contradiction.

Public health leaders often talk about reducing health disparities, by which they mean achieving progress in health by eliminating the differences between the worst health performance and the best, which provide benchmarks of good practice. "Health equity" or "health justice" is replacing this idea as a conceptual framework by asserting that health and the health care required to achieve it is a human right and a public good. Health justice calls upon society to provide the best attainable medical care and health protection to all people. The constitution of the World Health Organization (WHO) states that "the enjoyment of the highest attainable standard of health is one of the fundamental rights of every human being . . ." Conceptually, health justice is therefore different from other rights, which mostly free persons from restrictions or limits so that they can reach their own potential. This concept of health justice is consistent with the WHO constitution and several covenants of the United Nations and with the Universal Declaration of Human Rights and was particularly influential in the 2006 Convention on the Rights of Persons with Disabilities, which led to major breakthroughs in opportunity for persons with handicaps. Health justice is still controversial in the United States, as demonstrated by the bitter controversies surrounding the Affordable Care Act.

The right to a safe and healthy environment is largely a restatement of the right to social security (meaning safety), quality of life, and the material necessities of life, which are recognized as natural rights under the United Nations Charter (1945) and the International Covenant on Economic, Social and Cultural Rights (1966). Material rights are widely recognized as natural rights and a cornerstone of the rights debate in Europe and are written into many instruments of the United Nations. The great Indian economist and theorist of justice Amartya Sen (b. 1933) emphasized that economic development is about freedom to make choices without constraint, not only of force (coercion) but also of want (privation). He was led to this conviction not only by his economic studies but also by his personal experience having grown up in a poor developing country and witnessing famine firsthand as a child.

It is perhaps surprising that societies representing so many different cultures and histories have adopted similar points of view, at least at the national level, on whether there is a legal right to a clean and healthy environment. There has been surprisingly little controversy over the issue around the world. Of the 193 member states of the United Nations, 173 have rights to a clean environment written into their basic law, established by legislation, determined by court opinion, or committed by international agreement. (How effectively these rights are honored in some countries is another matter.) The major relevant exceptions are the United States, Canada (although six provinces do recognize such a right), Australia, New Zealand, Malaysia, and Japan (which has a postwar constitution that was American in origin). This strongly suggests that there is a philosophical divide on the issue marked by differences in tradition.

The American tradition places little emphasis on material rights other than property rights, although it values personal liberty as paramount. This may be because it is assumed in the United States that a person should be able to provide for him- or herself and family in a country where such opportunity abounds. American history, law, and tradition reflects the attitude (derived from Britain), and shared by other large English-speaking countries that have experienced similar circumstances of economic advantage, that "freedom" implied above all "freedom of opportunity" and therefore the freedom to fail. By comparison, European history had its laws, traditions, and attitudes shaped by its terrible legacy of war and want, leading up to the 1950s, when many of these provisions were adopted, the recent memory of war and post–Second World War suffering. In postwar Europe and in other societies with a long history of poverty and even starvation, "freedom" implied "freedom from privation."

This is reflected in the provision of social support services. Until recently, the welfare model for western Europe was to fund social services directly. (Europe is now going through a wrenching adjustment because this model was quite expensive.) Employment was a secondary consideration. The welfare model for the United States, on the other hand, has been to encourage the poor and disadvantaged to find work and lift themselves up out of poverty. This works well enough when there are jobs and when people actually have the capacity to lift themselves up but results in a permanent underclass when these conditions are lacking.

The American tradition converged with the social democracy model of material rights once, when President Franklin Delano Roosevelt articulated the "Four Freedoms" in his 1941 State of the Union address and specified "freedom from want," which was tied to a "healthy peacetime life" through economic agreements. (The other freedoms were freedom of worship, freedom of speech, and freedom from fear.)

A right to a safe and healthy environment confers a social benefit and promotes progress toward the goal of sustainability. It is also consistent with the philosophy that conscientious people, being aware, are custodians, not owners, of the planet.

Health Rights

Is there a right to good health? The United Nations believes there is. The Universal Declaration of Human Rights (1948) stipulates that there is a right to health, including health care. Of course, concerns over health are a major driver of activism and environmental justice. This connection links the drives for "environmental justice" and for "health rights." It also confers urgency and priority beyond actual public health impact on the "toxics" arm of the environmental movement, which motivates a great deal of activism on environmental issues. (See figure 11.1.)

The idea that there is a right to health is a widely held view in other developed countries, as demonstrated by the priority placed on national health-care systems, especially in Canada, Europe, and Japan. As countries grow richer, usually one of the early items on their social agenda is a universal health-care system for their citizens, usually introduced in phases (as in Mexico and Turkey).

Advocacy for health care as a human right has been going on much longer than the dialogue on environmental justice, but it has also been mired in ideological opposition and practical details regarding implementation of health-care reform. And in the case of the United States, partisan politics has also been a major factor.

In the United States, Franklin Delano Roosevelt also agreed that there was a basic right to health; in his 11 January 1944 State of the Union address he called for a second Bill of Rights that would include, among a variety of other material rights, "the right to adequate medical care and the opportunity to achieve and enjoy good health." Just a year later his successor, President Harry Truman, launched an initiative to operationalize this right for access to medical care with a national health insurance system for the United States. Opposed by the American Medical Association and not supported by American industry, the proposal died. Comprehensive insurance coverage for medical care did not become a reality until the Affordable Care Act (ACA) of 2010. The ACA barely passed, was ferociously opposed on a partisan basis, and has been repeatedly attacked since it became operational in 2014. As of late March 2014, the US House of Representatives (which converted to a Republican majority in the two elections after the ACA was passed) had voted to repeal the ACA fifty-four times, but the Senate (which retained a Democratic majority) did not, so the law stands, as of the time of this writing. Obviously, in the United States there is no general agreement about the existence of a right to health care, let alone good health. Equally obvious is that those who disagree with the existence of this right do so vehemently and as a matter of principle.

In the United States there is a wide range of opinion on this issue. These opinions are deeply held but not often stated explicitly, not least because it requires courage to take a public position. A gradation of views has emerged at various times during the debates surrounding the fifty-year national conversation about US health standards that culminated in the ACA. They have formed a sort of subtext, unstated but real in their influence, of underlying attitudes toward the law. Many, probably most who count themselves politically progressive, agree that access to health care and the means to live a healthy life (good food, decent shelter, clean air and water, low levels of environmental hazard) should be available to all Americans, but there would be disagreement as to whether the government should make this happen or whether it is the individual's obligation to manage his or her own health.

FIGURE 11.1 Cultural factors shape the expression of concern for health and sustainability and bring together many disparate themes. Health is not only a principal concern for sustainability and environmental protection but is also used as a powerful metaphor. This demonstration by supporters of the Green Party in Germany brings together potent symbols of life (the sunflower, children, the color green) in opposition to symbols of death (the skull, the gray color of the nuclear reactor cooling towers, the shrapnel like flying particles from the atom being split). Note that the skull is laughing, which implies that proponents of nuclear energy are irresponsibly arrogant. The casual dress, laid-back manner, and informality of many environmental advocates, including most Greens, is also an implicit critique of formal hierarchy, trust in officialdom, and reliance on engineering; it also emphasizes their roots in the community.

Centrists agree selectively, in the sense that elderly and disabled people have a right to health care and accommodation of their disabilities proportionate to their needs. Perhaps most people believe that adults in disadvantaged circumstances should get emergency care. Americans, in general, may agree that children deserve health care but are not so sure about a right to care that goes beyond that.

Vocal advocacy for health as a human right today (as this is written in 2014) have been almost exclusively a progressive or "liberal" (in the contemporary sense of "left" on the political spectrum) position in in the United States. Thus it would seem on the surface that support for health-care rights is narrow and confined to the left of the political spectrum. However, this is somewhat misleading. Other politicians have accepted the need for reform of the health-care system in ways that imply acceptance of a right to health.

If advocacy of health care and health insurance reform is counted as actualized support for health care as a human right, whether explicitly advocated or implied, then the political spectrum in agreement is much broader. Every American president after Herbert Hoover, except Ronald Reagan, advocated some form of health-care reform and most of them expanded government-sponsored or funded health-care reforms at some time during their administrations, including conservative presidents Dwight Eisenhower, Richard Nixon, Gerald Ford, George W. Bush, and George H. W. Bush, as well as "liberal" (in the popular sense, not the traditional sense) Franklin Roosevelt, Harry Truman, Lyndon Johnson, Bill Clinton, Jimmy Carter, and, of course, Barack Obama (who actually achieved broad-based health-care reform). The current political polarization in attitudes and opposition to health-care reform obscures the fact that for much of American history there was a leadership consensus that change was needed; this consensus somehow did not translate into broad recognition of health-care as a human right.

A large part of the debate is how the question is framed. If what is meant is a right to health care at whatever cost it takes, then even societies with generous medical insurance or national health services have limits. If what is meant is a right to be protected from harm, there might be few arguments, especially if children were under consideration. If what is meant is access to free services to keep people healthy, many people would object, on the grounds that personal responsibility and self-interest should govern rational behavior, or that the means to stay healthy should be available to everyone but that taking advantage of those means is each person's responsibility or that of parents and guardians (in the case of children). These nuances have seldom been reflected in poll questions put to the American people. It is therefore somewhat unclear where the limits are in support of health rights in the United States (and its various "publics," to use the terminology of risk perception).

Many people frame the question differently, however, and believe that there is no natural right to a state of being. Some people, especially those who identify themselves as libertarians, believe that there is no right to health care because there is only a right to act, never a right to demand from others; thus, it is up to the person involved to have the means to obtain health care. If someone cannot afford health care then, following the strict libertarian argument, that is the problem of the individual because society has no obligation to support them. They may receive charity, but only if someone wishes to bestow it on them. In the marketplace, the argument continues: Is health care a commodity to be bought and sold like any other product or service and subject to supply and demand? Should there be marketing to attract the lower-cost, higher-paying patients and diagnoses? Despite many years of debate, these questions are still not resolved.

Environmental Justice

"Environmental justice" is the right to a clean and safe environment with an emphasis on ending discrimination or disparities due to race, social class, income, or location. It has emerged as a foundational principle for sustainability and equity in health risk. "Health justice" and environmental justice embrace a similar set of principles for achieving sustainable health and links healthy living with human rights, fairness, and equality. Both advance the argument that there is a right to a healthy living and commit to reducing disparities and eliminating discrimination. Compared to health justice and the presumed right to health care, environmental justice has been more widely accepted, even though they are closely related. Environmental justice is more accepted in the United States and has some recognition in law, largely through a mandate of the Environmental Protection Agency.

Environmental justice is commonly defined in the United States as the right to a safe, healthy, productive and sustainable environment for all, with a broad biological, physical, and social definition of environment. The US EPA, which has been at the forefront of translating environmental justice into action, defines it as "the fair treatment and meaningful involvement of all people regardless of race, color, national origin, or income with respect to the development, implementation, and enforcement of environmental laws, regulations and policies." Fair treatment means that no group of people, including any racial, ethnic, or socioeconomic group, should bear a disproportionate share of the negative consequences resulting from industrial, municipal, and commercial operations or the execution of federal, state, local, and tribal programs and policies.

It has long been known that pollution tends to be worse in disadvantaged communities, usually consisting of people of color and characterized by low socioeconomic status and low income, with resulting political disadvantage.

Environmental hazards may have been "sited" (located) where they are because richer and better connected communities opposed them and because property was less valuable. Or the hazard may have been imposed because the community was unable to resist it politically; there may have been neglect or less stringent enforcement or regulations.

Environmental justice is a dimension of social justice as applied through environmental equity, which is the idea that people should not have disproportionate environmental risk, no matter where they live, who they are, or what their income status may be. It also requires fair treatment and environmental protection regardless of class, race, income, or location. As a movement, environmental justice encourages community empowerment for the purpose of health protection and environmental equity and advocates the elimination of health disparities—and therefore is closely linked to health justice. It pursues equal justice and protection under the law for all environmental laws. In practice, this means equal application of regulations, reducing disparities in risk that arise from environmental conditions, and correcting historical patterns of discrimination (sometimes called "ecoracism" or "ecoclassism") and opportunism (typically when a facility is sited in a location because the community is not able to resist it).

The action agenda of environmental justice is community empowerment, such that residents are able to resist unwanted development and insist on the same standards of environmental protection as elsewhere. In practice, this also often means documenting and pursuing legal actions for toxic torts (environmental lawsuits over alleged injury from chemical exposure). This is often (but incorrectly) called "ecopopulism." (This term is something of a misnomer because "populism" as a political description usually implies not only that the common people are favored over elites but that the public is also manipulated by politicians who play on their values and beliefs; this is not at all what "ecopopulism" is supposed to be about.)

Environmental justice, as a movement, arose out of concern for environmental equity that was documented in a seminal study originating with Walter Fauntroy (b. 1933), who was the elected but nonvoting representative of the District of Columbia in Congress. (DC has no vote in Congress, but its representative has certain privileges, including sitting on committees and voting as a member of them.) Fauntroy requested that the US General Accounting Office conduct a study of landfill "siting" (choosing a location) and enforcement measures and minority (African American and Hispanic) communities in the eight Southern states that comprised EPA Region 4. (Landfill approval is a state responsibility, not a federal responsibility.) It found that about 75 percent of landfills in the region were located near predominantly low-income minority communities. About 60 percent of these minority residents lived in communities with toxic waste sites, and the US EPA took 20 percent longer to cite (take legal notice of) abandoned sites in minority communities. And companies convicted of pollution violations paid fines that were 54 percent less than in predominantly white communities. The report concluded in 1983 that "race was in fact the most significant factor in determining the siting of hazardous waste sites." This, of course, raised the issue of whether the lives of people of color (African American and Hispanic) were thought to be worth less or whether the assumption was that these communities would not resist, perhaps because of discriminatory assumptions about education, intelligence, and capacity to understand the risk. These communities invariably have a history of discrimination and have many other associated problems that may impede the expression of community will and make it hard to organize. These include poverty, lack of economic and educational opportunity, a paucity of models for young people in the community, generational conflicts, social issues (often including drugs), and often dysfunctional politics and corruption in local government. In immigrant communities, Hispanic or otherwise, language and acculturation barriers complicate these problems.

The narrative emerged, and was widely accepted, that owners or corporate interests behind unwanted facilities such as polluting factories or landfills lack respect for residents. In turn, the environmental injustices and inequities interfere with the healing of social problems in these communities because the industries or facilities are often exploitive and employ few residents. These injustices also tend to drive out alternative economic development options, perpetuate poor community self-esteem, and perpetuate the marginalization of communities across generations. The real issue is seen to be marginalization and the lack of both economic and political power.

The reality turned out to be somewhat more complicated. There was a past legacy of racial and ethnic discrimination, poverty, and lack of opportunity, combined with economic depression that attracted marginal industry that was undesirable to other communities. Historically, communities often grew up around a polluting industry or contaminated site that had employed local residents. Sometimes communities had welcomed polluting industries for jobs and economic development. Low land prices also attracted unwanted and marginal land uses and facilities. Racial discrimination, low income, low educational levels, lack of mobility (and inadequate transportation in the community), and political disenfranchisement not only affected the siting of new undesirable facilities but kept residents from moving away from the old ones. A community would be forced to accept a new unwanted siting because the community had no political clout to resist or to insist on remediation. Community residents lacked direct participation in planning and approval process. Then, a different narrative emerged on the issue—one of community empowerment.

Around the same time, an attempt to site a hazardous-waste landfill in a predominantly impoverished African American community sparked a protest movement.

Warren County in North Carolina is a rural county with a highly segregated population. The landfill would have been issued a permit to receive soil contaminated by polychlorinated biphenyls (PCBs), one of the classes of POPs (chapter 5). The successful opposition to this landfill drew additional attention to the siting of unwanted facilities in impoverished communities consisting mainly or entirely of people of color. As a direct result, in 1994 President Clinton signed Executive Order 12898, which committed the EPA to a policy of environmental justice.

A special case also came to light, especially in the western US, in Native American communities and on tribal lands. These are often geographically isolated and financially dependent on the federal government or on a limited range of resource-based industries, mostly mining. Many are deeply impoverished (especially in the north-central Great Plains) and most of those that have an economic base experience "boom or bust" cycles in their primary economy, which is one reason so many tribes have turned to gambling (which is legal on tribal lands).

Since the 1980s, the narrative has grown a bit more elaborate. It is still clear that the real issue is marginalization and the lack of power, but it is also clear that even disadvantaged communities can organize and successfully oppose unwanted land uses. "Ecopopulism" (the term is used advisedly) thrives in communities rich or poor and of any racial or ethnic makeup when residents are aware and educated and there is a strong sense of self-determination and democracy in the community. Opposition to a particular project is generally most effective when there is a past or present grievance that has sensitized the community, the object of the opposition is either novel (such as a hydrogen refueling station) or stereotypically negative (such as a landfill). Opposition, in turn, is most effective when it is rooted in evidence and can be delivered by articulate spokespersons for the community.

Because much initial concern had centered on hazardous waste sites and landfills, the development of environmental justice was closely connected to the Superfund program. This program was mandated by the Comprehensive Environmental Response, Compensation and Liability Act (CERCLA) of 1980, which created broad federal authority in the United States to respond directly to present or threatened releases of hazardous substances that pose a risk to human health or the environment. Superfund mostly dealt with prioritization, interventions, and cost recovery for cleaning up (remediating) the worst hazardous waste sites. The Superfund Amendments and Reauthorization Act (SARA) of 1986 encouraged the use of innovative technologies, added additional enforcement tools, created a duty on industry to supply information for medical treatment, and encouraged citizen participation in making decisions about Superfund sites. Although the Superfund program was enormously expensive (funded by a trust fund that held allocations from the chemical industry) and painfully slow in individual cases, it did establish the principles of environmental justice and right to information and thus facilitated progress of the entire movement.

New tools became available for community representatives to use, including legal action under Title VI of the Civil Rights Act (1964), under the authority of which other federal agencies were bound to the same principles. This effectively extended the reach of environmental justice policies beyond the mandate of EPA to cover energy, transportation, agriculture and forestry, and other infrastructure, including the important issue of highways that divide communities. EPA also launched a grants program, Community Action for a Renewed Environment. Precedents were set in early cases involving the relocation of a tank farm in Texas and managing brownfields (contaminated industrial sites), establishing that the principles were not limited to landfills. California, Minnesota, and New York passed state legislation that covered community risk assessment and siting of facilities such as power plants.

Much of the power that has since shifted to communities comes from a break in the monopoly on information about hazards and risks, which began on the community level and really broke open on a national level with SARA in 1986. Prior to about 1980, information about what chemicals and other hazards may be present in the community was considered proprietary. It might be shared with a fire department but was otherwise unavailable. Even local public health agencies were often in the dark about hazards that might cause catastrophic incidents, such as tanks of flammable chemicals, hazardous industrial processes, and sources of contaminant release, a series of federal, state, and local laws were passed collectively called "community right to know" that established access for citizens to information on the location and magnitude of mainly chemical hazards, in their communities.

This movement accelerated after the horrendous disaster in Bhopal, India, in 1984, in which half a million people were exposed to a highly toxic gas (methyl-iso-cyanate) and at least 2,259 died (estimates vary to as high as 8,000) as a result of a leak from a plant owned and operated by an American-owned chemical company. The shock of the Bhopal incident led to fears that something similar could happen in the United States and drove pressure to adopt protective legislation.

The major new law in response to Bhopal was the Emergency Planning and Community Right-to-Know Act (1986), which mandated creation of a Toxic Release Inventory, which led directly to the National Environmental Monitoring Initiative; this led in turn to the National Public Health Tracking Program, which extended the federal role from observing releases to the environment to monitoring trends in actual human body burden. The new availability of information on potential exposure was matched with greater capacity for the assessment of human health risk, for example by the Agency for Toxic Substances and Disease Registry (now absorbed into the Centers for Disease Control and Prevention). Accompanying these federal programs were a myriad of state initiatives that went even further, particularly in California.

A side effect of the new access to data is that some activist groups have been tempted to try and match release information with disease prevalence in their communities, usually without taking into account the many other factors that influence health. This practice, called "community epidemiology" by some, has been applied mostly to cancer clusters and almost never yields a valid picture of the health status of the community. These studies, often called "quick and dirty" even by their proponents, are too often uninterpretable or intentionally biased. However, people have a right to information and to use it as they wish, as well as to express their views on what it means. Since 11 September 2001, the right of access to much community-level data on chemicals has been curtailed for security reasons, since knowledge of where chemicals are stored could be used by terrorists.

Since its origins, the idea of "environmental justice" became less narrowly attached to chemical safety and hazards and has extended to concerns over land use and media quality, such as ambient air pollution. It has also been applied to disparities in health outcomes, particularly asthma, on the assumption that these disparities are likely to be driven by environmental factors.

Today, most communities in which environmental threats are present, at least in the United States, are not nearly as defenseless as they were in the 1980s. Since the 1960s, almost three generations of community organizers and activists have worked with most of these communities, at one time or another and on various issues and for community development. Many of them now have extensive networks with other, similar communities and they trade strategies and case studies freely. It is now rare to find a community that lacks local leadership on controversial local issues that is well-informed on tactics and strategy. Knowing this, and knowing that these communities, once engaged, are quite capable of embarrassing them with demonstrations or voting them out of office, local politicians are usually far more responsive in dealing with such issues than they used to be. On the other hand, proponents and economic interests who may not have roots in the community often misjudge it, with predictable consequences. Consumer power is much greater than it used to be and the reputation damage to an oblivious or wrongheaded company can translate into bad investor advisories, difficulty with lines of credit, enforcement actions, tax scrutiny and many other headaches. Media has also changed; local newspapers are no longer the lapdog of local business, as they have been so often in the past, because they are in a fight for survival with the Internet. That does not mean that there is not still a power imbalance but rather that communities are empowered to a degree that was difficult to imagine in the 1980s.

In some parts of the United States and in many other parts of the world, communities are often still at the mercy of a small power elite, usually consisting of elected officials of the county government, the important employers in town, the leaders of the chamber of commerce, and, sometimes but less often than in the past, the publisher of the local newspaper (who is losing power and influence by the day and more likely than not is responsible to a distant owner with no stake in local affairs other than to seem to be relevant to support circulation). In the past, this power elite, or "establishment," largely settled matters among themselves and acted in what they felt was the best interest of the community, without necessarily involving local residents in affairs that directly concerned them. This oligarchy, a form of parallel government, is more difficult to maintain today mainly because of the erosion of trust in elites in American life over the last decades. In the United States and other developed countries, the means of social control are now usually much more subtle than having all the important decisions made by a self-selected and self-perpetuated oligarchy of "the great and the good."

"Not in My Back Yard" (NIMBY)

NIMBY is an acronym for a local-level movement whose objective is the prevention of an action involving the placement of a facility in a local area by the residents, all without regard for overall need or necessarily benefits to the community. This phenomenon is well known as the "not in my backyard" (NIMBY) syndrome, in response to "locally undesirable land uses" (LULUs). The NIMBY syndrome does not imply that residents do not see the overall need for the facility or the potential benefits to the community. Rather, opposition centers on the placement of the facility in their own immediate area. The notion of a trade-off between individual rights and benefits and the collective is not often perceived as meaningful in small communities or neighborhoods with accessible local government, where individuals are known by name and their fates are obvious to all and local residents do not see or think of themselves as part of a bigger whole. Local residents simply see that a decision that does not suit them and that may lead to risk, annoyance, or discomfort is being imposed on them.

A LULU may be unwanted public facility or a private development. The most common LULUs that precipitate a NIMBY response seem to be industrial facilities, university campus expansion plans, halfway houses (residential centers for soon-to-be-released or paroled prisoners or persons on probation), drug treatment centers, housing developments, zoning changes, and, above all, landfills. Box 11.2 presents the case of Aurum, a large-scale landfill issue that played out in Alberta in the 1990s. Aurum ably represents the issues inherent in NIMBY and also illustrates some of the specific issues associated with landfills.

Landfills are lighting rods for communities because of their association with trash and other people's waste. Unmanaged landfills (trash dumps) do present a variety of hazards and nuisances. Decomposing trash produces methane, which can catch fire (and is also a potent greenhouse gas). Free material can blow around with the wind. Rainwater and snow melt can run off the trash, carrying dissolved chemicals, and water that percolates through the trash can penetrate the floor of the landfill and carry chemicals and dissolved chemicals into groundwater, thus potentially contaminating the aquifer. Exposed trash attracts seagulls and rats. However, these problems should never occur with modern "sanitary landfills."

Sanitary landfills are engineered to prevent these problems. The purpose of a sanitary landfill is to dispose of trash in a location that is selected to cause the fewest problems, to control leachate, run-off, airborne emissions, and to prevent fires, nuisance, and blowing litter. Trash is compartmentalized into "cells" and covered with soil. A plastic liner is installed to prevent migration of "leachate" (chemicals and dissolved materials that are carried by water) into the ground and then down to groundwater. To be safe, groundwater is regularly monitored for evidence of contamination. A storm-water drainage system keeps water out of the landfill. The methane formed by trash decomposition is vented or, in large and advanced landfills, captured and used for energy production. Access to the landfill is restricted to prevent illegal dumping and to keep hazardous materials out. Liquid disposal is kept to a minimum. The most advanced landfills are usually built close to recycling centers and are engaged in programs to educate people to reduce their waste. However, no matter what the technological sophistication or promises of management, nobody wants to live next to a landfill and anyone who is told they have to, or that one will be built near their home, is likely to feel that they are being, literally and figuratively, dumped on.

LULU developments may be resisted because they present a threat or because residents are concerned about property values. The community may become or be seen as a less desirable place to live. Most often, however, communities empowered to resist unwanted sitings seem to do so because they have a vision of their community as supportive, a sense of security and feel imposed upon and even violated if asked to compromise their interests, as they see it. In most cases, the struggle against a LULU reflects at some level a fear of loss of local control, inequitable distribution of risk, and threats to the integrity of the community's structure. Underlying the concern over health issues in many if not most NIMBY situations may be a fear that the previously shared interests of the community will begin to diverge and that the unity and solidarity of the community itself may unravel. The despised target of the NIMBY movement forges unity in the community through consolidated opposition to the LULU.

The NIMBY phenomenon is characterized by a spirit of self-defense among community residents, often well out of proportion to the risks as they might seem to an observer without a stake in the community. "Concerned residents" (whose opinion and will is regularly invoked by activists) may vehemently oppose the local siting of unwanted facilities, using every practical political,

BOX 11.2

A Case Study in the Politics of NIMBY

The Aurum project was a landfill to serve the City of Edmonton, Alberta (Canada). It was proposed in 1989 to be built on an isolated and mostly vacant site nominally within city limits but situated across a river from the rest of the city on land owned by Edmonton but intruding into the political boundaries of adjacent County of Strathcona. The siting was bitterly opposed by the residents of and government of the country, on grounds that the site was unsuitable on geological, engineering, and public health grounds, and it was not approved. In the end, the landfill site was rejected on a technicality in response to massive community opposition. The story is not so simple, however, and a close examination reveals complicated social dynamics.

Edmonton is a major city, Canada's fourth largest at the time of these events. Most of the city is contiguous and is set almost entirely on the west bank of the North Saskatchewan River. However, there are two relatively isolated parcels of land in the northeast that lie east of the river: Clover Bar, formerly a significant community with an economy based on coal mining, and Aurum, which had always been farmland. Both are sparsely populated. At the time of these events, Clover Bar was the site of the city's only landfill and Aurum was agricultural, with very few permanent residents.

Sherwood Park is a planned community on the east bank of the river designed as a planned community to be the population center of Strathcona County, which covers a total area of 1,172 km^2, most of which comprises farmland, residential acreages (some of which are second homes), and open space. Sherwood Park was settled in the 1950s as residential community for workers in the petrochemical industry. In the 1970s, it became a planned residential community on a larger scale and grew rapidly into a major suburb catering to relatively affluent families. In 1991 Sherwood Park had a population of 56,559. In order to diversify its economic base, which was initially closely tied to jobs in Edmonton and a nearby petrochemical complex (third-largest in North America), Sherwood Park developed an industrial park and attracted light industry. By the time of these events, Sherwood Park featured its own substantial town center that was beginning to attract concert-goers and shoppers from Edmonton. Sherwood Park housing prices, while lower than for comparable homes in Edmonton, were high relative to other nearby towns and the housing stock was new and of high quality.

Strathcona County generally and Sherwood Park, in particular, have had a long history of controversy related to environmental health risks, many with little or no justification. Strathcona County is generally unremarkable in its risk profile save for being distantly downwind of a refinery complex, but these issues still tarnished its reputation as a healthy and safe place to live and eroded trust in health authorities. As a result, the county has a great deal of experience in defending itself against perceived environmental risk.

THE ISSUE

In 1988 the City of Edmonton determined that the existing landfill, at Clover Bar, was nearing capacity. Diversion of waste through intensive recycling had already been achieved, and further capacity was thought to be impossible or impractical. Edmonton already had one of the highest rates of recycling in North America but had experienced problems finding markets for recovered plastics and paper due to distances from larger markets.

It was decided that a new landfill would be needed within five years. Because groundwater levels in the entire region tend to be close to the surface, relatively few sites were considered feasible for excavation and development of a new secure landfill. A few sites were identified within the City of Edmonton, among them two in northeast Edmonton, on the west (north) bank of the North Saskatchewan River, not far from Clover Bar. Local residents, citizens of Edmonton, objected to the siting of all four sites. These sites were then ruled out of further consideration. However, another site was identified nearby: Aurum.

Aurum was owned by the city of Edmonton and so was considered to be within the boundaries of the city but it was an enclave bounded an all sides but the river by Strathcona County. The population of the Aurum site and the immediately surrounding area was approximately eighty at the time so very few Edmonton voters would be inconvenienced. The land supported a farming operation, a mushroom farm, and very few homes, most of them old farmhouses. The land was on a bluff overlooking the river, upstream and separated by a buffer of farmland from Sherwood Park, which one might have assumed would soften opposition from residents of the County.

Politically, Aurum therefore appeared to be ideal from the standpoint of Edmonton's interests. The Edmonton City Council ordered planning to begin for a landfill on the site, without early consultation with Stratchcona County.

Strathcona County disposed of its waste at the Edmonton municipal landfill in Clover Bar, under a contract dating from 1978. Strathcona County itself would therefore be affected by the closure of Clover Bar and so would also need an alternative for its own waste disposal but this fact was not much discussed at the time.

The site at Aurum is not naturally suitable for a landfill. Preferred sites for landfills are distant from bodies of water and downstream from drinking water intakes and ideally sit on nonporous strata that impede leaching and migration of waste products. The Aurum site was above and upstream from the City of Edmonton and its water intake. It was possible, if things went badly, that leachate from a landfill could percolate relatively rapidly through the permeable soil and drain into the river, where it could enter Edmonton's water supply. The landfill at Aurum would therefore have had to have been extensively engineered and constructed to rigorous specifications to ensure that it would have maintained its integrity and held back leachate for prolonged periods. Further technical problems included traffic on rural roads, as the trash would have to be trucked in and, unlike Clover Bar, the site was not near a bridge or major highway.

In 1989 the City of Edmonton released its plan for the site. The plan for the landfill featured redundant liners, an active system to monitor and recover leachate, and an integrated waste management facility on site that would divert recyclable products and industrial waste from the landfill itself. A decommissioning plan was drawn up to manage the site after the useful life of the landfill expired in 2060. However, these precautions would have required ongoing effort for decades beyond the lifetime of the landfill, with attendant investment, active monitoring, and a plant on the location with outbuildings. The plans for the Aurum site were assuming the dimensions of a major industrial site.

The government of Strathcona County advanced six major objections:

1) The site was unsuitable, requiring extensive engineering to achieve minimal acceptability. (Edmonton countered that the engineered landfill was secure and did not depend on the site characteristics and so the unfavorable site characteristics were therefore irrelevant.)
2) Leachate control could be problematic. (Edmonton countered that leachate would be actively monitored and pumped.)
3) Land use was committed far beyond the operations of the landfill itself. Specifically, a new industrial park would have had to have been created to support the waste facility and its attendant recycling, diversion, and management operations. Plans for these specialized facilities after the landfill closed, and after the waste stream ceased to be, were not spelled out in detail. (Edmonton countered that the support facilities were necessary for efficient operation, would create jobs and would create a model system for ecosystem and human health protection.)
4) Hazardous waste could enter the waste stream. (Edmonton countered that industrial waste would be identified and diverted from the waste stream more reliably than before.)
5) Nuisance issues were not resolved (e.g., transportation, birds). (Edmonton countered that this would be mitigated and that the proper operation of a sanitary landfill presents few nuisances to local residents, although it was conceded that road traffic would increase.)

6) The proposal did not adequately address health issues. (Edmonton countered that sanitary landfills, properly operated, present few documented health hazards.)

From these positions, the public policy debate played out in media and hearings. The City of Edmonton participated mostly at the level of City Council and its Department of Environmental Services. The County of Strathcona participated mostly on the level of the County Council and the office of the reeve (the chief executive of the county). A citizen's group (SAD, or "Stop Aurum Dump") was formed, and the local residents made their opposition known. The provincial government stayed neutral as a matter of policy on issues of municipal relations and land use, but the ministry responsible for reviewing the environmental implications, Alberta Environment, declared the site technically unsuitable.

The Aurum issue was complicated by changes in legislation on regulation and governance in public health and environmental management at the time, which will not be described here. Planning approval ultimately became contingent on a mandatory review for health protection and compliance by the Edmonton Board of Health. The City of Edmonton treated this review as a technicality and submitted voluminous technical reports, which nevertheless had little content directly relevant to health. Strathcona County, on the other hand, recognized the strategic significance of the review and prepared by hiring environmental health consultants to prepare detailed reports, including a two-volume health risk assessment. Opponents of the landfill called for a comprehensive "health impact study" to be conducted by the city, certainly to document possible health issues but also perhaps to cause delay in order to raise the cost of the project substantially.

Highly charged language appeared at times, with opponents accusing the Edmonton City Council of "arrogance." Accustomed to thinking of itself as environmentally minded and progressive (and having a reputation in Canada for environmental awareness and responsibility) the Edmonton City Council and municipal government did not understand the significance or appreciate the magnitude of the opposition. In short, it was "blind-sided" (caught unawares) by the opposition.

In 1990 the Board of Health completed its evaluation and unexpectedly rejected the application, making it impossible for the city to issue "planning approval." The grounds for the rejection were actually narrowly technical, specifically that the text of the report was deficient on issues of health, but the city withdrew the plan and abandoned the proposal to site the landfill at Aurum.

Case Analysis

Although the primary arguments for opposing the landfill were tied to health, it is not clear that the health effects would have been significant or that the operation of the plant would have been disruptive to more than a few dozen people at most, given its remote location. Most large municipal sanitary landfills operate with fewer controls than were proposed for this site. Siting of landfills on the prairies is almost always a major challenge because of geological and hydrological characteristics. Leachate would migrate away from the county, not toward it, and the quantity generated would be small compared to surface run-off, even with a liner leak. Thus, the concerted and effective opposition mounted by the County of Strathcona appeared to be disproportionate to the objective benefit to be gained by stopping the project. Certainly, the rhetoric of the damage that the Aurum landfill might have caused cannot have helped local economic development and property values.

In addition to the issue of health risk, there were many other issues swirling in the debate. Among them were

- rural location and preservation of local lifestyle
- location of the site on prime agricultural land
- lack of confidence in implementing the city proposal (in particular, there had been a series of failures in the installation of clay or plastic liners for local landfills of much smaller dimensions)
- concern over nuisance issues
- wish to avoid an undesirable facility in the neighborhood that might affect the reputation of the County as clean and attractive

However, these factors do not satisfactorily explain the intensity of the opposition. Strathcona County itself has approved building on farmland, the area in question had very few residents, it was already owned by Edmonton, and the site was buffered from Sherwood Park. Leaving aside the technical question of geological and siting suitability, it was an ideal site from a political perspective.

On the other hand, the disconnect between the opinions of "experts" and community residents follows a pattern well documented in other studies, in which technical authorities fail to perceive issues of loss of control, loss of community civic sense (which would have been paradoxical in this almost vacant area), loss of equity, and environmental injustice.

Because of this unexpectedly hostile attitude toward the city, an investigative journalist hired as a research assistant for the project started digging into the basis for this apparently deeply held and concealed (but pervasive) sense of distrust, going back to reinterview some of the key players. She uncovered a political agenda that had been forgotten by residents of the city: Edmonton's repeated attempts to annex the population center of Strathcona County, three times, the most recent just a decade earlier.

This history had not been mentioned in the press or at hearings during the debate on Aurum. It is likely that few participants in the debate on the Edmonton side were aware that it was in the background. Edmonton civic leaders were in fact surprised when this was mentioned to them, because almost none of the current generation of leaders remembered the attempted annexations. From the point of view of Strathcona County, Aurum was perceived at least as much as an insult as a health threat.

In this case study, the dynamics of the NIMBY were conditioned by a preexisting issue, more than a decade old and by then decidedly one sided, that contributed to risk perception and the political motivation to oppose the proposal. Like a diamond that is "cut" by cleavage along the preexisting fracture planes of the crystal, stresses such as LULUs fracture communities and political interests along preexisting planes of division.

A few years after the Aurum issue, Strathcona County faced another NIMBY issue, this one related to rights-of-way for electric transmission lines. Needless to say, after Aurum the community was well positioned to fight that battle.

social, legal, and economic weapon available to them. The opposition may take the form of demonstrations, testimony at hearings convened by local government, advocacy of local government action (usually to suspend or ban the proposed development), litigation, and appeals to higher levels of government. NIMBY battles tend to become "all or nothing" fights in which there is little incentive for compromise or for reengineering the project. For that reason they tend to drag on and aggravate existing schisms in the community. Once the NIMBY battle is over, the community and the leaders who have emerged during the conflict often move on to a successor issue which keeps engagement and the apparatus of opposition alive.

In extreme cases, resistance to a LULU even involves residents who may stand to benefit from the proposed development. At times, the NIMBY movement may be so strong that residents damage the reputation of their own communities and the future value of their property, to their own apparent detriment. Until the final, tragic aftermath to the Ghostpine landfill project (see box 3.1, in chapter 3), that would have been a fair evaluation of the attitude of residents of that small, doomed rural resort.

Environmental justice has provided NIMBY, as a movement, with a way of analyzing such issues, a vocabulary with which to discuss them, and a set of principles that are followed today in any community-level environmental issue. For every new development, one should ask before the project is publicly proposed: who gains and who loses? Too often the answer is cast in terms of jobs, property values, tax revenue, and investment and such—not in self-perception, respect, and how the community sees itself. When LULUs are imposed on a community in the face of opposition, it is perceived by many as a violation. No benefit will ever erase the perception of disrespect and brutality toward community sensibilities, and it usually takes a long time after losing the battle for the community to accept the LULU as something they have to live with. In the meantime, passive aggressive behavior, opportunistic exploitation, distrust, and even harassment interfering with local operations may diminish the value of what the proponent or developer initially thought was a good investment.

There is also a broader strategy at work with respect to constraining the generation of waste, emissions or pollution, or unwanted activities through environmental restrictions in order to force more fundamental change. Ideally, the solution to emissions, landfills, and waste disposal, would be to stop pollution and waste generation at the source. Many activists believe that the way to do this is to deny siting unwanted facilities and to deny disposal of waste, so that the system "backs up" with no place to go and forces fundamental change. In this way, NIMBY is often a part of a much larger activist strategy to deny local options, but it would be a mistake to assume this is the case in every local issue.

The literature on the phenomenon of community resistance to local siting of undesirable facilities, the so-called not in my backyard syndrome (NIMBY), emphasizes community action, civic unity, and motivation of residents to oppose what they feel is an imposition. NIMBY issues tend to play out in a relatively predictable way, as outlined in box 11.3. This reflects in part the resourceful use of the legal and community organizing tools available and in part the interaction and sharing among community groups that takes place, trading strategies and lessons learned.

The study of NIMBY as a phenomenon has been based mainly on detailed case studies and is therefore largely anecdotal. Efforts to frame the issue in sociological terms have concentrated on the dimension of environmental justice and the imposition of unwanted facilities on a disempowered community. However, from the selected case studies that are in the literature, it is clear that the NIMBY phenomenon is reported most often in communities that are politically and socially empowered, often affluent, majority, and middle class and that enjoy access to the resources required to mount an effective opposition. These advantaged communities use much the same logic and rhetoric of environmental justice as disadvantaged communities.

BOX 11.3

How a Typical NIMBY Issue Plays Out

Vocal residents in the community unite in predictable ways. A NIMBY issue is similar to other contentious environmental conflicts but has the added dimension of being hyperlocal. NIMBYs often pit neighbors against neighbors.

A typical scenario for a NIMBY issue follows a certain script that is the result of politics, organizational dynamics, shared community organizing methods, and the perceptions of the players about the roles they occupy in the community. There is an initial announcement of a proposed new development, such as a landfill. Most such projects eventually require detailed plans that come with analysis of risks and benefits, but these are usually not complete or available at the very beginning. The issue is therefore subject at once to allegations of withheld information or lack of detail.

A small group of local residents may decide to oppose the development. Local likeminded residents might then quickly organize to oppose the development, usually forming an association with an evocative name or abbreviation, such as "Desperate to Stop Pollution in Our Neighborhood" ("Don't SPIN"). Opposition emerges in public primarily by expressing "concern," laying the groundwork for demonstrating why the project cannot satisfy these concerns.

Once the specifications or details of the plans are revealed, the organization analyzes the proposal (often with the help of outside experts) to identify weaknesses, gaps, mistakes, and potential risks. The opposition then goes directly to the public with a critique, the battle is carried forward in meetings and media, and within the community the issue becomes a litmus test of partisanship and neighborliness.

The strategies available to oppose LULUs vary with the local situation but usually involve one or a combination of the following claims:

- The siting is technically deficient.
- The siting conflicts with other land uses.
- The siting is not secure; or, the technology required to ensure that the siting is secure is either flawed or untested.
- In populated areas, the proposed development places people at too great a risk.

- In less densely populated areas, the proposed development is unfair to locals who must bear the burden for others.
- The siting is out of scale with the modest scale of the community.
- The siting poses a risk for children.
- The environmental impact statement (if one is required) is deficient.

The power relationships of the players often shift and can invert entirely during a NIMBY process. Ordinarily, players with a financial or political interest in seeing the project go forward will actively promote and advocate for it. However, in the hyperlocal world of NIMBY, those who stand to gain from the project are often reluctant to be overly vocal for fear of being accused of promoting their self-interest at the expense of the community. Local government officials variously play the role of mediators or proponent but are perceived as weak by the community because they usually try to avoid making a definitive decision for political reasons.

Activist opponents of the project typically present themselves as amateurs or grassroots representatives of the community when in fact they devote themselves as a vocation to mastering the details of the process and learn at least as much about the background as most of the professionals involved. In their focus on this issue, the activists risk becoming far removed over time from the opinions of the community as a whole, which they claim to represent. For many activists, a local NIMBY issue becomes a springboard for a political career. For others, their role is informal but no less influential as community leaders.

Unless the proponents of the project have near-authoritarian powers (for example, in a "company town" where one employer exercises overwhelming influence) and unless they are prepared to deal with the fallout of community outrage for years to come, the opposition usually wins in time. It is unwise to underestimate the power of an organized community.

In the end, the community has in effect institutionalized a posture of resistance and is well organized to protect its interests. Typically there is a successor issue that comes along soon after. The same leaders and community groups who were successful the first time around will move to another issue: partly to exercise their new empowerment and partly to keep the organization viable. But they move on to other issues mainly because every community is under some type of threat and there is no end of community problems to address.

As the Aurum case study demonstrates, there may also be deeper agendas behind a NIMBY. The full history and motivation may not be known or clearly understood by the stakeholders, even while it motivates or constrains their actions. The presence of other political and social agendas does not invalidate the health or environmental concerns of the project, of course, but may explain the evolution of the issue. In some cases (but definitely not Aurum) this history may even hold the seeds of an eventual solution based on compromise, compensation, or accommodation.

12

Spirituality and Sustainability

RELIGIOUS, MORAL, OR spiritual sustainability directs believers toward a more profound moral case for treating the earth and its creatures responsibly, whether this is through stewardship that arises from a mandate from God, moral responsibilities, sensing a "one-ness" with the natural world, or through an ethical obligation that transcends relationships among people and creates duties and rights with species and ecological communities (e.g., "deep ecology"). The connections between spirituality, sustainability, and health are many and complicated. (See table 12.1.)

The literature of environmental activism is replete with references to spiritual values, aesthetics, stewardship, and the psychological and cultural value of wilderness and nature. Are these subjective and intangible, vested with importance only because people believe them to be important? Or is there a transcendent connection between faith, spirit, and both person and the human community that connects to health as wholeness and to sustainability and a responsible way of living?

To many people of faith and to those who have reached a level of self-actualization (described below), and also to those with a deep sense of personal attachment to the world, these questions have deep meaning and emotional resonance. To others who believe in a materialist rather than transcendent world, the answer lies in the person and the emotional response that is uplifting for that individual. In other words, to those who believe and who feel deeply spiritual, the connection between spirituality and sustainability is paramount. To those who are aware but do not live guided by faith or with a sense of transcendence, the question is not meaningless but one of personal values. The majority of people, who have no strong conviction either way (aside from a cultural affinity for the religion of their heritage) are commonly assumed to be unaware and unbothered by the problem—but this may be an illusion. People from all walks of life think deeply about such things but are not always able to articulate their feelings, or inclined to share them.

Table 12.1 **Table of sustainability values: Spirituality**

	Sustainability Value	Sustainability Element
I.	Long-term continuity	A community with a sense of stewardship and responsibility for the future is supportive of sustainability and providing for future generations; the sense of stewardship traditionally arises from adherence to faith and recognition of an obligation to others that transcends practical concerns and is rooted in the spiritual.
II.	Do no/minimal harm	A community that respects creation, either as the work of God or as a wonder of nature, is less likely to harm the environment and more likely to care for it as a sacrament or duty; likewise, a community that sees the human body in a spiritual tradition (which could be as created in God's image, as a sacred creation of God, as the temple of God, or as the inviolate vessel of humanity) is more likely to strive actively for good health and to avoid habits or practices that threaten good health.
III.	Conservation of resources	A community that sees the world as God's creation or a finely balanced clockwork may conserve resources, especially for the benefit of future generations; a community that believes that resources were put on earth for mankind's unrestricted use and that God will provide, regardless of human consumption, will exploit and waste resources.
IV.	Preserve social structures	A community that respects traditional religion is often motivated to preserve constructive social structures that support sustainability and continuity but risks becoming a society that is rigid in its beliefs.
V.	Maintain health, quality of life	A community of faith or secular spirituality that respects life and human dignity will protect the environment that supports life and human growth.
VI.	Performance optimization	A community of faith or secular spirituality will usually (not always) seek harmony, justice, and charity, which support stability, sustainability, and continuity, and which minimize threats to security and health.

(continued)

Table 12.1 Continued

	Sustainability Value	Sustainability Element
VII.	Avoid catastrophic disruption	A community of faith or secular spirituality will usually (not always) seek to avoid catastrophic disruption; however, a community an apocalyptic vision, that surrenders all control to a deity, or that believes that the only purpose of earthly life is to worship God, come what may, may not have the motivation to avoid catastrophe and may even welcome it as the fulfillment of a divine plan.
VIII.	Compliant with regulation	A community of faith or secular spirituality is likely to support the values of sustainability and health and to be accepting or neutral toward compliance with regulation; regulation will usually be perceived as a secular issue, irrelevant to faith except as a means to a positive end.
IX.	Stewardship	A community of faith or secular spirituality will usually (not always) accept the idea of stewardship, but in the tradition of some faiths may believe that the world was created for human exploitation.
X.	Momentum	A community of faith or secular spirituality will usually (not always) tend to support forward momentum for sustainable development and continuity but there is also a risk that the faith will become rigid and thus profoundly conservative, retarding progress.

Concepts of the relationship of man to God's creation, stewardship, and oneness with all things come from belief systems that are not provable or demonstrable scientifically. For some people they arise from religion and for some they arise from personal growth and humanism. However, in a pluralistic society some people believe, others do not, and some do not pay attention. Sustainability rooted in a particular belief system risks becoming irrelevant or partisan if it depends on faith and that faith is challenged.

This is a topic that has preoccupied thinkers since ancient times. The world's first literary epic has as its central theme the relationship between humankind and the natural and spiritual world, with a dramatic and startlingly violent example of the violation of spiritual and ecological sustainability. (See box 12.1.)

No discussion of religion or spirituality can be comprehensive. Not only are there many recognized religions in the world (and so many sects, denominations,

BOX 12.1

The Epic of Gilgamesh

"Gilgamesh" is an epic poem that was already centuries old in 2100 BC, when it was written down, in cuneiform on clay tablets, in the version that has survived. It is the earliest work to be recognized as a masterpiece of world literature. Fragmented elements of the ancient epic can be found in the Bible (the earliest books of which were written centuries later), in Greek myths, and in the worship of Baal (the golden calf). The poem was lost in ancient times and was not rediscovered until the 1850s, when ten tablets from a Babylonian king's library were deciphered. It was recognized immediately as a significant writing because of its description of the Flood and the presence of a character corresponding to Noah.

The story has a political theme and a surprisingly contemporary feel, highly relevant to sustainability and health. Its protagonist, Gilgamesh, was an actual, historical king in the first dynasty of royalty that governed Uruk, which was a great city in the land of Sumer (later Mesopotamia, contemporary Iraq). He ruled at a time when the Great Flood was almost within living memory. Among other achievements, he built a wall around Uruk for security. In the tradition, he is half man and half god (on his mother's side).

This is the story of Gilgamesh:

Gilgamesh is bold, strong, and intelligent, but he is very ambitious and self-centered. Gilgamesh thinks he is a good king with the potential to be so great as to rival the gods themselves. One of his achievements is to build an invincible Great Wall around Uruk. He sees himself as the protector of Uruk's citizens. However, he is actually cruel, selfish, ruthless, and arrogant—but he does not realize his true nature. He relies on the knowledge that he is half god to feel superior and invincible. Despite his seeming iron control of Uruk as a despot and his overweening self-confidence, he has inexplicable nightmares. In them, he is struck by a mysterious sudden weakness and dies. He does not understand these dreams.

The gods fear his growing power, so they decide to create a rival for Gilgamesh. They call on a nameless sacred prostitute (forerunner of the later ancient Greek *hetairai*, who had a religious mission) in the temple of Ishtar and ask her to bring fully to human form a wild half man, half animal in the forest named Enkidu. Enkidu is beloved by the animals of the forest, who consider him to be one of them. He sets animals free from traps and snares set by the hunters of Uruk. Through sensuality and human passion, Enkidu is pulled out of his animal instinct and brought into human self-awareness.

He gains understanding and acquires language. He sees the plight of the people of Uruk and of all creatures in the kingdom. Saddened by what he sees with his newfound awareness, he then bravely goes to Uruk to confront Gilgamesh. There, he tells the absolute ruler that the people of Uruk are suffering and tries to make Gilgamesh realize that he is actually a bad king.

Gilgamesh resents this message, of course. He could easily have had Enkidu killed by his soldiers, but he knows that if he did, he would appear weak to the people of Uruk and before the gods. He decides that in order to show his superiority he must vanquish Enkidu in a test of strength. They engage in a wrestling match. However, the two turn out to be equally strong, and they exhaust each other. Impressed by their physical equality, in that each is unable to dominate the other, they come to respect one another. In a short time, they become pals and drinking buddies. (This relationship puts a half god and a half animal on an equal footing, which would be contrary to the natural order of things except for their shared half-human sides.) As the relationship between the two develops into a deep friendship, Gilgamesh learns respect for other people, which he is experiencing for the first time.

However, Gilgamesh dreams of glory and wants to perform heroic deeds so that he will be remembered by history. He proposes a number of exploits that Enkidu, with his natural wisdom, initially resists. However, Enkidu's loyalty to his friend and king makes him feel that he cannot desert Gilgamesh even when Gilgamesh is wrong.

Tragically, Gilgamesh's ambition pushes the reluctant Enkidu into a violation of everything Enkidu stands for and a betrayal of sacred law and nature itself. Together, they destroy the sacred Cedar Forest. Gilgamesh and Enkidu kill the fearsome forest guardian (Humbaba, or Humwawa) for no better reason than to show how tough they are. Ishtar, the goddess of sex and wealth, is impressed by their manly actions and attempts to seduce Gilgamesh. But the king, who seems to be growing a little more more aware and sensible, refuses. He knows that she is insatiable and that the world does not contain enough to make her happy. Ishtar, deeply offended, appeals to the Sun God to allow her to send the great Bull of Heaven god to avenge the insult. Gilgamesh. Enkidu kills the bull god, who dies for no good reason other than the vanity of Ishtar and Gilgamesh. This makes the gods even angrier, more so at Enkidu than Gilgamesh (presumably because Enkidu knew better the ways of nature), and a council of the gods condemns Enkidu to death.

The two friends return to Uruk and boast of their deeds as if they were heroic, not the destructive and cowardly acts they were. Gilgamesh does not realize, nor does Enkidu recover his awareness, to see that they have made the gods angry. One of them must die for the sacrilege.

Despite their high spirits and celebration, Gilgamesh continues to be bothered by his terrifying dreams. Enkidu had previously sworn to Gilgamesh's mother that he would protect Gilgamesh from harm. Although Enkidu had agreed to protect Gilgamesh primarily in a physical way, he ends up protecting him in a psychological or spiritual way as well. Enkidu takes on himself Gilgamesh's nightmares of weakness and death and internalizes Gilgamesh's fears as his own. Gilgamesh, now relieved of his psychic burden, sleeps soundly for the first time since he became king, but Enkidu is now the one who is troubled.

After one of these nightmares, Enkidu falls ill. After several days, he dies. The animals of the forest grieve for Enkidu. The are sad not only because he was one of them (being half animal) and that in the end he betrayed them but also because they understand that Enkidu was led into error and deceived by pride.

Gilgamesh is distraught by grief over the loss of his friend, recognizing at last Enkidu's superior character and virtue. He falls into a deep depression, which deepens into a long psychic darkness that paralyzes him into inaction. Gilgamesh has now learned firsthand the meaning of death and now has empathic emotions, feeling for something outside of himself.

Eventually, because he feels responsible for the death of Enkidu, Gilgamesh sets out on a journey to find a cure for death. The only cure is a legendary flower known only to the one person in the history of the world who has cheated death: he who kept his family and the animals of the earth alive during the Great Flood, Utnapishtim (the biblical Noah). Desperate to bring Enkidu back to life, Gilgamesh sets out on an emotional and perilous journey to find Utnapishtim. He experiences many trials, overcomes many fears and temptations but finally finds the legendary Utnapishtim, who lives with his wise wife in the Underworld by the River of Death, within which the flower lies.

Utnapishtim's wife, Siduri, bakes Gilgamesh seven loaves of bread, which he is allowed to eat, one each day, on the condition that he does not sleep. Gilgamesh cannot stay awake that long. Rising to another level of self-awareness, Gilgamesh, who had previously exulted in his dominance, despairs of his lack of discipline over his own body and his inability to control time. However, he is still devoted to his mission of finding Enkidu and returning him to the world of the living. Taking pity, Utnapishtim tells Gilgamesh where the flower lies and how to get it.

Gilgamesh barely survives the dive into the River of Death to find the flower. Only because he is half god does he succeed in bringing it to the surface. He is tempted to eat some of it himself, in order to ensure everlasting life (which is not guaranteed to a half god), but he does not because he has learned empathy and the love of others, and so he feels that he must save all of it for Enkidu. Gilgamesh heads back to Uruk with the precious flower. Then, in a surprise accident, a snake eats the flower when it is left unattended for a moment while Gilgamesh stops to bathe. (This is why snakes shed their skin—because by eating the flower they have become immortal.) Gilgamesh learns from this experience that it is not possible to go backward in time or to influence events that occurred in the past.

After all his travails, he must now return to Uruk having failed, despite an effort that this time was genuinely heroic. However, he returns having learned empathy. He now realizes that to be a good king he must listen to the part of him that is human and not just to the godlike part. He commits himself to the present and renounces the search for immortality (eternal life), as it is not appropriate for a human being to want or to expect this. He even rejects seduction by no less than the goddess Ishtar herself, since he now wants to live entirely as a human being and not as a god apart from his subjects.

He declares to the people of Uruk that he now wishes to live as if he had no name and to serve others, as well as to be a good king at last through love. After his death, the people of Uruk feel sad because, in the end, he became a good king and protected them after all.

The gods think so highly of him and his hard-won wisdom that in the shadow world they make his spirit a judge and a counselor to the other spirits.

Figure 12.1 is an evocative artistic impression of Gilgamesh, capturing moments in his journey through life.

Obviously this story presents many narrative problems to the modern reader, not the least of which is that in the end Gilgamesh becomes immortal after all and dwells among the gods, which seems to contradict the point of the story, which is that he chose to be human. Myths often do not make a lot of sense in their details.

Gilgamesh is an attractive protagonist because he is capable of change and insight. His early faults are callow and adolescent, unworthy of a king. It is the consequences of Gilgamesh's self-centered acts and the journey that follows that matter. The story is powerful and represents deep thinking on issues of health, sustainability, the relationship of humankind to the natural world, and the origins of self-awareness.

It hardly matters that there actually was a king of the ancient civilization of Uruk named Gilgamesh, at a place in the succession of kings that matches the story and who is also said to have built a wall. Whatever actually happened was elevated to another level and became a story for the ages and a precursor to the biblical story of Noah. It was celebrated before the first pyramid was built and was then lost to civilization, long before Homer's *Iliad* and *Odyssey* were written.

FIGURE 12.1 "The Progress of Gilgamesh," by Jonathan Mayer (Scapegoat Studio, Seward, Nebraska), follows the life of the hero: first a precocious child, then a brave and clever youth, a brash and overconfident young king, and a just and wise but saddened monarch in old age.

and cults within almost all established religions), but there is huge variation within most of the significant traditions and in personal faith among observant believers. Meaningful discussion of religion and health and sustainability requires engaging specific tenets or problems of theology and exploring how particular beliefs motivate behavior or create social controls. This naturally causes problems of which religions to omit, which theological tradition to consider mainstream for a given religion, and how the practice of a religion among the nominally faithful may differ from that religion's mainstream theology. However, the topic cannot be avoided. Discussions of spirituality can at least take a relatively detached and secular approach in describing the psychological and social dimensions of spiritual beliefs and how they relate to health and sustainability.

Religion is too important to ignore in this discussion, and some examples are necessary to convey how important these issues are among the communities of believers. Approaching the problem of spirituality from an exclusively secular point of view misses the critical importance of faith. Approaching it from a confessional point of view, emphasizing one religion or tradition's belief system, channels the discussion in a direction that many readers may not share and risks becoming partisan and sectarian. An inclusive society in which religious freedom is celebrated needs to have a common belief system based on ethics, and values but one independent of the faiths of its members. For these reasons, this chapter will at least touch on selected religious issues and how they affect attitudes, values, and practices in health and sustainability without attempting to explore any particular tradition in detail.

A fundamental problem of sustainability and health is to find a common way of living in a sustainable manner that respects spirituality and individual faith but does not impose a particular religion or belief system. By definition, that means that in a pluralistic society, a collective or public philosophy that supports sustainability and promotes good health has to be secular.

The Moral Case for Sustainability

The value placed on health, helping others in need, and protecting others (especially children and the vulnerable) from harm runs deep in morality and religious belief. Embedded within the definition of sustainability, particularly the value cluster used in this book, is an undeniably moral argument that values sharing, equity, justice, empowerment, and freedom from want. This moral argument is recognized as either deriving from moral beliefs about what is right or from religious belief. It does not derive from practicality, although strong practical arguments for sustainability and health protection can be made and are obvious throughout this book.

Although closely related, there are important distinctions between morals and ethics. Both describe rules to live by and define what is right and wrong by a covenant with a higher authority. Morals relate to good behavior in an absolute sense, regardless of interactions with others and what constitutes a good and just life. Moral behavior is judged against an ideal. Morals are internalized in every individual by belief, which may be religious or a personal conviction, or the lack of belief. A moral life would be lived in exactly the same way by a believer whether it involved interaction with others or not. It would not matter whether the person living the good and just life was a hermit (e.g., in the early ascetic Christian tradition) or a rich merchant constantly engaged in commercial transactions with other people but given a gift to see a higher plane. (This example from Islam is chosen because the Prophet Mohammed was a rich merchant, not to mention a political and military leader.)

Ethics, on the other hand, is a set of rules that govern relationships between people and within a social system. Ethics are accepted as part of living in and belonging to a community; in that sense they are imposed externally rather than from internalized belief. Therefore, in the context of sustainability, ethics has more to say about how we distribute wealth, protect one another from harm, and treat our fellow human beings, while morality has more to say about what kind of society we are building and our relationship to the natural world. Ethics may be informed by belief and tradition, but ethical principles are secular and presumed to be general.

Religion always carries with it a distinct and usually highly explicit moral code as part of the belief system. The Ten Commandments and Christ's teachings from the Sermon on the Mount are familiar examples from the Judeo-Christian tradition, but there is much convergence of moral thought across the major world religions. In secular traditions, morality derives from what the philosopher Immanuel

Kant (b. 1724–d. 1804) called "the categorical imperative," which is a requirement for behavior that is absolute, unconditional, universal, beyond reason or rationalization, and not subject to any form of personal judgment or choice. Whether derived from religion or the moral imperative, a moral system is considered by believers to be universal. If a person does not adhere to these beliefs, the implication is that they are outlaws within their community. Those who believe otherwise and act without regard for others or for sustainability may be deemed immoral, selfish, or even dangerous; however, in a pluralistic society with personal freedoms one moral code cannot be applied to everyone and so has to be replaced by more inclusive, coercive, and ultimately cruder means of regulating behavior, such as a set of laws.

Kant, in formulating the categorical imperative, was saying that there exists a morality derived from being human beings, as a thinking species, that is inherent in any form of human community and that in a secular context serves the same guiding purpose as morality drawn from religious belief. In either case (or both cases in a pluralistic society), there is an absolute right and a wrong quite separate from maximizing "utility" (the best outcome for the largest number). Examples relevant to sustainability might include the beliefs that every species has a right to be protected from extinction, that there is a responsibility to future generations, that every person has an obligation beyond his or her own comfort and the benefit of family and relatives, that the planet should be preserved for its own sake, and that suffering should be alleviated both among people and animals.

Moral persuasion is powerful. Recently, a study conducted in the Netherlands suggested that people were more easily persuaded of the need to address climate change when the argument was framed in moral terms rather than as a practical necessity. This is contrary to the common assumption that it is more convincing to show people that there is a practical or monetary benefit in sustainability and health protection. It is an illustration of how people think about right and wrong even if they cannot easily express their thoughts in moral terms.

One way of exploring morality in sustainability or health is to ask, "What would it take for me not to care?" For example, "What would it take for me not to care about climate change?" One would have to answer that it would take not caring about other people, not caring about future generations, not caring about the sacredness of the world, and not caring about the fate of one's own children and loved ones.

Because morality and conviction require identification with a community and an appreciation for coherency of the moral code, faith-based organizations, communities of believers, and secular communities committed to sustainability on moral grounds have certain features in common, regardless of the tradition they follow. They emphasize connectedness, seeing things in terms of linkages rather than separation. They emphasize the collective good rather than individual gain, although they are not scornful of personal success and are not opposed to profit unless it is associated with exploitation. They see a higher plane than is experienced by humanity and a higher mission than mere survival.

However, the response of such communities to evident contradictions between the moral argument and scientific or objective findings varies according to their belief. The most obvious example, to which this chapter will return, is the question of whether the natural world was "God given" for the benefit of human beings to take from freely: in this case, exploitation may not be perceived as immoral at all. There is also the question of whether the natural world is a sacred creation in its own right: if so, then exploiting it is sacrilege. Much depends on whether the belief system and moral code are inseparable (in other words, morality defines the religion) or the moral code is interpreted or inferred from religious teachings (religion sets out the moral code), and whether it can be interpreted or has to be adopted in its totality and followed literally. Even so, thoughtful believers may accept that what is occurring today demands action simply because "God helps those who help themselves."

Some communities of faith are adamant that the moral argument, being religious and handed down by the Deity, has primacy over facts, which could be just an illusion. For example, the creationist point of view in fundamentalist Protestant Christianity is that the evidence of biology and paleontology only appear to support evolution—it only looks that way because anomalies were created within and together with the natural world, perhaps as a way of testing faith but for reasons unfathomable to human beings. For these communities, reconciling the contradiction means recasting the understanding of observation and deduction to fit first principles. For example, if one considers the world as created by intelligent design (meaning literally by creation within seven days, as described in the Old Testament), then the idea that the world is only 8,000 years old does not need to be reconciled with the demonstration of broad climate changes over a much longer period. Some scientists of faith are devoted to reconciling science and religion in literal terms and look for examples of historical validation, interpretations of prophecy that fit objective reality, and convergence between scientific theories and doctrine. One of the most influential examples was the Jesuit priest Teilhard de Chardin (b. 1891–d. 1955, French), who believed in a divinely guided form of evolution that was aimed at an ultimate goal of perfection.

On the other hand, communities in which the faith is less literal and more theistic or diffuse may take the approach that religious and moral belief exists on a different plane entirely from verifiable scientific fact and reconciliation is unnecessary. In between is the position known in theology and philosophy as "God of the gaps," which is the notion that divine will is evident in those things that are unexplained and where there are gaps in knowledge. The problem with that line of thinking is that when gaps in knowledge are filled or an explanation is provided, the space available for crediting divine intervention shrinks.

Religious Belief and Sustainability

Religious beliefs about sustainability have at least three dimensions:

- First, and usually most important because it is foundational, is the relationship between God and the creation of the natural world and the question of omnipotence.
- Second is the relationship of the believer to God or another deity (or divine equivalent) and the implications of that relationship for worship and sacrament which carries a duty to protect creation.
- Third is the relationship between people and the problem of moral behavior, which usually includes charity and equity in distributing the abundance of nature or alleviating poverty (giving alms) and privation during times of scarcity (as in the biblical story of Joseph and the lean and fat years) through sharing and careful management.

Each of these dimensions has direct relevance to motivating believers to support a society that values sustainability.

Religious beliefs about health are varied, but most religions seem to assert that there is a duty to respect the body as the creation of God (or the divinity) but to accept God's will in what happens to it and the outcome of illness or injury. In the Christian tradition, especially, the perfectibility of the body is considered ultimately impossible but to be cultivated as a vocation; in other words, one tries to be perfect knowing one will fail because the process of trying is worthy in itself. The concept of health as "wholeness," in body and spirit (see chapter 1 on the definition of "health"), seems deeply rooted in most religions. St. Basil, a patriarch of the Orthodox Church, wrote that the "the medical art has been granted to us by God . . . as a model for the cure of the soul." Many religions (for example Roman Catholicism, Buddhism, Judaism) consider ill health to be a trial or natural part of life; others (Hinduism, especially) consider it the consequence of personal deeds in the past (karma). Almost all religions seem to have some thread or tradition that considers ill-health to be a punishment from God for personal and collective misdeeds, but most also seem to have a tradition, usually not as prominent, that rejects such a punitive reprisal on mortals as unworthy of an omniscient deity. Many religions are highly prescriptive in their health practices, particularly Judaism and Islam (the dietary laws of both probably derived from health practices), and newer religions such as Mormonism ("herbs and mild food"), Seventh-day Adventism, and Jehovah's Witnesses. A few, such as Christian Science, renounce theories of medicine and health altogether and believe only in divine intervention, although the actual practice of believers varies.

The idea of purity as a sacrament is paramount in many religions such as Orthodox Judaism. Cleansing rituals before prayer are central to many traditions such as Islam and Buddhism.

Although it is easy to assume that religions with a strong sense of predestination or God's will would always resist action taken to achieve sustainability, these same religions (including Islam and, historically, Christianity) have also had strong traditions that the believer is expected to act on God's will and to help and protect themselves.

Traditional Religious Beliefs

Belief in a creator is inextricably intertwined in issues of equity and respect for the natural earth. Because treating the environment respectfully (together with propitiating the gods) is about all that human beings could do to sustain the yield of renewable resources, it is not surprising that most indigenous cultures have religious or philosophical beliefs to the effect that the earth is a provider, a sentient being, and an ancestor. To act in ways that do no harm to the planet is a highly adaptive and desirable code of behavior when a relatively static population depends on a sustained yield of food and fiber. In the absence of an articulated theory of resources and material distribution, a traditional society naturally incorporates these adaptive biocentric ways of thinking into religious belief systems, including the concept of earth (or "Gaea," "Gaia," or "Erda") as nurturing mother. In such a society, sustainability would mean (if it were meaningful as a separate thematic strand in the culture) sustaining the earth as well as all its creatures, including its human communities.

When early religions emerged they may well have had a purpose (of which the ancient shamans may well have been aware) and an adaptive role in maintaining tribal or clan awareness of their environment and the limits on what they could exploit. Something like this may survive in the rites and beliefs of peoples such as Australian Aborigines, who survived for 20,000 years by intensively managing an environment that produced as much as was needed but never in abundance and for whom religion is now a source of social cohesion. Likewise, the creation myths and spirit beliefs of Aboriginal peoples in North America (Native Americans, and First Nations in Canada), especially the widespread tradition in hunting of honoring and spiritually respecting the animals that gave their lives to sustain the people, undoubtedly played an important role in maintaining ecological stability.

Perhaps it did not matter much, because, it is often assumed, aboriginal tribes tend to be small, scattered, and in the earliest times unlikely to have much impact on the land. However, this was not necessarily true, then or now. Aboriginal populations were quite capable of causing large-scale ecological change, primarily through the use of fire and by intensive hunting of species (in the case of mammoths, to extinction). It is also now clear that the aboriginal population in the Americas, especially, was much larger than previously assumed and that relatively large settlements existed before European contact, with specialization of labor and trade. These settlements are, of course, well known as full-scale cities in Mexico, Meso- and South America, but large permanent settlements were also present on a smaller scale in the eastern Mississippi and Ohio River valleys (sometimes called "mound builders") and Pueblo cultures (in what is now the southwestern United States). Such large settlements undoubtedly put pressure on local resources, particularly during drought, and would have had to deal with problems of local ecological impact, water and food supply, and distribution. There is even evidence that what is now rainforest in the Amazon once was savannah and supported a much greater human population. Clearly, these societies had to manage their use of resources carefully, and undoubtedly religion and traditional knowledge, which for most traditional societies is inseparable, guided them for long-term sustainability.

Modern people, particularly introspective city-dwellers, tend to project onto aboriginal populations whatever traits of wisdom, knowledge, or nobility they believe is lacking in their own flawed society. These ideas are based, primarily, on stories, ideals, and second-hand accounts, including objective descriptions by detached anthropologists, distortions from popular culture, and only occasional voices within the culture from those who have some personal reason or motivation to communicate with the majority culture. The authenticity of received wisdom about aboriginal cultures can be questionable and is often manipulated. Not long ago, American Indians ("First Nations" in Canada) were talked of as "noble savages" ("savage" here meaning "wild," as in the French "sauvage"), an idea derived from the philosopher Jean Jacques Rousseau (b. 1712–d. 1778, Geneva), who believed that humankind had a grace and moral superiority in its original, primitive state that was lost with the illusory refinement of civilization. The unintended consequence of this seemingly sanctifying idea was to simultaneously deny respect for aboriginal civilization when the aboriginal community stayed apart and to consider them to be further degraded when they assimilated, but always to marginalize them as different and maladapted. What followed was centuries of persecution and genocide.

Perhaps the most bizarre example of how influential aboriginal thought has been on matters of the environment, as well as how easily it could be manipulated, was the case of Grey Owl (styled Wa-shon-qua-asin, 1888–1938). Grey Owl was an immensely popular writer and lecturer in Canada. His books were read throughout the British Empire and the United States in the 1920s and 1930s. He was an early advocate for sustainable harvesting and conservation of nature. Handsome, and with classic "Indian" features, he claimed to be half Scottish and half native, not an unusual combination in Canada at the time. Grey Owl spoke especially for the protection of beaver and other fur-bearing animals. He denounced the destruction of the natural environment and depletion of natural resources, showcasing traditional ways of the Ojibwe for living with minimal impact. The only problem was that Grey Owl's real name was Archibald Stansfeld Belaney, and he was thoroughly English by birth, as he had emigrated to Canada as a young man after the First World War. Toward the end of his life, his physical decline from alcoholism made him erratic and unreliable in performances, which reduced his appeal and his popularity. The newspaper in the area of Ontario where he lived knew that his personal story was fraudulent but did not divulge the truth until after he died, after which the story of his deception was reported across North America. The exposure of his fictionalized persona destroyed his legacy, damaged his credibility as a defender of Nature, and deeply hurt his conservationist causes. However, by modern standards he could be considered a performance artist who went too far rather than an outright fraud. He genuinely knew the wilderness, had lived among Ojibwe trappers and knew their ways, and had apparently been "adopted" in some fashion by the local Ojibwe band who mentored him. His inspiration also came from his relationships with aboriginal women, especially his Ojibwe wife, Anahareo, who taught him to see trapping from the animal's point of view. He is remembered today as a deeply flawed but prescient figure who was inauthentic but correct in the essentials of his teachings.

Ancient Religious Beliefs

The first religions, insofar as is known, were animist beliefs worshiping natural phenomena, sacred places, and animal spirits. Early religions were traditional knowledge (see chapter 2), repositories of collective belief and experience that were indivisible from other knowledge and wisdom. It was thus natural that these religions, as well as the animistic religions that survive today, have a strong sense of place and of naturalism that today is interpreted as an ecological consciousness. At the time of origin, however, it is safe to infer that religion and belief in the supernatural, history, empiricism, and explanation of natural phenomena were not differentiated, as they are now.

In examining religions of the past, there is a natural human tendency to pick and choose elements for emphasis, with a bias toward emphasizing what feels consistent with contemporary values. The "Mother Goddess" of Paleolithic Europe, Anatolia, and Asia Minor (Cybele, later Artemis in Hellenic and Roman times) was indeed an important deity, representing motherhood, fertility, abundance, and creation. The idea appeals to the notion of sustainability. The idea that there was once a worshiped maternal deity is attractive in an age and society when feminism and gender equality are still relatively recent concepts (and undergoing constant reinterpretation). However, there were other deities in the region, including the cow god (later Baal), the deer god, and the king gods (such as Gilgamesh).

The primacy of a female influence was not universal, of course. Most religions are and remain highly patriarchal, not matriarchal (although some, such as Judaism, are matrilineal). Emphasizing a Mother Goddess is a choice, based on current cultural impulses, not necessary an accurate representation of a phase common to all Western culture. The point is not that these ideas are meaningless or had no enduring cultural influence. It is that ancient beliefs are mostly of inspirational value, rarely providing contemporary society an authentic living narrative relevant to today. It is also clear that the literal significance of these ancient beliefs is less important to society today than the sense of reverence and tradition awe comes from believing that these traditions are still current and once had literal power.

Similarly, the ancient Celtic religion, dating from the Iron Age, is often considered a model for a sustainable belief system. The reality seems to be that nobody really knows if this is true because historical sources on Celtic civilization are second hand (mostly written by their conquerors), confusing, and contradictory. The narrative surrounding Celtic culture has been elaborated to legend by literature, deeply distorted by politics and wishful thinking, and later appropriated by the propaganda of nationalism, especially in their popularized forms. The rituals and hermetic beliefs of the Druids, the priests of the Celtic religion, have not survived intact because there was no written tradition. Of course, esoteric knowledge of the Druids themselves was kept secret. Inevitably, this vacuum of sure knowledge has led to the modern tendency for people to project onto the Celts and their Druids whatever it is they want to believe that the Celts believed. That the Celts were ultimately defeated and their civilization extirpated by the technologically advanced and more organized Romans creates a poignant narrative that resonates with modern beliefs about technology overcoming and threatening ecology, but it is not at all clear that this narrative has anything to do with what really happened in the time of Julius Caesar and of his Celtic adversary Vercingetorix.

The Celts may or may not have been spiritually enlightened and sensitive peoples deeply in tune with nature. If they were anything like the Norse, about whom much more is known, any superior ecological awareness and oneness with nature they may have had does not necessarily imply spiritual enlightenment and humane behavior. On the other hand, compared to the spirituality of the Romans almost any other contemporary culture might look good. (The Romans themselves, perhaps aware of their bad behavior, looked to the Etruscans as the font of ancient wisdom and tradition.)

The belief that these ancient cultures held mysteries of how to live in harmony with the natural world has become enormously influential in popular culture. (It has been the basic plot for hundreds of movies, most famous and successful to date being 2009's *Avatar*.) The idea permeates what is often called New Age philosophy (although "New Age" has mostly come to be used as a pejorative term in recent years) and much alternative religious belief. It is inextricably interwoven, both by design and by historical tradition, with contemporary alternative medicine and natural healing systems. The result is that these very different threads are often treated as if they were manifestations of a single belief system, either rejected together or embraced together.

This appropriation of ecological, sustainable thinking by New Age advocates and gurus, and its fusion with alternative medicine and romantic ideas about healing, has undoubtedly facilitated acceptance of sustainability ideas in many communities where the culture is receptive. However, it also presents practical problems for contemporary sustainability and for health advocacy in places where the culture is more socially and religiously conservative. Paradoxically, this is a threat to sustainability. Positive ideas that are important for mainstream cultural change may be marginalized as "crunchy" and rejected by the mainstream. (One can think of the difference in consciousness between Manitou Springs and Colorado Springs, which are separated by only six miles in Colorado but by a cosmic cultural divide.)

New Age philosophy, ecological awareness, a commitment to sustainability, personal awareness of health, and a taste for granola are not a single, indivisible belief system. It is quite possible and even responsible for those without a religious belief system to accept or reject various traditions or their elements as individual notions and to have a religious or secular view of values and responsibilities in the natural world (as with "deep ecology," explained below). For those with a strongly-held religious belief system, it is possible to have doubts and responsible to be self-aware, regardless of its teachings.

Contemporary Religious Belief

Religion can be discussed from the ecumenical point of view, looking for commonalities, or from the comparative point of view, looking at differences. The ecumenical point of view is attractive and peaceable, especially to those who believe that all religions share a common core of beliefs. However, people believe what they believe. It is constructive to search for agreement but not to pretend that all belief systems are fundamentally the same or can always be reconciled. The ecumenical approach begs the question of how people act on their beliefs and why and when they are willing to violate them, but it has the virtue, like sustainability, of emphasizing agreement and putting it in the foreground and not letting doctrinal differences impede progress toward a goal that uplifts everyone.

There are major and minor currents in every world religion that are more or less supportive of sustainability. In Christianity, for example, there are many Baptists who believe in stewardship (in the Christian tradition, often the metaphor of a shepherd is used) and a duty to protect God's creations in nature, but there may be more who believe in a literal interpretation of the verse: "God said, Let us make man in our image, after our likeness: and let them have dominion over the fish of the sea, and over the fowl of the air, and over the cattle, and over all the earth, and over every creeping thing that creepeth upon the earth." In Roman Catholicism there is also the strongly ecologically minded Franciscan tradition, but it has been a minor theme, only weakly influencing the church as a whole, never dominating the theological mainstream. The theological mainstream still emphasizes unrestrained dominion.

Actual practice of believers can be quite different from the theology and traditions of the religion. Roman Catholicism once emphasized unrestrained procreation, with its implication that the world was created to support human numbers, based on the literal interpretation of the biblical injunction to be fruitful and multiply. However for many years Catholics worldwide have taken their own paths on the issue of reproductive choice, resulting in today's low compliance with teachings about birth control. In Hinduism, there is great reverence for life and care of the earth. Even the great catastrophe that will end time, Shambhala, is not really the end of all time but will be followed by a new cycle and restoration of the natural world. On the other hand, there are Hindu practices, such as bathing people and animals in the River Ganges (not to mention defecating into it and using it to dispose of bodies), that are certainly not conducive to life, good health, and sustainability. The recently elected prime minister of India, Narendra Modi, a Hindu nationalist, has pledged to clean up the heavily polluted holy city of Varanasi.

Belief exists on a different plane than health and practical sustainability. With these few examples, it is apparent that religions, while almost always supportive of good health (sometimes more in theory than in practice), can be either supportive or critical of sustainability, depending on whether action taken to achieve it conforms to the idea of God's plan or predestination.

One important aspect of traditional religion and sustainability requires special mention: the influence of Judeo-Christian motifs on the activist "environmental" or "ecology" movement of the 1960s, especially in North America. Many observers have commented on similarities between the themes of environmental activism and Jewish and Christian traditions, with a heavy dose of John Milton's later poetical imagery. These include the idea of a "paradise lost," corruption of a perfect creation, expulsion from paradise as punishment for sin, the need for atonement as well as redemption, ascetic living, good in constant battle with evil, and the wisdom of innocents. On one level it would be surprising if, in a relatively religiously observant country such as the United States, there were not echoes of a familiar religious subtext. Framing the narrative in terms similar to religion serves to simplify what are often highly emotional, conflicting, politicized, and complicated issues. These similarities are often (but not always) quite apt. Urbanization and the Industrial Revolution did change things for the worse in many ways, even as they ushered in the modern world and its temptations. It is useful to recognize that these motifs do not subordinate environmental thinking and sustainability to a particular religion but actually represent an expression of ecological thinking using more familiar religious metaphor.

Of course, in reality there was no perfect world that existed before pollution and urbanization. Rural life in the nineteenth century, for example, was marked by brutally hard work and had already changed the landscape, destroyed habitats, forced some species to extinction, and was hardly free from pollution (especially the fecal kind).

"Deep Ecology"

Spirituality does not have to be religious. It does not necessarily have to be based on faith. One of the most important influences on sustainability arose from rigorous philosophical deduction, more inspired by than derived from spiritual conviction but still grounded in values and belief.

"Deep ecology" is a secular philosophy that concerns itself with the intrinsic worth of living beings. It holds that the earth and all life have value separate from their usefulness to humans. It considers the relationships among living things in nature to be a natural, interdependent order, to which people and human communities must adapt and accommodate. It then defines a right of this natural order to exist free of destruction and interference from human intervention. From this foundation, deep ecology builds a philosophy that stresses conservation of nature, preservation and nonintervention (as opposed to active protection) of remaining wilderness and habitats, control of human population so as to reduce threats to and stress on the natural order, and ways of living that are sustainable and consume the minimum of resources required for decent human life.

Deep ecology rejects the assumption that the world exists to supply the needs of human beings or that people are at the center of the universe. It replaces this "anthropocentric" view of the world with the idea that human beings have no greater priority on the earth's resources than any other form of life and that human beings carry a responsibility, because of evolved intelligence (sentience), to act in ways that have minimal or no effect on the natural world. It holds that human beings, as a species and as individuals, have the same right as any other species to live and thrive—but no more of a right. Deep ecology also does not admit a role for human beings in actively intervening in the natural world for the sake of its protection or improvement (so-called wise use), since to do so is to impose a human vision on a biological community that should be free to evolve and develop in its own way.

Foundations

Intellectually, deep ecology accepts the science of ecology and community biology as describing a reality of complex interactions that define the natural order. It then goes further to confer on the natural order a set of universal rights: most fundamentally, the right to exist. On this basis, it elevates protection of the natural order to a "categorical imperative" (a fundamental moral assumption, as per Immanuel Kant). From this basic tenet, it deduces a set of principles that should guide human relationships with the natural order. As a system of ethics, therefore, it is "deontological," deriving moral or ethical principles from a basic moral code. It does not attempt to justify its tenets in science or inductive logic, from religion or epiphany. Instead, it accepts the natural world on its own terms and does not attempt to assign moral value to relationships between organisms and within ecosystems. The moral code applies when sentient human beings encounter the natural world and applies to human behavior toward nature.

Deep ecology was formulated in 1972 by Arne Næss (b. 1912–d. 2006). (His middle names are Dekke and Eide, which is useful to know so as not to confuse him with several other prominent Norwegians with the name Arne Næss). Næss was already a well-established philosopher at the University of Oslo and well known as an activist on Norwegian environmental issues and as a mountain climber partial to hermetic revelations during climbs. His concept of deep ecology came to him as an alternative to utilitarian and business-oriented development, inspired at least in part while living in a mountain hut called Tvergastein, which became through his work an iconic symbol of nature as well as an implicit symbol of rising above the world to get a superhuman perspective. (He was also interested in the roots of philosophy and argumentation and in language, and in many ways uncannily resembled Ludwig Wittgenstein [b. 1989–d. 1951], who also famously did much of his most profound thinking in a hut in Norway.) Much of Næss's work apart from deep ecology was concerned with individual self-actualization and moral behavior.

Næss meant "deep ecology" to contrast with "shallow environmentalism," in which he felt the issues were parochial and superficial; the "deep" refers to probative questioning of essential truths, not profundity. Næss's thought constitutes a rejection of and challenge to Protestant attitudes toward property and production, in a sense constituting a moral critique of Martin Luther. The principal intellectual sources for the development of deep ecology were Rachel Carson [b. 1907–d. 1964]; Baruch Spinoza ([b. 1632–d. 1677], who said, "I make this chief distinction between religion and superstition, that the latter is founded on ignorance and the former on knowledge"); Martin Heidegger ([b. 1889–d. 1976], who was tainted by his Nazi associations and responsible for the quote, "Science is the new religion"); and Aldo Leopold ([b. 1887–d. 1948], who was an American naturalist and environmental philosopher who famously wrote in 1949, "[A] thing is right when it tends to preserve the integrity, stability and beauty of the biotic community. It is wrong when it tends otherwise").

The implications of deep ecology are profound. By rejecting the idea that human beings have a superior claim to the resources of the world, it pushes back on the assumption that economic and population growth are inevitable and proposes that the only moral forms of growth are either internal and transformational or population growth under the level at which natural resources close to home, using appropriate technology, can support human life (see "carrying capacity," as discussed in chapter 4). Deep ecology also rejects the idea of stewardship, in that human beings are not assumed to be guardians or caretakers but only a peer species which should not be controlling the natural world for any purpose.

Deep ecology is compatible with other philosophies that emphasize community-level control, as well as animal liberation and pacifism, but it is distinct. One difference is that it is "anti-humanitarian," meaning that it is intellectually indifferent to human welfare as such (although not in sentiment) when there is a conflict between human well-being and ecological sustainability. Thus, deep ecology does not concern itself with what is in the best interest for human beings but what is best for any and every species within its own context and in competition or cooperation with one another. It also does not concern itself with how the natural world got to where it is today. Although the assumption is that it evolved, consistent with the principles of biology, the philosophy would work just as well if the natural world were assumed to have been created in its totality by a deity.

Deep ecology has been criticized on many grounds, including the predictable perception that it constitutes a secular religion. This is not supportable on the face of it because deep ecology makes no assumption of a godlike creator and has no room for divine intervention; it explicitly confers autonomy and rights on living things and rejects authority and dominion. It has been criticized for failure to address social issues and conflicts, as they cause destruction of the environment. Some critics maintain that Næss projected human values such as free choice, autonomy, moral character, and the will to exist, on nonsentient life forms including plants and animals that have no consciousness or purposeful will. In other words, does a plant "care" what is in its self-interest and whether it thrives or withers, or is it the philosopher who projects onto the plant a will to live that may be more important to the philosopher than the plant? The ecosystem often requires that the plant be consumed or that it grow under suboptimal conditions, so does the plant exist in a state of freedom or subordinate to the ecosystem and other species? Are not human beings just another of these other species? Næss would most likely argue that the assumption is wrong and that the natural world is a given, that human beings are a different case from other species because of the degree of control they exercise over the environment, and that deep ecology is aspirational in describing what should be done (not what can be done) in the real world.

Some critics have suggested that deep ecology is antihuman because achieving its ideals would require a catastrophic reduction in the human population and that it implicitly justifies coercive restraint on human activity. Other critics go further and suggest a connection (through Heidegger, especially) between deep ecology and German naturism and nationalism, in its darker forms. This is somewhat prejudicial against a long and fairly benign history of German Romanticism and fascination with nature. It is also rather insulting to the long and proud traditions of Norway, which early (in the nineteenth century) achieved a unique blend of Romanticism and Enlightenment values distinct from German ideology. The imprecation begs the question of the validity of the philosophy of deep ecology and ignores the fact that it is compatible with many other philosophical viewpoints and religions but is not compatible with totalitarianism (since it requires human concessions to nature and release of control). Some scholars believe such criticism is irresponsible and libelous to Næss, who considered his work descriptive of a valid point of view, not prescriptive or coercive, and aiming for self-realization, which is at odds with authoritarianism.

The Gaia Hypothesis

Acceptance of sustainable development by society may depend on cultural values and even spiritual notions about the relationship of humankind to the earth. This is why the otherwise quasi-religious concepts often expressed in the environmental movement, such as the Gaia hypothesis, have deep value as metaphor even if they do not necessarily express literal fact. The Gaia concept is not part of deep ecology, but its ideas resonate with deep ecology teachings, like the epic of Gilgamesh.

Gaia is the concept of the earth as a self-sustaining organism of such complexity and adaptability that "she" can be said to be alive. The idea was first advanced and promulgated, famously as a "hypothesis," by chemist James Lovelock (b. 1919) and microbiologist Lynn Margulis (b. 1938–d. 2011) in the late 1960s. It is clear that the so-called Gaia hypothesis was not a hypothesis in a literal sense but a wise and compelling metaphor that made people think about the planet as though it were a living organism. This biomorphic, even "gynomorphic" (woman-shaped), view of the planet has been effective in motivating concern and advocacy.

Some people undoubtedly believe in Gaia literally, as the Goddess Cybele of early myth and worship who became more familiar in the cult of Artemis in the eastern Aegean. Gaia, now redefined as the earth itself, is personified and named after the ancient earth goddess, mother not only of all life but of the ancestors to the gods. (In Greek mythology, "Gaea," "Gaia," or "Ga," was the mother of the Titans, the race that gave rise to the gods, who then deposed them.) That this myth has found new resonance redefined in a scientific age is a testament to the power of its central idea.

Besides having merit as a literal construct, since the Earth, like an organism, does have intrinsic buffering, stabilizing and balancing mechanisms, the Gaia hypothesis is useful as a means of personalizing responsibility for environmental damage. One cannot "hurt" an inanimate object, although it can be damaged for a given purpose; one can "hurt" a living being, even of planetary dimensions. The animation of the idea of the earth makes sustainability personal. This idea recasts environmental stress on the earth into more evocative human terms.

If Gaia is alive, then Gaia can be sick. The idea that if Gaia, or "Mother Earth" is sick, then her children will be sick as well, is a compelling metaphor. If the Gaia metaphor for health concerns features Mother Earth becoming ill as a result of human activity, then catastrophic failure represents matricide. The Gaia myth would interpret this as poisoning Mother Earth, making her sick and incapable of nurturing and providing support for her children. Gaia may also be thought of metaphorically as an elderly or abused mother in poor health whose children are bickering over her medical care and finances.

The Gaia paradigm has clear resonance with the idea of "love for Mother Nature" (this term generated 96 million hits through Google). The Gaia myth or hypothesis finds new meaning in the idea that Mother Earth (or "Mother Nature") supports humankind and if treated badly will turn her back on her children. The Gaia metaphor that particularly applies to sustainability and health hazard from ecosystem change may be the notion in popular culture that "Mother Nature strikes back!" This is a phrase that has been searched for 26 million times on Google, demonstrating that the idea is very much a part of contemporary thinking. The Gaia concept, as originally defined, did not reflect this metaphor in popular culture for unsustainability but it is apt. Another dimension of the Gaia idea is the intimation that in the degradation of the environment and lack of attention to sustainability, humankind is showing a child's rebellion against a parent's authority, selfish to the point of self-injury. Yet another is failure on the part of her children to appreciate a mother's care and nurturing, with the result that ingratitude and insensitivity makes the family mean. Thus, Gaia provides a potentially powerful metaphor on many levels.

The idea of Gaia has romantic intensity—but it is also distracting from the reality that the world is now controlled by human agency. All parts of the world are affected and threatened by human activity now, and the planet is not capable of "healing" itself in the face of continuing pressure.

The scientists responsible for the Gaia hypothesis were not always consistent. By 2008 James Lovelock believed that climate change had gone too far to be reversed and that catastrophe is inevitable. He stated at that time that he considered mitigation efforts to be futile and efforts to live sustainably as insufficient and often deceitful in causing denial of the inevitable consequences of the past abuse. His attitude was that it does not matter what humankind does, so it does not matter how people live: "Enjoy life while you can. Because if you're lucky it's going to be twenty years before it hits the fan." He then reevaluated his position after surface temperatures appeared to remain relatively stable for a decade (although subsequent research has shown that this may not actually have been the case) and returned to advocating a commitment to mitigating technologies, including shale gas development (fracking) to reduce carbon loading. Lynn Margulis, although a scientist of great vision and insight, held some very idiosyncratic ideas about biology that provoked ambivalence about her teachings and so she left an even more confusing legacy.

The Spiritual Experience and Health

There is a body of work suggesting that immersion in (or even merely viewing) the natural environment is calming and reduces anxiety. Some evidence suggests that human beings naturally feel "at home" in an environment that resembles the savannah and woodlands that were the presumed environment in which the human species (plural, until the final domination of *Homo sapiens*) evolved. If so, there must be a strong cultural overlay to this, since inner-city urban residents are reported to have experienced initial anxiety when transported to camps in the woods. However, it makes sense that a species that evolved in a natural environment would retain characteristics, neurological responses, and instincts that are most appropriate to that setting.

Popular culture has often held that exposure to the wilderness or at least an experience in nature is needed for the full human experience. Deprivation of a wild experience, it has been asserted, leads to anxiety or even to mental illness. Whether such privation actually leads to psychological or spiritual harm, and if so in whom, is not so clear. The idea was given a boost in 2005 when the American writer Richard Louv (b. 1949) published *Last Child in the Woods*, which dealt with the alienation of children from the natural and wilderness experience and coined the term "nature deficit disorder." (This condition is not a formally recognized psychiatric disorder.)

If there is a relationship between the wild experience and mental health, then one would expect there to be more anxiety associated with the perception that the natural environment is being lost. No doubt there is such anxiety among individuals who care deeply about the environment, but it is not clear that this feeling is widespread. In comparison to the psychological distress caused by the threat of thermonuclear war in the 1950s and 1960s, global anxiety over climate change appears to be modest and (in the United States) even transient. This is the case even though there is much more that individuals can do to influence events and to protect themselves collectively from climate change than there was in the case of avoiding thermonuclear conflict, which depended on a handful of national leaders acting rationally.

It is tempting to consider the idea of "nature deficit disorder" and the calming influence of nature to be nostalgia and sentiment. On the other hand, there is undeniably a subjective feeling of pleasure and a calming influence of natural beauty felt by most people, whether it is a neurological requirement or culturally programmed or a conditioned response to being less stressed while able to take time to be in such settings. Another interpretation, however, is that people want to connect to something and nature represents home.

It is true that people want to be connected to their natural environment; however, at a level as deep or deeper, they are also expressing in that longing a loss of the sense of community. Many deeper environmental concerns are expressions of disconnectedness and isolation. The reverse side of society's emphasis on individualism and its resistance to collective action—except in business organizations—is that people lose out on the family, neighbourhood, and society. Society, it is felt, needs to move back toward a balance. Much of the critique of industrial society that supported the ecology movement and that now supports sustainability is a call for an examination of personal values and the return of small-scale communities with the opportunity to matter as an individual person rather than as an interchangeable part.

13

The Professionalization of Sustainability

MOST LARGE, PUBLICLY visible companies have committed to sustainable practices because it is now the expected way for responsible companies to do business. Many smaller companies are doing the same either to create a market niche or out of genuine commitment, or both. To operate sustainably, they need managers who know something about their business, a lot about sustainability, and a very great deal about how to be an effective manager.

Sustainability is maturing as a management field. As sustainability matures and becomes expected of every organization's management, it is opening full-time career opportunities for those who can understand its principles, turn policies into strategy, develop business plans for achieving targets, evaluate products and services by quality and effectiveness, monitor compliance with policy and regulations, and manage the details of sustainability in organizations. (See table 13.1.) This has created a demand for people who know how to run sustainability-based services, can manage operations, and have good communication skills.

The biggest growth in sustainability careers and jobs, however, is now occurring in the business sector. The professionalization of sustainability is creating job opportunities that add value to organizations and society, as well as employment opportunities for the perceptive and well prepared. Just as the environmental movement institutionalized concern for the environment in politics, enterprise sustainability is institutionalizing sustainable practices in the conduct of business. Sustainability also plays an increasing role in government operations and the assessment of public policy, and environmental protection has long been a government function in all developed countries. Government continues to be a major employer of individuals with sustainability expertise and increasingly seeks out and rewards those with strong quantitative skills and those with expertise in risk assessment.

Table 13.1 **Table of sustainability values: Professionalization**

	Sustainability Value	Sustainability Element
I.	Long-term continuity	Progress in sustainability is likely to be faster and more consistent if there is a group of qualified, experienced professionals available to manage the process in important organizations, such as companies and institutions, and to disseminate best practices.
II.	Do no/minimal harm	Professionalized sustainability managers are likely to make fewer and lesser mistakes with training and experience. The selection of qualified managers would be easier with a professional credential.
III.	Conservation of resources	Managers of sustainability are likely to be more effective in conserving resources and reducing environmental impact if managers are well trained and share experiences.
IV.	Preserve social structures	Effective managers have to understand social institutions and work within complicated organizations.
V.	Maintain health, quality of life	A key aspect of professionalization is to define the core expertise required to practice. Managers should have an understanding of relevant issues in health and how sustainability measures affect the quality of life.
VI.	Performance optimization	Well-prepared and professionalized sustainability managers are more likely to enhance and balance economic, social, and environmental performance.
VII.	Avoid catastrophic disruption	Emergency management, business continuity, and continuity of operations require thoughtful planning by management and linkage with sustainability; well-prepared managers are in a better position to do this.
VIII.	Compliant with regulation	Monitoring of compliance with regulations is often a technical function requiring professional preparation and certification.
IX.	Stewardship	Professionalized sustainability managers are in a position to promote the values of sustainability and its inclusion in strategic planning.
X.	Momentum	A professional sustainability manager must know how to communicate the value of sustainability so that it becomes part of the culture of the organization.

Sustainability is taking on some of the characteristics of a profession and is becoming sufficiently technically complicated to require specialized expertise. This requires preparation and mastery of a particular skill set. What this skill set consists of and what expertise is required for a career in this field is becoming increasingly clear. Environmental management is also a professionalized function, and many aspects of it already require special training and certification of skills. In all probability, sustainability will soon be professionalized as well. The predominant models for professionalization are health care (traditionally, medicine) and law, with all the trappings of licensure or admissions to the bar and formal qualification. Sustainability is not likely to go so far, but it can be expected to go down the same road toward some form of certification as have other fields in which there is variable competence among practitioners, the expertise required is fairly clear, and the stakes are high (in money, legal compliance, and risk).

Students interested in health careers usually think too narrowly: not every health career involves medicine or nursing. As much as they contribute in their own way, these traditional health professions are not on the forefront of sustainability or environmental protection. Careers in public health are often, perhaps usually, completely overlooked by students interested in sustainability, and yet they are fundamental to sustainability (as this book should have demonstrated by now), well-established, rewarding, and filled with substantial opportunities.

Professionalization of Sustainability

Because of the increased demand for expertise relevant to sustainability, combined with the unevenness of preparation and the complexity of the issue, there has been growing interest in setting standards, creating career pathways, and identifying core competencies required to do higher-level work in implementing sustainability measures. This process may be called "professionalization," although the term is inexact and does not necessarily imply formal licensure or the other characteristics of a traditional profession. It does imply quality standards for performance in what might be called sustainability "practice" and encourages employers to look for certain characteristics, elements of preparation, and types of experience in candidates before entrusting sensitive or costly decisions to an applicant or self-declared expert in sustainability.

Who can best provide these sustainability management services is not clear. Sustainability practitioners may claim to have expertise with no training, experience, or credentials, while well-prepared, experienced, and qualified professionals are not always recognized in the field because their qualifications do not come with the label "sustainability." Employers, clients, and customers need to know why a sustainability "manager," "expert," or "consultant" thinks he or she can do the job and how he or she has earned the right to be recognized as an expert.

At present, there is no easy way to demonstrate this mastery, as by citing a set of initials after one's name or a common qualification. There are certification programs that apply to sustainability but no universal or business qualification. It is not clear that this would even be desirable in such an interdisciplinary, rapidly developing field such as sustainability. There are still few academic programs that identify themselves as emphasizing sustainability, although there are many in closely related and overlapping fields such as environmental studies and environmental sciences that embrace sustainability. A clear, agreed-upon definition of sustainability (something that this book attempted in chapter 1) will lead to recognition that sustainability management logically requires a certain skill set, deep understanding of certain principles, insight into how to apply principles in a particular context and place, and mastery (as in health) of a relevant body of knowledge. However, as a practical matter the lack of standards is likely to hold the field back for the near future. This is well recognized, and so the field is currently moving in the direction of establishing standards and promoting professionalization, especially for business practice. Professionalization of sustainability has the positive effect of institutionalizing favorable environmental and health practices in business and the material culture. It has the potential danger of removing or marginalizing sustainability practice from society as a whole and making it the domain of a restricted professional group.

Characteristics of a Profession

To provide consistency and quality assurance in sustainability, some practitioners have urged its professionalization. "Professional" status is based on similarities to the traditionally recognized professions of law, medicine, and the clergy, to which have been added, since the Middle Ages, numerous other recognized professions such as architecture and engineering that share:

- a defined, specialized, and technical body of knowledge
- standards of practice and mechanisms for quality assurance
- codes of ethics specific to the profession that control relationships with clients, colleagues, and the public, with mechanisms for reviewing and identifying malpractice
- autonomy, including the delegation to the profession of power to handle discipline for its own members
- work that is primarily cognitive (thought derived) rather than manual or physical
- work that requires a confidential relationship between the practitioner and the client
- work that assumes a substantial public obligation

- the means within the profession for improvement in practice and advancement of knowledge
- a mission of service in the public interest, deriving income through fees instead of profit
- regulation, usually by the government at some level, through registration, licensure, or issuance of a controlled identification number

Obviously not all of these characteristics apply literally to sustainability-related positions or consulting services. It is extremely unlikely, for example, that society would create a board to issue licenses to sustainability professionals or grant them the right to create the equivalent of a bar (for law) or a board or college (for medicine) to discipline one another. Even so, sustainability is clearly professionalizing in a broader sense.

Responsible practitioners with careers in sustainability are expected to act professionally, know what they are talking about, be honest with clients, and work in the public interest. In the future, they will be held to these standards more consistently and with greater professional liability than they are now. Figuratively, one can speak of future "sustainability professionals" as highly skilled, disciplined, honest, and committed to continuous improvement in the science and application of sustainability.

Efforts to create a professional credential for sustainability professionals are currently focused on the International Society of Sustainability Professionals (ISSP) (http://www.sustainabilityprofessionals.org/about-issp), which began in 2008 and has rapidly built its membership up to almost a thousand members at the time of this writing (2014). ISSP has the potential to become hugely important as sustainability practice and consulting grows and diversifies. ISSP has been developing a certification mechanism based on standardized competencies that would apply to all aspects of sustainability, while sidestepping issues of special expertise for each modality.

Professionalization of sustainability has the positive effect of institutionalizing favorable environmental and health practices in business and the material culture. It has the potential downside of removing or marginalizing sustainability from the mainstream of business practice, creating barriers to entry and therefore reducing the supply of qualified professionals, and making sustainability the domain of a restricted professional group.

Codes of Ethics

Codes of ethics are particularly important for professionals. Codes both commit and protect sustainability professionals and must be taken seriously. The most obvious reason to have a code of ethics is to hold practitioners accountable and to ensure that they uphold standards and do not harm clients or the environment and do not discredit the profession. This is not just a question of the practitioner's own performance. In individual cases and in a number of settings, there may be political pressure on environmental managers to compromise integrity. In these situations, a code of ethics is also useful in establishing standards of professional practice. Codes of ethics can be incorporated by reference into contracts and define a minimum standard of practice for a contractor. If the contractor does not observe the code, the individual or firm can be declared in violation of the contract. This is very important in the sustainability sector because so many environmental services are outsourced and handled by contractors and subcontractors.

Codes of ethics also place a responsibility on the professional who is governed by the code to avoid dealing with providers who do not act with integrity. The environmental services industry has had a long history of shady practices and corruption, including penetration by organized crime in some areas. The waste disposal industry is particularly attractive to organized crime because the visible part of the business can be entirely legal, the margins are very high (especially with inflated contracts), the business itself is easy to get into, the opportunities for bribing city officials and politicians to gain advantage are numerous, and the risks are low. Companies that were contracted to pick up liquid hazardous waste to take to a secure disposal area have been known to dump it by the side of the road or down a municipal sewer. Waste-hauling companies have been known to take hazardous waste to a regular landfill and falsify records on where solid waste goes. Corrupt business practices are not rare, such as bribery, kickbacks, extortion, price fixing, and collusive bidding (where bidders on a contract together rig the bids in order to take turns getting contracts at a higher price than competition would warrant). Many of these cases, especially those that cross state lines (for example, between New York and New Jersey) are prosecuted by the federal government in the United States under the Racketeer-Influenced and Corrupt Organizations Act of 1970, which applies to conspiracies to commit criminal acts; one of the more famous cases in the mid-1990s involved Mafia involvement in garbage hauling. Some communities have taken a principled stand and have not allowed corrupt businesses to take over their service contracts. In 1996, for example, the city of Thousand Oaks, California, refused to allow their contract for waste hauling to be transferred to a company that allegedly had repeatedly violated health and safety regulations and whose officers had pending criminal charges.

The United States is not alone in facing this problem. It is multinational, in the sense that it exists as a feature of the domestic industry in many European and Asian countries. It is also a global problem, in the sense that there are both cross-border and international networks of illegal waste disposal operations.

Jobs and Careers in Sustainability

Sustainability consists of innumerable sensible acts that together form a sensible strategy. This means that a sustainability career rests in large part on knowledge and skills in four areas: knowing what needs to be done and how to do it (the operational level), knowing in what order and combination and when to do what needs to be done (the strategic level), being sensible about doing it and knowing how to do it better (the leadership level), and understanding how doing it fits in the big picture (visionary level).

There are obvious definitional problems regarding a "sustainability career." First, it has to be a career and not just a job. Careers are commitments to a certain field or pathway and have a logic to them. Entry into careers is usually by climbing the "ladder" within an industry or via recognized academic preparation and special training. Sustainability has an advantage as a field in that it is building on the existing structure of environmental protection, which already has defined career pathways; however, as a field sustainability has not yet built its own career ladder or industry. The career pathways are emerging, however, as entrants into this relatively new management field progress through promotion and career transitions.

Employment in sustainability ranges widely from task-oriented jobs in the operation of environmental protection or necessary functions such as recycling and energy conservation, to management positions, to leadership positions. Relevant and satisfying jobs depend importantly on experience and a practical skill, for which training may be needed. Careers require an element of preparation and continuity. Careers at a management level are salaried positions and usually require special preparation in addition to relevant experience.

The US Bureau of Labor Statistics (BLS) tracks jobs in a variety of sectors and has paid close attention to the emergence of careers in sustainability. They have observed growth in many fields related to sustainability, but as yet there is no standard definition of the field they can apply to monitor trends in sustainability as such. Operationally, BLS equates sustainability with being "green" and defines a "green job" as "one whose primary duty is related to the use of environmentally friendly production processes" and in which more than half the time is spent researching, maintaining, installing, and/or using technologies of practices to lessen the environmental impact of their establishment" or training others to do so. BLS also notes transportation, storage, and distribution management, which are general categories that may incorporate sustainable operations not easily recognized in the job description. Categories of careers related to sustainability include, but are not limited to:

- *Executive functions.* These careers include the positions called "chief sustainability officers" or vice presidents in charge of sustainability. For a leader at this level, leadership skills are most important, with the ability to balance priorities, encourage creativity, financial management, and to present a vision that engages talented people who have their own visions and ideas of how the world works.

- *Management.* For managers, communication and administrative skills are essential, including the ability to make and follow a budget. In-depth knowledge of operations is required to manage sustainability at the operational level and to supervise the people who are doing the work. An excellent manager will spot opportunities to achieve greater savings or better sustainability performance and will motivate cooperation on the part of employees.
- *Consultants.* The need for expertise in sustainability has resulted in many individuals with more or less appropriate credentials offering their services as consultants. One level of consultant focuses on particular issues, such as the indoor environment or energy use, and are primarily problem solvers for their clients. Another level specializes in business management, such as strategic planning, financing for green projects, or project management. Not infrequently, they lay the groundwork for the eventual hire of a permanent executive position in sustainability for the company.
- *Scientists and science technicians.* The careers identified as sustainability related are in environmental sciences or sciences applied to environmental issues. They also have a variety of levels, including science managers. Most of these professions deal in one way or another with environmental protection. A few are dedicated to innovation and research directly applicable to sustainability.
- *Green architecture.* The design of structures to be sustainable and low impact has entered the mainstream of architecture, but some architects and firms specialize in sustainability, both because of its importance and because the client wishes to make a statement. Note that "green building" *construction* is not included in this list. That is because construction is well behind design when it comes to sustainability. (See chapter 9.)
- *Engineering.* Environmental engineering (a branch of civil engineering) dominates in direct application to sustainability, but industrial process engineering and industrial or occupational hygiene (defined in chapter 9) are also noted by BLS.
- *Loss Prevention.* Working in a specialized area of management, loss prevention officers typically review insurance coverage, preventive measures, occupational health and safety, the risk of being sued, security, and fire and disaster management in order to ensure that the enterprise is at low risk of a major adverse event and that any harm and loss will be mitigated.
- *Compliance officers.* Compliance officers, often embedded in loss prevention departments, are charged with monitoring the enterprise's performance and

ensuring that no laws are broken or standards violated; although it is good practice, compliance officers are not necessarily responsible for achieving better performance than the law requires.

- *Urban and regional planning.* Positions as planners generally require extensive academic preparation and graduate study. The positions are mostly in city and county government and academics, but some large private developers have need for planning expertise.
- *Occupational health and safety professionals.* These are highly responsible positions that involve the management of the workplace environment (see chapter 9). They have many characteristics in common with environmental health professionals (see below) but are focused on the health of workers and working populations, and on the specific environment of the workplace.
- *Operations supervisors or coordinators.* This is a category recognized by BLS but not usually treated by employers as an occupation in itself. It is a broad grouping that includes those occupations that involve overseeing and coordinating operations such as recycling, that relate to sustainability but are not an integral part of production. It also includes inspectors and auditors who monitor energy or water use or other aspects of sustainability performance.
- *Operators.* Operators do the work. For operators, skill and experience are primary but may not require special preparation or education. When the skill level is low, pay is usually low also, and these jobs usually do not make viable careers. When the skill level is high, operators have the potential to move up to supervisory positions and to develop consulting opportunities. The strategy is to master a particular area, such as building management services, waste management, or recycling operations and to know such details as equipment models, performance specifications, and how to integrate services and then build a career taking on increasing responsibility. (One pathway to success in this category is to become expert in dealing with something that other people do not want to worry about, such as trash; such positions often pay better because they are less attractive.)
- *Environmental protection.* These positions may involve water quality monitoring, air quality monitoring, laboratory analysis, auditing compliance with environmental regulations, or performing many other services for governments, usually at the state or municipal levels, or for large companies. They are similar to scientific technician positions and can lead to senior management careers. The key to success in these jobs is dedication to quality assurance.
- *Public health professionals.* The field of public health is rich in career opportunities related to sustainability (as explored in this book) and dedicated to the protection of health in entire communities or groups, rather than individual medical care. A particularly noteworthy category of public health professionals specializes in environmental health services. That field is itself so broad that it is treated separately below.

- *Environmental health professionals.* Environmental health professionals are a subset of public health professionals devoted to environmental health risks. These careers may be involved in direct services at the municipal level, which could be inspecting food services, vector control, investigating events of contamination or pollution, injury prevention, protecting children from lead and other serious hazards, inspecting housing quality, water quality monitoring, laboratory analysis, and reviewing planned construction projects to avoid health problems. A partial list of the areas of practice for environmental health, mostly at the local and state level, is presented in box 13.1. Environmental health specialists were ranked 22nd of the 100 best jobs in America by CNN Money in a story in 2010.

BOX 13.1

Specialty Areas in Environmental Health

1. air quality management (monitoring and control of local sources of air pollution)
2. consumer safety
3. control technology (engineering and control technology)
4. emergency preparedness (disaster management)
5. environmentally related chronic disease epidemiology
6. food protection (toxicology and microbiology)
7. housing conservation and rehabilitation
8. institutional environmental health (sanitation in schools, health facilities, jails, etc.)
9. noise control
10. outbreak investigation (of cases possibly related by environmental exposure)
11. radiation health
12. recreational health and safety (control of injuries and health problems associated with sports and recreation)
13. risk assessment
14. solid waste management
15. toxic substances management (including hazardous waste control)
16. toxic substances control
17. traffic safety
18. vector control
19. drinking and recreational water supply and treatment
20. water quality management (wastewater treatment)

Preparation for Careers in Sustainability

Academic programs in "sustainability" on the graduate level vary considerably in content and emphasis. For an employer, this is a problem. The label on the degree may not be informative, and the definition of sustainability used by the institution may not be explicit or may not agree with the employer's point of view. This forces human resources to investigate further to determine what the candidate actually did study, which is a lot of trouble unless they already consider the candidate to be an attractive hire. This places the new graduate at a considerable disadvantage in job hunting when the job market is weak. Human resources departments do not have to think twice when an applicant has a degree in a particular, named field. The question for sustainability is, prepared to do what?

The great majority of programs with "sustainability" in the title combine the label with environmental sciences (interdisciplinary study in scientific disciplines with environmental applications) or environmental studies (interdisciplinary studies in the context of environmental issues, usually emphasizing policy or the humanities) or a relevant profession or technical field (such as the MBA or a law degree).

The Association for the Advancement of Sustainability in Higher Education (http://www.aashe.org) has registered 1,403 academic programs emphasizing sustainability at 462 campuses, in all states and Canadian provinces, and some in other countries. Of these, 450 programs in their database are at the level of master's degrees (which is the most appropriate level for professional preparation because sustainability is a rather narrow academic focus for an undergraduate education) and doctoral programs (which mostly prepare students for academic careers). Certain trends are obvious.

Academic programs in sustainability generally fall into one of the following categories:

- Programs carrying the name "sustainability" alone, of which there are currently very few (as of 2014)
- Programs emphasizing sustainability but identifying themselves as environmental sciences or a closely related discipline, usually emphasizing ecosystem protection and environmental management
- Programs emphasizing sustainability but identifying themselves as environmental studies, usually emphasizing policy and business management
- Programs adding a sustainability emphasis to a professional or disciplinary field, such as law or business

From the university's point of view, it usually makes more sense to start a track within an existing program than to create an entirely new major from environmental sciences or studies. Inevitably, however, this may bury the identity of a program on the Internet and may obscure its relevance to "sustainability" as a whole. The absence of "sustainability" on the diploma or in the department name may make it less attractive to prospective students who are narrowly focused on sustainability as a career.

The time to commit to a program bearing the label of sustainability may be when a school identifies a "sweet spot" where it can do well because of location or prior positioning. Programs distinguish themselves from one another by an unusual emphasis or by presenting a competitive advantage. For example, sustainability education, rather than management, is emphasized at Oregon's Portland State University. Progressive curricula in a small institution is the appeal of Goddard College in Vermont and of Arizona's Prescott College. Relative newness and motivation distinguishes northern Virginia's George Mason University. The institutional advantage of a major gift (from the Margaret A. Cargill Foundation) for development of the area allows Bard College in New York, a relatively small institution, to develop programs with coherency and to develop them over time. The resources of a large, well-established university are part of the appeal of Arizona State University. Some differentiating areas of emphasis are logical extensions of existing academic strengths (the University of Southern California's program in sustainable regional planning), while others are not intuitively obvious (the University of Texas at Arlington's program in real estate and sustainability). Programs that already have an established reputation for excellence in environmental studies (such as the University of Western Ontario, in Canada), in which sustainability studies are embedded, may face a marketing problem if they introduce "sustainability" as a new field of study, since the addition of a new but similar program could compete with the existing offering for students.

Environmental Sciences and Studies

Environmental sciences and studies programs have been studied intensively by the National Science Foundation and by the National Council on Science and the Environment (NCSE), an association which monitors trends closely and is the administrative home of the Council of Environmental Deans and Directors, as well as the Association of Environmental Studies and Sciences. NCSE conducts regular surveys and produces a series of reports on the state of environmental sciences and studies education in the United States. These sources are data-rich, timely, and have been consistent for the last several years in describing overall trends. They provide an overview of the field at the time of writing (in 2014).

For a student entering sustainability studies, which by definition is interdisciplinary, the quandary is whether to major in a recognized discipline (such as biology, physics, chemistry urban planning, or economics) and then gain knowledge of sustainability applications through further study or experience, or to seek interdisciplinary education in sustainability or environmental management per se. The advantage of the disciplinary approach is that it is clear what the degree represents, but the disadvantage is that it may be a long wait to apply the knowledge to sustainability. The advantage of the interdisciplinary approach is that the student emerges with a thorough grounding in the relevant knowledge, but employers may not recognize the degree, and there is a risk of being overly specialized if the job market is challenging.

In recent years, interdisciplinary programs related to environmental sustainability have consolidated into two types of environmental curricula: (1) "environmental sciences," which features education in the scientific disciplines that support environmental analysis, research, and protection and has a high technical content; and (2) "environmental studies," which emphasizes social context, public policy, and education and typically has some introductory science but without much depth. As of 2013 environmental sciences degree programs were more popular than environmental studies for undergraduate students; it also had a much larger proportion of graduate students. Environmental studies degree programs overwhelmingly consisted of undergraduate majors, but there were some master's programs and a handful of PhD opportunities. The two together constituted about 60 percent of all so-named environmental curricula. None of the other degree programs accounted for more than 10 percent of the total identified as incorporating sustainability, and most for less than 5 percent: natural resources management, coastal and marine studies, earth and environmental sciences, environmental policy, energy studies, environmental management, humanities and social sciences, technology studies (not technological fields), and planning. The emergence of a small number of programs labeled as "sustainability studies" was evident, mostly undergraduate. Among the fields of science, technology, engineering, and mathematics (STEM), and engineering accounted for very few programs in the survey, but this may be misleading because civil engineering programs almost always have environmental engineering tracks (but may not consider them interdisciplinary). Likewise, planning has a long and distinguished history of interest in sustainability as a cornerstone of the field but may not consider their programs to be interdisciplinary.

It is noteworthy that none of these interdisciplinary programs are based in the health sciences. This should not be surprising, since medicine and public health are traditionally graduate and professional disciplines and require undergraduate preparation in a basic disciplinary field, such as biology. This may change in the future, however, as epidemiology and public health are becoming increasingly popular as undergraduate majors. Epidemiology excels at teaching analysis and problem solving, as demonstrated initially at Swarthmore College (Pennsylvania) where David Fraser, a president of that institution in the

1980's, was a former epidemiologist with the Centers for Disease Control and Prevention and realized the discipline's academic potential as a highly effective but not resource-intensive alternative for teaching science and analytical thinking to undergraduates. Since then other institutions, led by The Johns Hopkins University (Baltimore, Maryland), have initiated undergraduate majors in public health generally, which further expands the scope of science education that can be achieved through health studies.

Interdisciplinary environmental sciences and studies have, of course, become a major trend in higher education generally. A survey of 804 responding program leaders at 652 institutions representing undergraduate and graduate programs was conducted as a doctoral thesis project by Shirley Vincent at Oklahoma State University (she is now with NCSE) and reported in 2008. Preliminary analysis showed that almost two-thirds had full-time faculty devoted to environmental curriculum, and one-third had autonomous standing as a department or similar academic unit; and half reported having a smaller budget than comparable programs at their institution. Factors believed to be associated with success include use of real-world examples, program leadership, and strong administrative support. Exploratory factor analysis identified many clusters, including wide support for rigor in natural sciences and quantitative skills. Contrary to market information, communications skills were less highly valued, which is odd because they are repeatedly mentioned as critical by employers (but such skills may be taken as a given by the educators). Highly valued characteristics associated with competence included management skills, research skills, social sciences/humanities content, cognitive skills. It is unlikely that these relationships have changed in the ensuing years.

Environmental sciences and studies appear to be particularly attractive to women as a gateway to scientific and environmental management careers. The broad field therefore plays an important role in achieving diversity and parity in the STEM fields (Science, Technology, Engineering, and Mathematics). An alumni survey has been administered every two years by the National Bureau of Economic Research at Harvard since 2005, tracking the annual intake of graduates at baccalaureate, master's, and doctoral levels and tracking career paths longitudinally. There are slightly more women than men obtaining environmental curriculum–based degrees. Most decided on their career choice as an undergraduate and were heavily influenced in their choice by faculty members. Most were also motivated by an intellectual and aptitude match with the field, fewer by social benefit, and fewer by career attractiveness. Most have found jobs in the field and are satisfied with the salary. Career intentions on entry were overwhelmingly government related, but final career choices were equally divided between government and the private sector. The few who left the field did so because jobs were not available, presumably in the local market. The master's was the degree most highly valued. Retrospectively, graduates recommended more training in energy technologies, computer science, grant writing, more real-world experience, and more opportunities to present their work before peer and scientific audiences. Interpretation emphasized the creation of tracks between undergraduate and master's programs, good counseling during undergraduate years, and curricular enrichment in the areas noted.

Gender issues are highly significant in environmental sciences and studies. Women, as noted, outnumber men in environmental sciences and studies, and this is consistent with the observation that in general more women than men are in interdisciplinary studies (except in engineering). Environmental sciences and studies are a logical and welcome gateway for women into the STEM fields: together with the consensus that stronger preparation in math is essential, this suggests that an enhanced environmental curriculum could serve as a gateway for women into traditionally male-dominated STEM disciplines.

Because of the predominance of women in the field, faculty retention and promotion in environmental sciences and studies is therefore a gender as well as an academic issue. Models demonstrate that even small degrees of bias (1 percent to 5 percent) in making recruitment and retention decisions seriously degrade the "survival curve" (continuation in the field over time) of faculty retention of women and members of minority groups and leads to underrepresentation of women by as much as half.

Interest in a curriculum for climate change studies and solutions has developed during the same time period but not necessarily linked to other sustainability studies. The job market reflects growing employment opportunities in "carbon management." Climate mitigation-related jobs require innovation and engineering education, with a high priority placed on math and quantitative skills, environmental chemistry, social studies, and communications skills, a nontraditional but increasingly necessary combination.

Blending Health and Sustainability in a Career

The student or job aspirant who is looking for a career that combines sustainability and health is best advised to consider public health and especially environmental health, as mentioned previously. If his or her interest is primarily in public policy, another good option is risk science, which includes risk assessment.

This is advised not because the duties of a sustainability professional do not affect health in the big picture—far from it, as this book has argued—or because health professionals have no interest in sustainability—they do, or at least should. It is because there is not much rigorous content on health in sustainability-relevant education, such as environmental sciences and studies, and because professional practice in health requires mastery of other disciplines such as epidemiology or medicine.

Most people who work full time in health are licensed health professionals, most being nurses. There are no jobs for physicians or nurses in sustainability management and no career path. The great majority of academic programs in health are, of course, devoted to training that leads to professional licensure, as for physicians and nurses. Medicine and nursing, along with other licensed health professions, tend to be very applied, very narrowly focused to individual patient care, and extremely intense, given the amount of learning required and the high stakes in practice. As a result, there is little opportunity to combine training in the health professions with sustainability (or even environmental health) until training is completed. That can take years. Not surprisingly, hardly anyone has ever done it. Health care professionals with an interest in sustainability typically (and there are very few) learn about it after they are well along in their health careers and as a personal commitment.

On the other hand, preparation for a career in public health combines relatively well with preparation for a career in sustainability, particularly environmental health training (which dovetails well with environmental sciences). Health promotion would also seem to be a good career option for aspirants to sustainability careers and especially for environmental studies majors interested in behavioral change; however, again there is no defined career path.

Public health qualification typically relies on the master of public health degree (MPH), a PhD in a public-health-related discipline (such as environmental health), or the doctor of public health degree (DrPH), which is a generalist practitioner degree currently gaining in popularity. In the future there will no doubt be combined graduate programs, but they are not commonly available now.

A second pathway to combining health with sustainability-relevant training is, as noted, "risk science," which is a systematic approach to analyzing situations and making decisions involving uncertainty and planning how to deal with them. Risk science is a well-developed field at the graduate level, available at many major universities, usually in departments of public administration, psychology, or schools of public health. It rests on "risk assessment," which is a quantitative approach to analyzing the potential for harm or for benefit of a particular hazard or policy option and is heavily used in government for framing and supporting regulatory policy decisions. Risk assessment is heavily grounded in statistics and often involves mathematical modeling. Risk science as a whole also includes "risk management," which is a broader field of selecting policy options to implement the decision, together with "risk communication" and "risk perception," which is a heavily cognitive science–oriented field that examines how people and communities ("publics" is the term of art) understand and process information about risk. These programs should not be confused with programs in loss management in business, insurance, and finance, which also use the terms "risk assessment" and "risk management" but in a different context.

Today, the field of sustainability is just coming together on the scientific foundation of environmental science and studies and the management foundation of business and environmental protection. The connection with health, so obvious in the rest of this book, has not manifested itself in actual job opportunities. The enterprising student or aspirant has virtually no job definition, career path, or template to follow.

The potential to create a career that combines health and sustainability is there for the energetic and enterprising. The way forward will be highly individualistic; it is also unlikely to be a straight path and will prove unpredictable for the foreseeable future.

The only certainty is that anyone attempting to create such an opportunity needs to be well prepared in both health and sustainability management and have the best qualifications possible. An aspirant also needs to be an excellent manager with well-developed communication skills, which will be the key to success. Such a person cannot be reckless in gambling their career on a long shot but should not be overly timid in pursuing their vision. Beyond that not-very-helpful advice, there is no guidance from experience.

Selected Sources

The sources listed below have been especially useful or relevant to the topics covered in each chapter. No attempt has been made to reference every passage in the text or to provide a comprehensive bibliography. In the age of the Internet, cited facts and statistics can be found more quickly online than tracing them back through reference files that can become dated quickly.

Guidotti, T.L. 2010. *The Praeger handbook of occupational and environmental medicine*. 3 vols. Santa Barbara, CA: Praeger. This medical and public health textbook has several chapters elaborating on the ideas in this book. See especially the chapters on toxicology, epidemiology, risk assessment, prevention science, and environmental health.

Yassi, A., T. Kjellström, T. de Kok, and T.L. Guidotti. 2001. *Basic environmental health*. London: Oxford University Press. Produced for the World Health Organization, our textbook was structured with an approach consistent with the present book. It remains one of the most popular introductory textbooks of environmental health. However, more recent sources should be consulted (as always) for time-specific information.

1. HEALTH AND SUSTAINABILITY

Health

Bourdelais, P. 2006. *Epidemics laid low: A history of what happened in rich countries*. Baltimore: Johns Hopkins University Press.

Coggan, J. 2012. *What makes health public: A critical evaluation of moral, legal, and political claims to public health*. Cambridge, MA: Cambridge University Press.

Evans, R.G., M.L. Barer, and R. Marmor Evans. 1994. *Why are some people healthy and others not? Determinants of health in populations*. New York: Aldine de Gruyter.

Guidotti, T.L. 1996. Of blindmen, elephants, and environmental medicine. *New Solutions* 6(4):25–30.

Guidotti, T.L. 1996. Preventive medicine: Notes toward an agenda for change. *Am J Prev Med* 12:165–171.

———. 1997. Why are some people healthy and others not? A critique of the population health model. *Annals RCPSC* 30:203–206.

———. 2010. *The Praeger handbook of occupational and environmental medicine*. Santa Barbara, CA: Praeger.

———. 2011. The literal meaning of health. *Arch Environ Occup Health* 66(3):189–190.

Institute of Medicine. 2013. *Public Health Links with Sustainability: Workshop Summary*. Washington, DC: National Academies Press.

McMichael, T. 2001. *Human frontiers, environments and disease: Past patterns, uncertain futures*. Cambridge, MA: Cambridge University Press.

Rose, G. 1992. *The strategy of preventive medicine*. London: Oxford, 1992.

Sustainability

Brower, David. 1960. *The meaning of wilderness to science*. San Francisco: Sierra Club.

Brundtland, G.H. 2002. *Madam prime minister: A life in power and politics*. New York: Farrar, Straus and Giroux.

Charter, S.P.R. 1962. *Man on earth: A preliminary evaluation of the ecology of man*. New York: Grove Press.

Commoner, Barry. 1963. *Science and survival*. New York: Viking.

———. 1971. *The closing circle: Nature, man, and technology*. New York: Alfred A. Knopf.

———. 1977. *The poverty of power*. New York: Bantam, 1977.

Dansereau, P. 1966. The 27 laws of ecology. New York: Botanical Garden. http://www.jackmtn.com/PDF/27_laws_of_ecology.pdf.

Garver, K.L., and B. Garver. 1991. Eugenics: Past, present, and the future. *Am J Hum Genet* 49(5):1109–1118.

Hengeveld, R. 2012. *Wasted world: How our consumption challenges the planet*. Chicago: University of Chicago Press.

Lawton, J.H. 1999. Are there general laws in ecology? *Oikos* 84(2):177–192.

Lockwood, D.R. 2008. When logic fails ecology. *Q Rev Biology* 83(1):57–64.

Milbrath, L.W. 1988. *Envisioning a sustainable society: Learning our way out*. Albany: State University of New York Press.

Sen, A. 2009. *The idea of justice*. Cambridge, MA: Belknap/Harvard.

World Commission on Environment and Development. 1987. *Our common future*. Geneva: Center for Our Common Future.

2. KNOWING

Bell, S., S. Morse. 2008. *Sustainability indicators: Measuring the immensurable?* Styterling, VA: Earthscan.

Camilli, R., A. Bowen, C.M. Reddy, J.S. Seewald, and D.R. Yoerger. 2012. When scientific research and legal practice collide. *Science* 337:1608–1609.

Corvalán, C., T. Kjellström, and K. Smith. 1999. Health, environment and sustainable development: Identifying links and indicators to promote action. *Epidemiology* 10:656–660.

Eccles, R.G., and M.P. Krzus. 2010. *One report: Integrated reporting for a sustainable strategy*. Hoboken, NJ: John Wiley and Sons.

Elkington, J. 1998. *Cannibals with forks: The triple bottom line of 21st century business*. Stony Creek, CT: New Society Publishers.

Environmental Protection Agency. 2000. *America's children and the environment: A first view of available measures*. Office of Children's Health Protection and National Center for Environmental Economics.

Epstein, M.J. 2008. *Making sustainability work: Best practices in managing and measuring corporate social, environmental, and economic impacts*. San Francisco: Berrett-Koehler.

Goldman, A.I. 2004. *Pathways to knowledge: Private and public*. Oxford: Oxford Univ. Press.

Guidotti, T.L. 1994. Critical science and the critique of technology. *Pub Health Rev* 22:235–250.

Guidotti, T.L., and S.G. Rose, eds. 2001. *Science on the witness stand: Scientific evidence in law, adjudication and policy*. Beverly Farms, MA: OEM Press.

Jasanoff, S. 1995. *Science at the bar: Law, science and technology in America*. Cambridge MA: Harvard University Press.

Meufeld, P.J., and N. Colman. 1990. When science takes the witness stand. *Scientific American* 262(5):46–53.

Michaels, D. 2008. *Doubt is their product: How industry's assault on science threatens your health*. New York: Oxford University Press.

Pastides, H. 1995. An epidemiological perspective on environmental health indicators. *World Health Stat Q* 48:140–143.

Quental, N., J.M. Lourenço. 2012. References, authors, journals, and scientific disciplines underlying a sustainable development literature: A citation analysis. *Scientometrics* 90(2):361–381. http://www.slideshare.net/nquental/references-authors-underlying-the-sustainable-development-literature.

Savitz, A.W., and K. Weber. 2006.*The triple bottom line*. Hoboken, NJ: John Wiley and Sons.

Shapin, S. 1994. *A social history of truth*. Chicago: University of Chicago Press.

Suppe, F., ed. 1977. *The structure of scientific theories*. 2d ed. Urbana, IL: University of Illinois Press.

World Bank, and L. Segnestam. 1999. Environmental performance indicators. Environmental Economic Series, paper 71.

3. CATASTROPHE

Diamond, J. 2005. *Collapse: How societies choose to fail or succeed*. New York: Viking.

Ehrlich, Paul R. 1968. *The population bomb*. New York: Sierra Club/Ballantine.
Funk, M. 2014. *Windfall: The booming business of global warming*. New York: Penguin.
Kolbert, E. 2006. *Field notes from a catastrophe*. New York: Bloomsbury.
Matthews, H.D., and S. Solomon. 2013. Irreversible does not mean unavoidable. *Science* 340:438–439.
McMichael, A.J. 2012. Insights from past millennia into climate impacts on human health and survival. *Proc Nat Acad Sci* 109(13):4730–4737.
Muller, R.A. 2012. Conversion of a climate change skeptic. *New York Times*, 30 July.
Pinker, S. 2011. *The better angels of our nature: Why violence has declined*. New York: Penguin.
Pratt, A.G., Klein R.J.T. Schrõter, and A.C. de la Vega-Leinert. 2009. *Assessing vulnerability to global climate change*. Washington DC, Earthscan.
Shrady, N. 2008. *The last day: Wrath, ruin, and reason in the great Lisbon earthquake of 1755*. New York: Viking.

Population

Friedman, T.L. 2008. *Hot, flat, and crowded*. New York: Farrar, Straus and Giroux.
Engelman, R. 2008. *More: Population, nature, and what women want*. Washington DC, Island Press.
Hardin, G. 1968. The tragedy of the commons. *Science* 162(3859):1243–1248.
———. 1993. *Living within limits: Ecology, economics and population taboos*. New York: Oxford University Press.
Homer-Dixon, T.F. 1999. *Environment, scarcity, and violence*. Princeton, NJ: Princeton University Press.
Malthus, T., J. Huxley, and F. Osborn. 1960. *Three essays on population: Thomas Malthus, Julian Huxley, Frederick Osborn*. New York: Mentor.

Climate Change

The literature on climate change is, of course, vast. The definitive source is the Intergovernmental Panel on Climate Change and its series of publications. For the nonspecialist reader wishing to monitor the science closely, *Science* (the journal/magazine of the American Association for the Advancement of Science) is the indispensable source. Here are some accessible book titles.

American Geophysical Society. 2011. Climate change looking much the same, or even worse. Summary of the 2011 American Geophysical Society meeting. *Science* 334:1616.
Archer, D. 2009. *The long thaw: How humans are changing the next 100,000 years of earth's climate*. Princeton, NJ: Princeton University Press.
———. 2010. *The global carbon cycle*. Princeton, NJ: Princeton University Press.
Calvin, W.H. 2008. *Global fever: How to treat climate change*. Chicago: University of Chicago Press.

Dressler, A., and E. Parson. 2010. *The science and politics of global climate change: A guide to the debate.* 2d ed. Cambridge, UK: Cambridge University Press.

Flannery, T. 2005. *The weathermakers: How we are changing the climate and what it means for life on earth.* Toronto: Harper Collins.

Guidotti, T.L., and J. Last. 1992. Implications for human health of global atmospheric change. *Trans Roy Soc Canada* 6(2):223–239.

Helm, D. 2012. *The carbon crunch: How we're getting climate change wrong—and how to fix it.* New Haven, CT: Yale University Press.

Hulme, M. 2009. *Why we disagree about climate change.* Cambridge, UK: Cambridge University Press.

Posner, E.A. 2010. *Climate change justice.* Princeton, NJ: Princeton Univ. Press.

Primack, R.B. 2014. *Walden warming: Climate change comes to Thoreau's woods.* Chicago: University of Chicago Press.

Ruddiman, W.F. 2005. *Plows, plagues and petroleum.* Princeton, NJ: Princeton Univ. Press.

Stern, N. 2006. *The economics of climate change: The Stern review.* Cambridge, UK: Cambridge University Press.

———. 2009. *The global deal: Climate change and the creation of a new era of progress and prosperity.* New York: Public Affairs.

Vallis, G.K. 2012. *Climate and the oceans.* Princeton, NJ: Princeton Univ. Press.

4. CONTAMINATION

The basic schema for differentiating natural cycles and anthropogenic cycles came to my attention in or around 1970 at a student environmental conference in Big Bear, California, where it was presented by an environmental planning student from Cal Poly at San Louis Obispo. The student was Steve Renfrew. I have never heard a definition for pollution as clear and as useful since and have used and adapted his formulation many times.

Guidotti, T.L. 2009. Emerging contaminants in drinking water: What to do? *Arch Environ Occup Health* 64(2):91–92.

Guillette, L.J. Jr., and T. Iguchi. 2012. Life in a contaminated world. *Science* 337:1614–1615.

Relyes. R., and J. Hoverman. 2006. Assessing the ecology in ecotoxicology: A review and synthesis in freshwater systems. *Ecology Letters* 9:1157–1171.

Truhaut, R. 1977. Ecotoxicology: Objectives, principles, and perspectives. *Ecotoxicol Environ Safety* 1(2):151–173.

———. 1976. Ecotoxicology: Objectives, principles, and perspectives. *Arch Belg Med Soc* 34(4):201–237.

5. CHEMICAL POLLUTION

The field of environmental sciences and environmental health, together with occupational and environmental health, suffers from the lack of a good, basic, easily accessible textbook of toxicology relevant to environmental rather than acute and medical or consumer exposure. The several chapters listed below that were written or edited by the present author were prepared specifically to fill that gap for teaching purposes, but none of them are long enough to support a full course.

Guidotti, T.L. *The Praeger handbook of occupational and environmental medicine.* Santa Barbara, Praeger ABC/CLIO, 2012. See chapter on Toxicology (2, pp. 55–200), Risk Science (7, pp. 379–458), Common Chemical Hazards (10, pp. 569–688).

Guidotti, T.L., and M.S. Moses. 2007. Toxicological basis for risk assessment. In *Risk Assessment for Environmental Health.* Edited by M. Robson and W. Toscano, 55–84. San Francisco, Jossey Bass/Wiley.

Guidotti, T.L., and L. Ragain. 2007. Protecting children from toxic exposures: three strategies. *Ped Clinics N Amer* 54(2):227–235.

Guidotti, T.L., J. Rantanen, S. Lehtinen, K. Takahashi, D. Koh, R. Mendes, H. Fu, R. J. Guzmán, and S.G. Rose, eds. 2011. *Global occupational health.* New York: Oxford University Press.

6. INFECTIOUS DISEASE

Dykhuizen, D. 2005. Species number in bacteria. *Proceedings of the California Academy of Sciences* 56.1(6):62–71.

Oppenheimer, G.M., and E. Susser. 2007. The context and challenge of von Pettenkofer's contributions to epidemiology. *Amer J Epidemiol* 166:1239–1241.

Ostfield, R.S., F. Keesing, and V.T. Eviner. 2008. *Infectious disease ecology: Effects of ecosystems on disease and of disease on ecosystems.* Princeton, NJ: Princeton Univ. Press.

Stallybrass, C.O. 1931. *The principles of epidemiology and the process of infection.* London: Routledge.

Walters, M.J. 2014. *Seven modern plagues and how we are causing them.* Washington, DC: Island Press.

Malaria

Malaria is used here as an example of an infectious disease with an environmental aspect to transmission. It is important to sustainability as a paradigm but also because of the enormous global burden it imposes. The description in the text is kept simple in order to keep the example accessible to readers outside the health sciences. For basic information on malaria itself, readers are encouraged to go online at the websites of the Centers for Disease Control and Prevention or the World Health Organization, or to consult textbooks of global public health rather than medicine (as texts on global public health are more relevant to sustainability). Textbooks of infectious disease or "tropical medicine" are more likely to emphasize the biology of the organism and the specifics of treatment.

Chiyaka, C., A. J. Tatem, J.M. Cohen, P.W. Gething, G. Johnston, R. Gosling, R. Laxminarayan, S., I. Hay, and D.L. Smith. 2014. The stability of malaria elimination. *Science* 339:909–910.

Davis, F.R. 2012. Deploying a powerful pesticide. *Science* 335:288.

Fidock, D.A. 2013. Eliminating malaria. *Science* 340:1531–1533.

Gyapong, M. 2014. Malaria vaccine technology roadmap update includes new targets. mviPATH. http://www.malariavaccine.org/

Heckel, D.G. 2012. Insecticide resistance after *Silent Spring*. *Science* 337:1612–1614.

Lemon, S.M., F. Sparling, M.A. Hamburg, D.A. Relman, E. R. Choffnes, and A. Mack, eds. 2008. *Vector-borne diseases: Understanding the environmental, human health, and ecological connections*. Workshop Summary, Forum on Microbial Threats. Washington, DC: National Academies Press.

Nájera, J.A., M. González-Silva, and P.L. Alonso. 2011. Some lessons for the future from the Global Malaria Eradication Program. PLoS, 25 January. http://www.plosmedicine.org/article/info%3Adoi%2F10.1371%2Fjournal.pmed.1000412. doi:10.1371/journal.pmed.1000412

Siraj, A.S., M. Santos-Vega, M. J. Bouma, D. Yadeta, D. Ruíz Carrascal, and M. Pascuat. 2014. Altitudinal changes in malaria incidence in highlands of Ethiopia and Colombia. *Science* 343:1154–1158.

Tanner, M., and D. de Savigny. Malaria eradication back on the table. Geneva: World Health Organization. http://www.who.int/bulletin/volumes/86/2/07-050633/en/.

7. SOCIALLY MEDIATED EFFECTS

The DPSEEA model was first brought to the author's attention by Tord Kjellström of the World Health Organization, during a collaboration in the 1990s. The theoretical basis was developed by Dr. Kjellström, together with many colleagues at WHO, including Yasmin von Schirnding and Carlos Corvalán, and has been extensively used in the EU as well as the UN system.

Bagehot. 2013. The parable of the Clyde: The devastation of a fishery shows the idiocy of much environmental politics. *Economist*, 31 August, p. 50.

Gilmore, J.S. 1975. *Boom-town growth management: A case study of Rock Springs–Green River, Wyoming.* Boulder CO: Westview Press.

Rosa, E.A., A. Diekman, T. Dietz, and C.C. Jaeger. 2010. *Human footprints on the global environment: Threats to sustainability*. Cambridge MA: MIT Press.

Soskolne, C.L., C.D. Butler, C. Ijsselmuiden, L. London, Y. von Schirnding. 2007. Toward a global agenda for research in environmental epidemiology. *Epidemiology* 18(1):162–166.

Cod:

Pershing, A. J., J. H. Annala, S. Eayrs, L. A. Kerr, J. Labaree, J. Levin, K. E. Mills, J. A. Runge, G. D., Sherwood, J. C. Sun, S. Tallack Caporossi. 2013. *The future of cod in the Gulf of Maine.* Portland: Gulf of Maine Research Institute, University of Maine, June 2013.

Hamilton, L.C., and M.J. Butler. 2001. Outport adaptations: Social indicators through Newfoundland's cod crisis. *Human Ecol Rev* 8(2):1–11.

Harris, M. 1999. *Lament for an ocean: The collapse of the Atlantic cod fishery.* Toronto: McClelland and Stewart.

Holloweed, A.B., and S. Sundby. 2014. Change is coming to the northern ocean. *Science* 344:1084–1085.

Kurlansky, M. 1997. *Cod.* London: Penguin.

Mason, F. The Newfoundland cod stock collapse: A review and analysis of social factors. *Electronic Green J* 1(7). http://escholarship.org/uc/item/19p7z78s#page-12

May, A. 2009. The collapse of the northern cod. *Newfoundland Quarterly (Memorial University of Newfoundland)* 102(2):41–44.

Myers, R.A., J.A. Hutchings, and N. J. Barrowman. 1997. Why do fish stocks collapse? The example of cod in Atlantic Canada. *Ecological Applications* 7(1):91–109.

Pikitch, E.K. 2013. The risks of overfishing. *Science* 338:474–475.

Seelye, K.Q., and J. Ridgood. 2013. Officials back deep cuts in Atlantic cod harvest to save industry. *New York Times*, 31 January.

8. ENVIRONMENTAL SERVICES

2014. Valuing the long-beaked echidna: Setting a price on nature is a useful exercise, up to a point. *Economist*, 22 February, p. 66.

Arriagada, R., C. Perrings. 2009. Making payments for ecosystem services work. Ecosystem Services Economics Occasional Papers Series, United Nations Environmental Programme, Nairobi (Kenya), August 2009, http://www.diversitas-international.org/resources/outreach/ArriagadaPerrings_2009_UNEPpolicybriezfESPayment.pdf.

Bakker, K. 2012. Water security: Research challenges and opportunities. *Science* 337:914–915.

Boyd, J., and S. Banzhaf. 2007. What are ecosystem services? The need for standardized environmental accounting units. *Ecol Econ* n.v.:616–626.

Costanza, R., R. d'Arge, and R. de Groot, et al. 1997. The value of the world's ecosystem services and natural capital. *Nature* 387:253–260.

Fisher, B., R.K. Turner, and P. Marling. 2009. Defining and classifying ecosystem services for decision making. *Ecol Econ* n.v.:643–653.

Guidotti, T.L., K. Teschke, E. Wein, and C.L. Soskolne. 1997. An agenda for studying human and ecosystem health effects in the boreal forest. *Ecosystem Health* 3(1):11–26.

Henry, D. *Canada's boreal forest*. 2002. Washington, DC: Smithsonian Institution.

Kinzig, A.P., and F.S. Chapin III, S. Polesky, V.K. Smith, D. Tilman, B.L.Tuner II. 2011. Paying for ecosystem services—promise and peril. *Science* 334:603–604.

Mburu, J., L.G. Hein, B. Gemmill, and L. Collette. 2006. Economic valuation of pollination services: Review of methods. Rome: Food and Agricultural Organization. http://www.fao.org/fileadmin/templates/agphome/documents/Biodiversity-pollination/econvaluepoll1.pdf.

Winfree, R., B.J. Gross, and C. Kremen. 2011. Valuing pollination services to agriculture. *Ecol Econ* 71:80–88. doi:10.1016/j.ecolecon.2011.08.001.

9. ARTIFICIAL ECOSYSTEMS

Beisner, B., C. Messier, and L.A. Giraldeau. 2006. *Nature all around us: A guide to urban ecology*. Chicago: University of Chicago Press.

Clewell, A.F., and J. Aronson. 2013. *Ecological restoration: Principles, values, and structure of an emerging profession*. Washington, DC: Island Press, Society for Ecological Restoration.

Guidotti, T.L. 1994. Comparing environmental risks: A consultative approach to setting priorities at the community level. *Pub Health Rev* 22:321–337.

———. 1995. Perspective on the health of urban ecosystems. *Ecosystem Health* 1(3):141–149.

———. 2012. Urban ecosystems. In *The Praeger handbook of environmental health*. Edited by R.H. Friss, 353–370. Santa Barbara, CA: ABC-Clio.

Listorti, J.A., and F. M. Doumani. 2001. Environmental health—bridging the gaps: environmental health assessments—rapid checklists. Washington, DC: World Bank Africa.

Takano, T., and K. Nakamura. 2001. An analysis of health levels and various indicators of urban environments for healthy cities projects. *J Epidemiol Community Health* 55(4):263–270.

Warner, P. 2010. *Through the end of nature*. Cambridge, MA: MIT Press.

Warren J., C. Lawson, and K. Belcher. 2008. *The agri-environment*. Cambridge, UK: Cambridge University Press.

Weber, E.P. 2003. *Bringing society back in: Grassroots ecosystem management, accountability, and sustainable communities*. Cambridge, MA: MIT Press.

Wimberley, E.T. 2009. *Nested ecology: The place of humans in the ecological hierarchy*. Baltimore: Johns Hopkins University Press.

Woodworth, P. 2013. *Our once and future planet*. Chicago: University of Chicago Press.

Yarnal, B., C. Polsky, and J. O'Brien. 2009. *Sustainable communities on a sustainable planet: The Human-Environment Regional Observatory Project*. Cambridge, UK: Cambridge University Press.

10. ENERGY

The energy sector is dense with statistics and data, and most of it is not well integrated into a comprehensive picture for readers. The author relies heavily on publications of the Energy Institute (London) to follow trends. Here are some accessible sources with an emphasis on current trends.

2013. Appalachian fall: New carbon regulations will make life harder still for a beleaguered region. *Economist*, 28 September, p. 30.

2014. Coal: The fuel of the future, unfortunately. *Economist*, 10 April.

2013. Extremely troubled scheme: Crunch time for the world's most important carbon market. *Economist*, 16 February, p. 75.

2013. How to lose half a trillion euros: Europe's electricity providers face an existential threat. *Economist*, 12 October, p. 27.

2014. Let the sun shine: The future is bright for solar power, even as subsidies are withdrawn. *Economist*, 8 March, p. 29.

2010. Lift-off: Research into the possibility of geoengineering a better climate is progressing at an impressive rate—and meeting strong opposition. *Economist*, 6 November, p. 93.

2013. Power struggle: The shadow of Fukushima, the world's worst nuclear power disaster after Chernobyl, hangs over Japan's energy future. *Economist*, 21 September, p. 45.

Burn-Murdoch J. 2014. Up in smoke: How efficient is electricity generation in the UK? *Guardian*, 6 July. http://www.theguardian.com/news/datablog/2012/jul/06/energy-green-politics

Cappa, C.D. 2012. Radiative absorption enhancements due to the mixing state of atmospheric black carbon. *Science* 337:1078–1081.

Clery, D. 2011. Step by step, NIF researchers trek toward the light. *Science* 334:449–450.

______. 2012. Report on future of fusion research says U.S. should hedge its bets. *Science* 335:1158–1159.

______. 2014. Laser fusion shots take steps toward ignition. *Science* 344:721.

Dunn, B., H. Kamath, and J.M. Tarascon. 2011. Electrical energy storage for the grid: A battery of choices. *Science* 334:928–935.

Ellsworth, W.L. 2013. Injection-induced earthquakes. *Science* 341:142.

Faiola, A. 2013. Europe's carbon market goes bust. *Washington Post*, 6 May.

Folger, P. Carbon capture and sequestration: Research, development, and demonstration at the U.S. Department of Energy. Congressional Research Service, Report 5-5700. http://www.fas.org/sgp/crs/misc/R42496.pdf

Gillis, J. 2014. A price tag on carbon as a climate rescue plan. *New York Times*, 20 May.

Gopstein, A.M., 2014. D. Arent, S. Wofsy, N.J. Brown, R. Bradley, G.D. Stucky, D. Eardley, R. Harriss. Methane leaks from North American natural gas systems. *Science* 344:733.

Hamilton, C. 2013. *Earthmasters: The dawn of the age of climate bioengineering.* New Haven, CT: Yale University Press.

Hindell, D., et al. 2012. Simultaneously mitigating near-term climate change and improving human health and food security. *Science* 335:183–188.

Kerr, R.A. 2012. Are world oil's prospects not declining all that fast? *Science* 337:633.

———. 2012. A quick (partial) fix for an ailing atmosphere. *Science* 335:156.

Kintisch, E. 2013. Dr. Cool. *Science* 342:307–309.

Levi, M. 2013. *The power surge: Energy, opportunity and the battle for America's future.* New York: Oxford University Press.

McNutt, M. 2013. Bridge or crutch? *Science* 342:909.

Mufson, S. The coal plant to end all coal plants? Overbudget and running late, much rides on Southern Co.'s carbon-capture experiment. *Washington Post.* 18 May 2014.

Nisbet, E.G., E. J. Dlugokencky, and P. Bousquet. 2014. Methane on the rise—again. *Science* 343:493–495.

Nocera, J. 2013. A fracking Rorschach test. *New York Times*, 5 October.

Parson, E.A., and D.W. Keith. 2013. End the deadlock on governance of geoengineering research. *Science* 339:1278–1279.

Prentiss, M. *Energy revolution: The physics and promise of efficient technology.* Cambridge MA, Harvard University Press (forthcoming).

Reuters. 2014. China to close nearly two thousand small coal mines. Reuters Online. http://www.reuters.com/article/2014/04/04/china-coal-idUSL4N0MW2OJ 20140404

Simon, P., Y. Gogotsi, and B. Dunn. 2014. Where do batteries end and supercapacitors begin? *Science* 343:1210–1211.

US Energy Information Agency. 2014. History of energy consumption in the United States, 1775–2009.http://www.eia.gov/todayinenergy/detail.cfm?id=10.

Williams, J. H., A. DeBenedictus, R. Ghanadan, A. Mahone, J. Moore, W.R. Morrow III, S. Price, M.S. Torn. 2012. The technology path to deep greenhouse gas emissions cut by 2050: The pivotal role of electricity. *Science* 335:53–59.

Warburg, P. 2012. *Harvest the Wind.* Boston: Beacon Press.

Yergin, D. 2012. America's new energy reality. *New York Times*, 10 June.

11. CULTURE AND RIGHTS

Attfield, R., and A. Belsey, eds. 1994. *Philosophy and the natural environment.* Cambridge, UK: Cambridge University Press.

Clayton, S., and S. Opotow, eds. 2003. *Identity and the natural environment.* Cambridge, MA: MIT Press.

Crosby, A.W. 1986. *Ecological imperialism: The biological expansion of Europe, 900–1900.* Cambridge, UK: Cambridge University Press.

Diamond, J. 2012. *The world until yesterday: What can we learn from traditional societies?* New York: Viking.

Guidotti, T.L., and Abercrombie, S. 2008. Aurum: A case study in the politics of NIMBY. *Waste Management Res* 26(6):582–588.

Guidotti, T.L., and P. Jacobs. 1992. Effect on a community of a perceived excess cancer risk: Case study of an epidemiologic mistake. *Am J Public Health* 83:233–239.

Jamieson, D. 2008. *Ethics and the environment: An introduction*. Cambridge, UK: Cambridge University Press.

Lane, M. 2012. *Eco-republic: What the ancients can teach us about ethics, virtue, and sustainable living*. Princeton, NJ: Princeton University Press.

12. SPIRITUALITY

The author is indebted to Ms. Allison Fisher, of Interfaith Power and Light of Maryland and Washington, DC, who presented the "What would I have to do for me not to care?" formulation at a conference on 28 January 2014.

Aitkenhead, D. 2008. James Lovelock: Enjoy life while you can, in 20 years global warming will hit the fan. *The Guardian* (UK), 1 March. http://www.theguardian.com/theguardian/2008/mar/01/scienceofclimatechange.climatechange.

Bellah, R.N. 2011. *Religion in human evolution: From the Paleolithic to the Axial age*. Cambridge, MA: Harvard University Press.

Biehl, J., and P. Staudenmaier. 2011. *Ecofascism revisited: Lessons from the German experience*. Porsgrunn, Norway: New Compass Press.

Dasgupta, P. 2001. *Human well-being and the natural environment*. New York: Oxford University Press.

Dominick, R. 1992. *The environmental movement in Germany: Prophets and pioneers, 1871–1971*. Bloomington: Indiana University Press.

Heidel, A. 1949. *The Gilgamesh epic and Old Testament parallels*. Chicago: University of Chicago Press.

Sessions, George, ed. 1995. *Deep ecology for the 21st century*: Readings on the philosophy and practice of the new environmentalism. Boston: Shambhala Press.

Gaia

Hickman, L. 2012. James Lovelock: The UK should be going mad for fracking. *The Guardian* (UK), 15 June. http://www.theguardian.com/environment/2012/jun/15/james-lovelock-interview-gaia-theory.

Johnston, I. 2012. "Gaia" scientist James Lovelock: I was "alarmist" about climate change. NBC News, 23 April. http://worldnews.nbcnews.com/_news/2012/04/23/11144098-gaia-scientist-james-lovelock-i-was-alarmist-about-climate-change.

Lovelock, J. 2001. *Gaia: A new look at life on earth*. New York: Oxford University Press.

———. 2009. *The vanishing face of Gaia: A final warning*. New York: Basic Books.
Ruse, M. 2013. *The Gaia hypothesis: Science on a pagan planet*. Chicago: University of Chicago Press.
Schaechter, M. 2012. Lynn Magulis (1938–2011). *Science* 135:302.
Watts, A. 2012. Breaking: James Lovelock backs down on climate change. MSNBC, 23 April. http://wattsupwiththat.com/2012/04/23/breaking-james-lovelock-back-down-on-climate-alarm/

13. PROFESSIONALISM

Most of the discussion in the section on environmental sciences and studies is derived from conferences sponsored by the National Council for Science and the Environment, Council of Environmental Deans and Directors, and the Association of Environmental Studies and Sciences, including presentations by Dr. Vincent and Dr. Pfirman. Exact figures vary from year to year and are not given here for any one year because the exact percentage is less significant than the overall trend (e.g., Vincent, S., S. Bunn, L. Stone. 2013. Interdisciplinary environmental and sustainability education on the nation's campuses 2012: Curriculum design. Washington, DC: National Council for Science and the Environment).

Association for the Advancement of Sustainability in Higher Education. 2014. http://www.aashe.org/resources/academic-programs/type/masters/
Best jobs in America. Number 22: Environmental health specialist. CNN Money. http://money.cnn.com/magazines/moneymag/bestjobs/2010/snapshots/22.html
Cairncross, F. 1995. *Green, Inc.: A guide to business and the environment*. Washington, DC: Island Press.
Cascio, J., G. Woodside, P. Mitchell. 1996. *ISO 14000 Guide: The New International Environmental Management Standard*. New York: McGraw-Hill.
Hamilton, J. 2012. Is a sustainable career on your green horizon? Washington DC, Bureau of Labor Statistics, 2012. http://www.bls.gov/green/sustainability/sustainability.htm
Rappaport, A., and S. H. Creighton. 2007. *Degrees that matter: Climate change and the university*. Cambridge MA: MIT Press.

Index

Lightning Source UK Ltd.
Milton Keynes UK
UKHW011822290319
340175UK00002B/99/P

9 780199 325337